Technological and Social Insights in Food Industry

Technological and Social Insights in Food Industry

Edited by **Lisa Jordan**

New York

Published by Callisto Reference,
106 Park Avenue, Suite 200,
New York, NY 10016, USA
www.callistoreference.com

Technological and Social Insights in Food Industry
Edited by Lisa Jordan

International Standard Book Number: 978-1-63239-591-7 (Hardback)

Printed in the United States of America.

Contents

Preface   VII

Part 1   Scientific and Technological Aspects   1

Chapter 1   The Application of Vacuum Impregnation Techniques
             in Food Industry   3
             A. Derossi, T. De Pilli and C. Severini

Chapter 2   A Discussion Paper on Challenges and Proposals for
             Advanced Treatments for Potabilization of
             Wastewater in the Food Industry   35
             D. B. Luiz, H. J. José and R. F. P. M. Moreira

Chapter 3   Microorganism-Produced Enzymes in the Food Industry   57
             Izabel Soares, Zacarias Távora,
             Rodrigo Patera Barcelos and Suzymeire Baroni

Chapter 4   Freezing/Thawing and Cooking of Fish   69
             Ebrahim Alizadeh Doughikollaee

Chapter 5   Novel Fractionation Method for
             Squalene and Phytosterols Contained in
             the Deodorization Distillate of Rice Bran Oil   83
             Yukihiro Yamamoto and Setsuko Hara

Chapter 6   Micro and Nano Corrosion in
             Steel Cans Used in the Seafood Industry   95
             Gustavo Lopez Badilla,
             Benjamin Valdez Salas and Michael Schorr Wiener

Chapter 7   Nanotechnology and Food Industry   111
             Francisco Javier Gutiérrez, Mª Luisa Mussons,
             Paloma Gatón and Ruth Rojo

Chapter 8   Characteristics and Role of Feruloyl Esterase from *Aspergillus*
             Awamori in Japanese Spirits, '*Awamori*' Production   145
             Makoto Kanauchi

Part 2   Social and Economic Issues   163

Chapter 9   The Industrial Meat Processing Enterprises
             in the Adaptation Process of Marketing Management
             of the European Market   165
             Ladislav Mura

Chapter 10   Facilitating Innovations in a Mature Industry-Learnings
              from the Skane Food Innovation Network   177
              Håkan Jönsson, Hans Knutsson and Carl-Otto Frykfors

Chapter 11   Functional Foods in Europe: A Focus on Health Claims   197
              Igor Pravst

Chapter 12   Organic Food Preference: An Empirical Study
              on the Profile and Loyalty of Organic Food Customers   241
              Pelin Özgen

              Permissions

              List of Contributors

Preface

This book has been a concerted effort by a group of academicians, researchers and scientists, who have contributed their research works for the realization of the book. This book has materialized in the wake of emerging advancements and innovations in this field. Therefore, the need of the hour was to compile all the required researches and disseminate the knowledge to a broad spectrum of people comprising of students, researchers and specialists of the field.

The rapid growth of processed food industry has led to new challenges and issues. Food is necessary for the survival of humankind and this book aims towards giving an overview of the food industry that runs the supply of food and thereby ensures health and comfort. The global food industry serves a population of seven billion people and has the biggest market. This book highlights the issues of world hunger and discusses how food shortages endanger water and energy supply. It covers two broad sections such as "Scientific and technological aspects" and "Social and economic issues". It further elaborates about how food production can be affected by climate change, droughts, global warming, floods, greenhouse emissions and storms. Food producers, industry experts, corrosion practitioners, academicians and designers of food processing equipment have made valuable contributions to this book based on their extensive knowledge and experience. They present different perspectives and approaches in the diverse aspects of food science and technology.

At the end of the preface, I would like to thank the authors for their brilliant chapters and the publisher for guiding us all-through the making of the book till its final stage. Also, I would like to thank my family for providing the support and encouragement throughout my academic career and research projects.

Editor

Part 1

Scientific and Technological Aspects

The Application of Vacuum Impregnation Techniques in Food Industry

A. Derossi, T. De Pilli and C. Severini
Department of Food Science, University of Foggia
Italy

1. Introduction

The interest of food scientists in the filed of microstructure is recently exponentially increased. This interest has raised after the recognition of the importance that chemical reactions and physical phenomena occurring at microscopic scale have on safety and quality of foods. This concept was well resumed from Aguilera (2005) which stated that "...*the majority of elements that critically participate in transport properties, physical and rheological behavior, textural and sensorial traits of foods are below the 100 μm range*". As example, Torquato (2000) reported that through microscopic observation it is possible to observe that only a portion of cells of the crumb bread solid matrix are connected, even though at macroscopic level it may appear as completely connected; so, the three dimensional spatial distribution of cell crumbs greatly affects the sensorial quality of bread. Before the scientific recognition of the above consideration, food scientists which focused their efforts on the effects of traditional and innovative industrial processes on food quality, analyzed only macroscopic indexes such as color, texture, taste, concentration of several nutritional compounds, etc. without to consider that they essentially are the result of chemical and physical phenomena occurring at microscopic level. However, with the aim to be more precise in the use of the term "microscopic" we may generalize the classification that Mebatsion et al. (2008) reported for the study of fruit microstructure. The authors considered three different spatial scales: 1. The macroscale which refers the food as a whole or a continuum of biological tissues with homogeneous properties; 2. The mesoscale which refers the topology of biological tissues; 3. The microscale which addresses the difference of individual cells in terms of cell walls, cell membranes, internal organelles, etc. Here, we would like focus the attention on the importance to study the foods at mesoscopic scale by considering them as mesoscopic divided material (MDM). On this basis the majority of foods may be defined as "*biological systems where an internal surface partitions and fills the space in a very complex way*". At mesoscopic scale the three dimensional architecture of foods may be studied analyzing the relation between void and solid phases where the voids (capillaries, pores) may be partially or completely filled with liquids or gases. The relations between these two phases and their changes during processing is one of the most important factors affecting the safety and the sensorial and/or nutritional quality of foods. Until few years ago only the porosity fraction of foods was reported in literature as experimental index of the internal microstructure but it gives us only low level of information. Instead, a second level of structure characterization may be reached by analyzing pore dimension, shape, length, surface roughness, tortuosity, connectivity, etc. So, one of the most important future challenge will be the precise characterization of the three dimensional architecture of foods, its changes during food processing and its relation with safety and quality. However, some pioneering researchers have focused their attention on this research field and the first results are reported in literature (Datta, 2007a; Datta, 2007b; Halder et al., 2007). For instance, the term "food matrix engineering (FME)" has been used to define a branch of food engineering that applies the knowledge about food matrix composition, structure and properties with the aim to promote and control adequate changes that improve some sensorial and/or functional properties in the food products as well as their stability (Fito et al., 2003). The authors reported that food gels and emulsions, extruded, deep fried and puffed foods may be considered as current FME. Instead, among the non-traditional processing, one of the most important and newest techniques based on the properties of food microstructure is the vacuum impregnation (VI). With this term are classified some technologies based on the exploitation of void fraction of foods with the aim to introduce, in a controlled way, chemical/organic compounds into the capillaries of biological tissues. VI is based on the application of a partial vacuum pressure which allows to remove native liquids and gases entrapped into the capillaries and to impregnate them with a desired external solution after the restoration of atmospheric pressure. At first, a vacuum osmotic dehydration (VOD) treatment was proposed and studied as method to accelerate water loss and solid gain during the immersion of vegetables into hypertonic solution. More recently, VI treatments have been studied as method to enrich food with nutritional and functional compounds, to introduce some ingredients with the aim to obtain food with innovative sensorial properties as well as to introduce compounds able to inhibit the most important degradation reactions and the microbial growth. This chapter has the aim to analyze and to discuss the application of vacuum impregnation techniques in food industry. At first, the theoretical principles and the mathematical modeling of the phenomena involved during the application of VI will be discussed. Also, the response of biological tissues to vacuum impregnation will be reviewed. In a second section, focusing the attention of the reader on different potential applications in food industry, the effect of process variables on both impregnation level and quality of final products will be analyzed.

2. Theoretical background and structural changes

The main phenomenon on the basis of vacuum impregnation treatment is the hydrodynamic mechanism (HDM) which was well described from Fito & Pastor (1994), Fito (1994) and Chiralt & Fito (2003). The authors, discussing the results of the water diffusion coefficients (De) obtained with both microscopic and macroscopic analysis during osmotic dehydration (OD), reported that the values were in general 100 times greater when measured at microscopic scale; in particular, De values were about 10^{12} m^2/s and 10^{10} m^2/s when relieved respectively at microscopically and macroscopically. These observations suggested that a fast mass transfer mechanism, in addition with the traditional diffusion, is involved during OD treatments. This mechanism is the hydrodynamic mechanism which is based on pressure gradients generated from the changes of sample volume and/or externally imposed at non-compartmented section such as intercellular spaces, capillaries, pores, etc. These phenomena play a key role in the solid-liquid operation increasing the rate of several processes during which mass transfer occurs. The action of HDM was well explained from Chiralt & Fito (2003). When food pieces are immersed in an external solution the surface of samples is washed and the solution partially penetrates into the open pores. After, in line with a deformation of cell membranes due to the loss of native liquids and gases, a pressure gradient is generated and HDM is promoted.

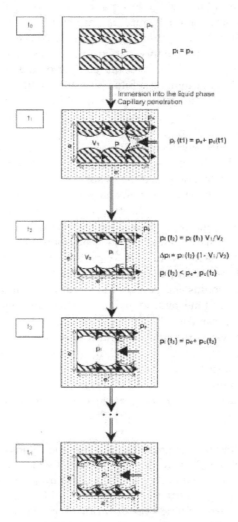

Fig. 1. Schematic representation of hydrodynamic mechanisms due to capillary action and pressure gradients as a consequence of internal volume changes (From Chiralt and Fito, 2003).

In figure 1 a schematic representation of HDM during osmotic dehydration is shown. Before the immersion of vegetable tissues into hypertonic solution the capillary pressure inside the pore is equal to the external (atmospheric) pressure. At t_1 capillary pressure promotes the initial gain of osmotic solution and the compression of gases inside the pores; so, internal pressure becomes greater than external one. In these conditions gases irreversibly tend to flow out and the cells in contact with hypertonic solution dehydrated due to osmotic pressure gradient. The water loss produces an increase of internal volume as a result of cells shrinkage during dehydration. Moreover, in line with the volume increase the pressure inside the capillary becomes lower to the external one promoting the suction of additional external solution. This process proceeds until the capillary is completely impregnated with the osmotic solution (Barat et al., 1998; Chiralt and Fito, 2003). All these phenomena becomes marked during vacuum impregnation treatments when a vacuum pressure gradient is externally imposed. The treatments are performed through two subsequent steps: 1. The immersion of foods into the solution and the application of a vacuum pressure (p) for a vacuum period (t_1) (also called vacuum time); 2. The restoration of atmospheric pressure maintaining the samples immersed into the solution for a relaxation period (t_2) (also called relaxation time). During VI in addition to HDM a deformation-relaxation phenomena (DRP) simultaneously occurs. HDM and DRP both affect the reaching of an equilibrium situation and their intensities are strictly related with the three dimensional food microstructure and mechanical properties of solid matrix.

Figure 2 schematically shows the phenomena involved during vacuum solid-liquid operations of an ideal pore. At time zero the samples are immersed into the external liquid and the internal pressure of the pore (p_i) is equal to external (atmospheric) pressure (p_e). After, a vacuum pressure (p) is applied in the head space of the system for a time (t_1) promoting a situation in which p_i is greater than p_e. In this condition, the internal gases expand producing the deformation (enlargement) of capillary and the increase of internal volume. Moreover, native liquids and gases partially flow out on the basis of the pressure gradient (step 1-A). At this time hydrodynamic mechanism begins and external liquid partially flows inside the capillary as a consequence of the pressure gradient. These phenomena simultaneously occur until the equilibrium is reached (step 1-B). In the second step the atmospheric pressure in the head space of the system is restored and the samples are maintained into the solution for a relaxation time (t_2). During this period, the generated pressure gradient ($p_i < p_e$) promotes both HDM and solid matrix deformation (compression) which respectively produce capillary impregnation and the reduction of pore volume until a new equilibrium is reached (Fito & Chiralt, 1994; Fito et al., 1996).

2.1 Mathematical modeling of vacuum impregnation and related structural changes

Fito (1994) and Fito et al. (1996) were the first scientists which translate in mathematical language the phenomena involved at mesoscopic scale during VI treatments. The model is based on the analysis of the contributions of both liquid penetration and solid matrix deformation of an ideal pore of volume $Vg_0 = 1$ during each step of VI. From time t = 0 to the step 1B a situation expressed by the following equation is obtained:

$$Vg_{1b} = 1 + Xc_1 - Xv_1 \tag{1}$$

Where Vg_{1b} is the pore volume at the step 1-B, Xc_1 is the increment of pore volume due to the expansion of internal gases and Xv_1 is the partial reduction of pore volume due to the initial suction of external liquid as a consequence of HDM.

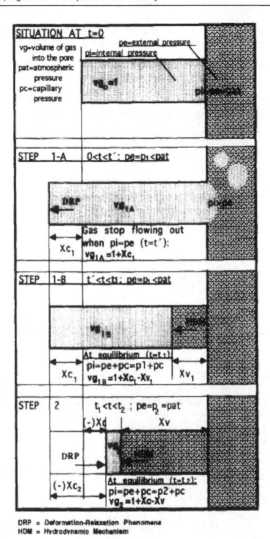

DRP = Deformation-Relaxation Phenomena
HDM = Hydrodynamic Mechanism

Fig. 2. Schematic representation of vacuum impregnation of an ideal pore. Deformation-relaxation phenomena and hydrodynamic mechanism occurring during vacuum period (t1) and relaxation time (t2) (from Fito et al., 1996).

At the end of step 2 the total liquid penetration and matrix deformation may be described respectively by the equations:

$$Xv = Xv_1 + Xv_2 \qquad (2)$$

$$Xc = Xc_1 + Xc_2 \qquad (3)$$

where Xv_1 and Xv_2 are the volume reduction due to liquid penetration respectively at the end of step 1 and 2; Xc_1 and Xc_2 are the volume pore changes as a result of solid matrix deformation (enlargement and compression) after the steps 1 and 2. Also, the total volume variation at the end of the process may be described as follow:

$$Vg_2 = 1 + Xc - Xv \tag{4}$$

Equations 2 and 3 may be used to calculate liquid penetration and solid matrix deformation of the total sample taking into account its porosity fraction value (ε_e):

$$X = e_e Xv \tag{5}$$

$$g = e_e Xc \tag{6}$$

As reported from Fito et al. (1996) when a pressure variation applied in a solid-liquid system and an equilibrium situation is reached, HDM assumes an isothermal compression of gas into the pores. So, the situation reached at the end of step 1-B may be mathematically express as:

$$\frac{Vg_1B}{Vg_1A} = \frac{1 + Xc - Xv}{1 + Xc} = \frac{1}{r} \tag{7}$$

Where r is the apparent compression rate ($\sim$ atmospheric pressure/vacuum pressure, dimensionless) (Zhao and Xie, 2004). From equation 7 may be obtained the following:

$$\frac{Xv1}{1 + Xc_1} = 1 - \frac{1}{r} \tag{8}$$

Also, by using the equation 5 and 6 it is possible to obtain:

$$X_1 = (\varepsilon_e + \gamma_1)\left(1 - \frac{1}{r_1}\right) \tag{9}$$

On these basis the equilibrium situation at the end of step 1-B may be mathematically expressed as:

$$X_1 - Y_1 = \varepsilon_e\left(1 - \frac{1}{r_1}\right) - \frac{\gamma_1}{r_1} \tag{10}$$

Furthermore, the same considerations may be extended for the phenomena involved from step 1 and step 2. So, between t = t_1 and t = t_2 the equilibrium situation may be expressed as:

$$X - \gamma = (\varepsilon_e + \gamma)\left(1 - \frac{1}{r_2}\right) - \gamma_1 \tag{11}$$

Starting from the above equations it is possible to calculate the porosity value (ε_e) at the end of VI process from the value of X, γ and γ_1 by:

$$\varepsilon_e = \frac{(X - \gamma)r_2 + \gamma_1}{r_2 - 1} \tag{12}$$

Where X is the volume fraction of sample impregnation by the external liquid at the end of VI treatments (m³ of liquid/m³ of sample a t=0), ε_e is the effective porosity, γ_1 is the relative volume deformation at the end of vacuum period (t_1, m³ of sample deformation/m³ of sample at t=0); γ is the volume deformation at the end of process (m³ of sample deformation/m³ of sample at t=0). However, although the measure of porosity fraction value is easily obtained from the apparent (ρ_a) and real density (ρ_r) values of the sample (Lewis, 1993; Gras, Vidal-Brotons, Betoret, Chiralt & Fito, 2002), the experimental estimation of X, γ and r are not easy to measure and some modifications of the equipment used for the experiments are necessary (Salvatori et al., 1998). Briefly, the experimental methodology is performed precisely weighting the sample at different steps of the treatment even if it is under vacuum condition (Fito et al., 1996; Salvatori et al., 1998). The authors defined a parameter called magnitude (H):

$$Ht = Xt - \gamma t = \frac{Lo - Lt - Mw}{Vo\rho} \tag{13}$$

Where L is the weight measured in the balance (kg), Mw is the liquid evaporated during the experiment, V is the sample volume (m³) and ρ is the liquid density (kg/m³). Instead the subscripts 0 and t refer respectively the time zero and any time t_i corresponding at each step of the process. Practically speaking the magnitude H is an overall index of the contribution of both liquid penetration (X) and solid matrix deformation (γ) on sample volume. A theoretical curve of H value as a function of time during vacuum impregnation is shown in figure 3.

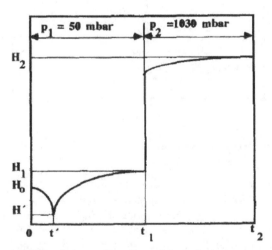

Fig. 3. Theoretical curve of H value as a function of time during vacuum impregnation treatments (From Fito et al., 1998).

From the figure it is possible to observe that as vacuum pressure is applied in the system, H value quickly decreases until H' at t = t'. This is because sample volume increases under the action of the gas expansion due to the positive pressure gradient (pi > pe). During the vacuum period (from t = t' to t = t_1) H value slowly increases because native liquids and gases flow out, solid matrix relaxes and external liquid begins to impregnate pores due to

HDM. H_1 is obtained when the equilibrium condition is reached at the end of step 1B. At time $t = t_1$ the restoration of atmospheric pressure (p = 1030 mbar) leads a sudden increase of H value as a consequence of liquid penetration. Also, during relaxation time (from $t = t_1$ to $t = t_2$) pore volume reduction due to capillary compression and HDM simultaneously occur. H_2 value is obtained when the new equilibrium situation will be reached. From the above consideration, the crucial importance of microscopic properties of foods such as dimension and shape of samples, their three dimensional architecture, the resistance of biological tissue to gas and liquid flow, solid matrix deformation, etc., is obvious. In particular, these factors affect the kinetics of all phenomena simultaneously involved during VI; so, the quality of foods will be a result of the rates by which each phenomena occur. For instance, if the solid matrix relaxation of food is slow, capillary impregnation could occur without significant deformation. On the other hand a fast deformation could significantly reduce capillary impregnation. Salvatori et al. (1998) studied the time evolution of H value of several vegetables submitted to VI at 50 mbar for different vacuum period (t_1) and relaxation time (t_2). As example, figure 4 shows the results obtained from apple samples (Granny Smith) submitted to VI for a $t_1 = t_2 = 15$ minutes.

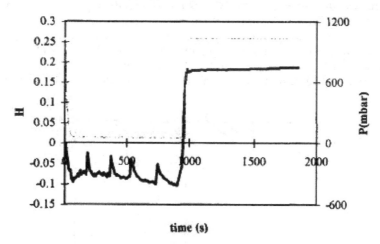

time (s)

Fig. 4. Experimental curve of apple sample submitted to VI treatment at 50 mbar for a t1 = t2 = 15 minutes (From Salvatori et al., 1998).

As expected, H value decreased quickly after the application of vacuum pressure in line with pores expansion. However, in disagreement with the behavior of an ideal pore reported in figure 3, at the end of vacuum period (~ 1000 s), H value was negative (H1 < 0) stating that pore volume deformation still was greater than liquid penetration due to the enlargement of pores. In particular, the authors reported that at the end of vacuum period the X_1 and γ_1 values (an average of all experiments performed in different operative conditions) respectively were – 4.2% and 1.7%. The negative value of the liquid penetration at the end of vacuum period was explained on the basis of native liquid release from the pores under the action of the negative pressure gradient ($p_e < p_i$). The peaks values observed in figure 1 are in line with the variability of both tissue structure characteristics and the surrounding fluid properties (Salvatori et al., 1998). After, the restoring of atmospheric pressure led to a significant increase of H value due to the compression of pores ($\gamma < 0$) and the suction of external liquid (X > 0). Indeed at the end of the process the authors reported a value of X and γ of 15% and -0.6% respectively. However, foods may show significant different trends as a function of their unique microstructure properties which are very different from the behavior of an ideal pore. For instance, mango and peach samples as well as oranges showed a positive solid matrix deformation (enlargement of the pores) also after the restoration of atmospheric pressure (Salvatori et al., 1998; Fito et al., 2001). Gras et al. (2002) reported that carrot, diced zucchini and beetroot, submitted to VI treatment in a sucrose isotonic solution, showed a pore volume increase (γ > 0) also at the end of process coupled with significant impregnation values of 16%, 20% and 7% respectively. These behaviors were explained with an overall situation in which the rate of liquid impregnation being very fast in comparison with the solid matrix deformation allowed to keep a residual capillary expansion.

2.2 Gas, liquid and solid matrix volume changes during VI

With the aim to have a complete theoretical analysis of the phenomena involved during VI it is important to consider the gases, liquid and solid matrix volume changes occurring during vacuum and relaxation times. Barat et al. (2001) proposed an experimental approach to study the volume changes during VI assuming food constituted of three phases: solid matrix (SM), liquid phase (LP) and gas phase (GP). On this basis the authors reported that total volume changes which occur during vacuum impregnation treatment may be mathematically express as:

$$\Delta V = \Delta V^{LP} + \Delta V^{GP} + \Delta V^{SM} \tag{14}$$

Moreover, since the variation of solid matrix volume may be assumed equal to zero equation 14 may be reduced at:

$$\Delta V = \Delta V^{LP} + \Delta V^{GP} \tag{15}$$

Here, ΔV may be obtained by pycnometer method of fresh and impregnated food whereas the volume changes of liquid phase (ΔV^{LP}) may be estimated by the following equation (Barat et al., 2001: Atares, Chiralt & Gonzales-Martinez, 2008):

$$\Delta V^L = \left\{ \left[m^t \left(x_w{}^t - x_s{}^t \right) \right] / r_{LP}{}^t - \left[m^0 \left(x_w{}^0 - x_s{}^0 \right) \right] / r_{LP}{}^0 \right\} / V^0 \tag{16}$$

Where m is the mass of sample (g), x_w and x_s are respectively the moisture and solid content of sample (water or solid, g/g of fresh vegetable), ρ_{LP} is the density of liquid phase (g/mL) and V is the volume of samples (mL). Moreover, the superscripts t and 0 refer respectively to each time treatment (t_i) and fresh vegetable.

The liquid phase density values may be estimated by the follow equation (Atares et al., 2008):

$$r_{LP} = \left(230 z_s{}^2 + 339 z_s + 1000 \right) / 1000 \tag{17}$$

where z_s is the solid content of sample liquid phase (g solid/g liquid phase). This method has been proved to give precise results during osmotic dehydration process (Barat et al., 2000; Barat et al., 2001). In particular, Barat et al. (2001) plotted the ΔV as a function of ΔV^{LP}

of apple (Granny Smith) samples submitted to traditional osmotic dehydration (OD) and pulsed vacuum osmotic dehydration (PVOD) performed applying a pressure of 180 mbar for a $t_1 = 5$ minutes.

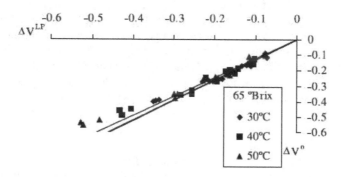

Fig. 5. Total volume changes as a function of liquid phase volume changes of apple samples submitted to OD and PVOD in different operative conditions.

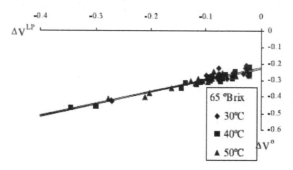

Fig. 6. Total volume changes as a function of liquid phase volume changes of apple samples submitted to OD and PVOD in different operative conditions.

Samples submitted to OD showed a linear trend with no significant intercept value stating that no significant total volume variation was observed before liquid penetration occurring under the action of HDM. By linear regression of experimental data it is possible to estimate the relative contribution of ΔV^{LP} and ΔV^{GP} by using the following equation:

$$\Delta V = s_1 \Delta V^{LP} \tag{18}$$

$$\frac{\Delta V^{GP}}{\Delta V^0} = 1 - \frac{1}{s_1} \tag{19}$$

In comparison with OD treatments, it is worth noting (Figure 6) that samples submitted to PVOD showed a significant intercept with an average value of 0.20. This may be considered as the result of the initial vacuum pulse which promotes the pores expansion coupled with the removal of native liquid and gases and the action of HDM. In this case the linear regression of experimental data shown in figure 6 assumes the following form:

$$\Delta V = i_2 + s_2 \Delta V^{LP} \tag{20}$$

Also, in the case of PVOD the general form of the equation 15 was modified with the following:

$$\Delta V = \Delta V^{LP} + \Delta V^{LP-VI} + \Delta V^{CR} \tag{21}$$

Where ΔV^{LP-VI} and ΔV^{CR} are the volume change due to liquid impregnation and volume changes due to compression-relaxation of solid matrix respectively. In equation 21 the two additional terms substituted ΔV^{GP}. Also, it is important to note that these two terms explain the intercept value of equation 20 and their sum is close to the porosity of food (for apples sample about the 0.22). In their conclusion the authors stated that osmotic dehydration in porous fruits may be explained in terms of LP and GP changes in line with the flow of gases and external liquid and the changes of pore volume as a consequence of the enlargement and compression of pores.

3. Process variables

As previous reported, vacuum impregnation is a technique that allows to introduce several chemical compounds and/or ingredients in the void phase of foods. The process is performed by applying a vacuum pressure in the head space of the system for a time t_1 and then by restoring atmospheric pressure for a relaxation time t_2. The numerous phenomena involved during the process are affected by several variables which may be classified as external and internal of foods. In the first class, vacuum pressure (p), time length of vacuum period (t_1), time length of relaxation time (t_2), viscosity of external solution, temperature, concentration of solution, product/solution mass ratio and size and shape of the samples may be introduced. Instead, the three dimensional architecture of food and the mechanical properties of the biological tissues may be considered as internal variables. However, it is worth nothing that the term "three dimensional architecture" refers to an ensemble of microscopic and mesoscopic characteristics such as porosity, size and shape of pores, connectivity, tortuosity, capillary curvatures, etc. which greatly affect the vacuum impregnation treatment. In this paragraph the effects of each variable above reported, considered both singularly and as synergistic effects will be discussed.

3.1 External variables

Among the external variables, vacuum pressure may be considered as the most important because it represents the force that produces the pressure gradient between the void phase of food and the atmosphere surrounding the external liquid. Briefly, vacuum pressure is the variable by which all phenomena previously reported may occur. In general, a vacuum pressure ranged between 50 and 600 mbar is reported in literature (Fito et al., 1996; Rastogi et al., 1996; Salvatori et al., 1998; Barat et al., 2001; Fito et al., 2001a; Fito et al., 2001b; Giraldo et al., 2003; Mujica-Paz et al., 2003; Zhao & Xie, 2004; Silva Paes et al., 2007; Corzo et al., 2007; Derossi et al., 2010; Derossi et al. 2011). Also, vacuum level is generally considered directly related to an increase of impregnation level (X) as a consequence of a higher release of native liquids and gases coupled with a greater HDM and DRP. Mujica-Paz et al. (2003a) studied the effect of vacuum pressure in a range of 135-674 mbar on the volume of pores impregnated from an isotonic solution of several fruits. The authors showed that for apple, peach, papaya and melon samples a greater impregnation was observed in line with an increase of vacuum pressure; instead, for mango, papaya and namey X values increased with the increasing of vacuum level until a maximum after that impregnation slightly decreased. Mujica-Paz et al. (2003b) studied the effect of vacuum pressure (135-674 mbar) on the weight reduction (WR), water loss (WL) and solid gain (SG) of apple, melon and mango slices kept in a hypertonic solution for a $t_1 = t_2 = 10$ minutes. For all experiments, melon and mango samples showed a positive WR, stating that fruits lost a significant fraction of their weight; instead, apples showed a negative WR values, stating a weight gain. In particular, for melon samples the authors reported a direct correlation between vacuum pressure and WR probably because as lower the vacuum pressure as greater the capillary impregnation, which decreased the weight reduction of the samples caused by both the osmotic dehydration and the removal of native liquids from the pores. Indeed the negative values observed for apple samples were explained on the basis of their high porosity fraction (~27.3%). In this conditions, the high free volume of apples increased the impregnation level more than water loss leading to an increase of weight of samples (figure 7a and 7b). Also, similar results were observed for water loss which assumed positive values for mango and melon and negative values in certain operative condition for apple samples (figures 8a and 8b).

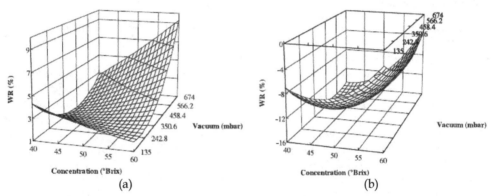

Fig. 7. Effect of vacuum pressure and osmotic solution concentration on weight reduction of melon and apple samples (from Mujica-Paz et al., 2003).

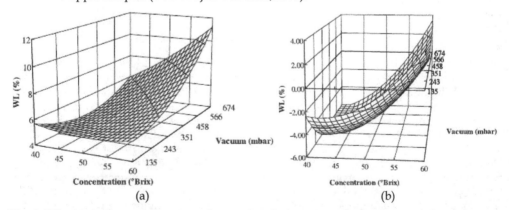

Fig. 8. Effects of vacuum pressure and osmotic solution concentration on water loss of mango and apples samples (from Mujica-Paz et al., 2003).

Derossi et al. (2010), studying the application of a vacuum acidification treatment on pepper slices, reported that pH ratio values (RpH) were lower when a pressure of 200 mbar was used in comparison with the results obtained applying a pressure of 400 mbar. Furthermore, some authors studied the effect of a decrease of vacuum pressure on the acidification of zucchini slices, showing that RpH values were directly correlated with vacuum pressure (Derossi et al., 2011). Nevertheless, the same authors did not observe a statistically significant variation of the porosity of the impregnated samples when vacuum level increased from 400 to 200 mbar. However, Derossi et al. (2010; 2011) showed that the use of a vacuum pressure significantly improved the rate of acidification in comparison with a traditional acidifying-dipping at atmospheric pressure. The authors attributed this result to the increase of acid-solution contact area due to the capillary impregnation. Hofmeister et al. (2005) studied the visual aspect of Mina cheese samples submitted to vacuum impregnation in different operative conditions. The authors reported a greater impregnation level when a pressure of 85.3 kPa was applied in comparison with the experiments performed at 80 kPa. In agreement with the theoretical principles this result was assumed as a consequence of a greater removal of air from the pores of cheese samples. Moreover, Andres et al. (2001) studied the effect of vacuum level on apple samples showing that impregnation level was affected by the applied vacuum pressure.

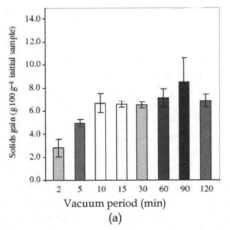 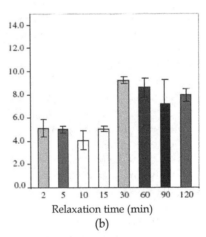

(a) (b)

Fig. 9. Solid gain values of apple cylinders submitted to vacuum osmotic dehydration at 40 mbar as a function of vacuum and relaxation times (from Silva Paes et al (2007).

Vacuum period and relaxation times represent two important variables affecting the results of the application of VI. Vacuum period (t_1) refers to the time during which food microstructure tends to reach an equilibrium situation after the application of the vacuum pulse. As previously reported, several phenomena such as deformation (enlargement) of capillaries, the expulsion of gases and native liquids from the pores and the partial impregnation of pores, simultaneously occur during t_1. Instead, relaxation time (t_2) is the period during which food structure moves toward an equilibrium situation after the restoration of atmospheric pressure. During t_2, capillary impregnation (by HDM) and deformation (compression) of pores occur under the action of a negative pressure gradient.

In general an increase of impregnation level of foods as a function of the time length of both t_1 and t_2 should be expected. Nevertheless due to the complex equilibrium occurring among the several phenomena involved the interpretation of the results is very difficult. Fito et al. (1996) and Salvatori et al. (1998) studied the response of several fruits (apple, mango, strawberry, kiwi, peach, banana, apricot) on both the X and γ values when submitted to vacuum impregnation at 50 mbar for a t_1 ranged between 5 and 15 minutes and a relaxation time ranged between 5 and 20 minutes. In both cases the authors did not observe significant differences of volume impregnated and solid matrix deformation as a function of t_1 and t_2 variation. Derossi et al. (2010) did not show a significant difference of total mass variation of pepper slices submitted to vacuum acidification at 200 mbar for a vacuum time of 2 and 5 minutes. On the contrary the results reported from Silva Paes et al (2007) stated a great influence of vacuum period and relaxation time on water gain (WG), solid gain (SG) and weight reduction (WR) of apple cylinders submitted to vacuum osmotic dehydration. As example figures 9a and 9b respectively show the solid gain of apple samples as a function of vacuum period and relaxation time.

It is worth nothing that SG values were directly correlated with an increase of t_1 in the range of 2 and 10 minutes. Instead, when a greater vacuum time was applied, no significant differences were observed. The authors suggested that when a short t_1 was applied, the removal of gases from the pores was not complete; so, their residues could have hindered the osmotic solution penetration. When vacuum period increased from 10 to 120 minutes, the complete removal of gases made the increase of t_1 unable to influence solid gain. In figure 9b it is possible to observe an increase of solid gain until a relaxation time of 30 minutes after that SG values are approximately constant. The authors hypothesized that 30 minutes is the time necessary to reach the equilibrium situation after which no differences were observed. Mujica-Paz et al. (2003a), studying the application of vacuum treatment to mango, papaya, peach and melon samples, showed that vacuum time directly affected the pore volume impregnated from external solution in a range of 3-25 minutes. Instead, the authors reported that vacuum time did not influence the X values for banana and apple samples. In the same way Salvatori (1997) reported that t_1 in a range of 5 and 15 minutes did not affect the impregnation level for mango and peach samples. Guillemin et al. (2008) studied the effect of sodium chloride concentration, sodium alginate concentration and vacuum time on weight reduction, chloride ions concentrations and pectinmethylesterase activity (PMEa) of apple cubes. The results showed that the effect of vacuum period was less significant and the more difficult to explain. Hofmeister et al. (2005) reported that an increase of vacuum time significantly affected the impregnation level of Mina cheese submitted to intermittent vacuum impregnation at 85.3 kPa. Chiralt et al. (2001), reviewing the scientific literature concerning the application of vacuum impregnation in salting process, reported that for meat fillets as longer the vacuum time as smaller the weight loss, in accordance with a greater pore volume impregnated from brine solution. Derossi et al. (2010) showed a significant positive effect of vacuum time on the pH reduction of pepper slices submitted to VI at a pressure of 200 mbar; instead a not clear behavior was observed for the experiments performed at 400 mbar. Moreover, in all cases the authors showed a positive effect of the increase of relaxation time on acidification level of the samples in a range of 10 and 30 minutes. Hironaka et al. (2011), studying the enrichment of whole potato with ascorbic acid through a vacuum impregnation treatment showed that an increase of vacuum time ranged between 0 and 60 minutes was effective in the improvement of ascorbic acid concentration of different varieties of whole potatoes. Apart the three above variables which are much strictly related to the vacuum process, the ensemble of the external solution characteristics with particular attention to its viscosity and temperature significantly affect the quality of vacuum impregnated foods. As reported from Barat et al. (2001), which studied the structural changes of apple tissues during vacuum osmotic dehydration, the impregnation level may be inhibited from a high viscosity of the hypertonic solution. The authors speculated that the relaxation of samples could occur when sample is taken out from the osmotic solution leading to an impregnation (regain) of gas if a liquid with very high viscosity is used. Also, the authors correlated the volume changes caused by capillary impregnation, $\Delta V^{LP\text{-}VI}$, and the volume changes due to compression-relaxation phenomena, ΔV^{CR}, as a function of viscosity and temperature of the osmotic solution (figure 10).

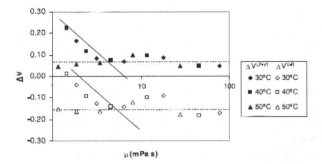

Fig. 10. Relationship between viscosity and temperature of osmotic solution and volume changes due to impregnation and compression-relaxation phenomena of apple samples submitted to vacuum osmotic dehydration (from Barat et al., 2001).

It is worth noting that for the experiments carried out at 40°C and 50°C the increase of viscosity did not show effects on impregnation and deformation. Instead, for samples treated at 30°C and for a viscosity of osmotic solution ranged between 0.5 and 3.5 mPa it is possible to observe that as viscosity increased as $\Delta V^{LP\text{-}VI}$ and ΔV^{CR} respectively decreased and raised. This behavior may be explained with a slow flow (because hindered from the high viscosity) of the osmotic solution into the pores; in this way more time for the relaxation process of the solid matrix is available which becomes the major contribution to the total volume changes of food. In line with this consideration, Barat et al. (2001) stated that: *"when viscosity is very low, impregnation tends to the theoretical value of a non-deformable porous matrix, and the deformation tends to zero"*. Mujica-Paz et a (2003), studying the effects of osmotic solution concentration on water loss of apple, melon and mango samples, reported similar results. In general they showed that WL values increased in line with an increase of hypertonic solution concentration due to the greater osmotic dehydration; nevertheless, apple samples showed negative value of WL (a gain of water) for solution concentration lower than 50° Brix. In this condition the penetration of osmotic solution probably was the most important phenomenon which produced an increase of water content; also, it is important to take into account that the experiments performed from the authors were constituted from a vacuum period and a relaxation time of 10 minutes that represent a very short osmotic treatments to generate a high dehydration. However, for the experiments performed with a solution concentration > 50°Brix, WL values were positive because the OS has difficulty to penetrate into the pores and the pressure gradient promoted the outflow of water from the apple microstructure. Similar effect of the osmotic solution concentration was observed from other authors (Martinez-Monzo et al., 1998; Silva Paes et al., 2007).

The temperature of solution directly affects the mass transfer of process like osmotic dehydration, acidification, brining, etc., but also affects the viscosity of external liquid and the viscoelastic properties of solid matrix of foods. The latter properties which are on of the most important internal variable will be reviewed in the following paragraph; in general, as foods are soft as higher is the contribution of relaxation phenomena on the total volume changes which, in turn reduces the impregnation level.

3.2 Internal variables

Vacuum impregnation is a technique strictly dependent from the characteristic of the food structure both at mesoscopic and microscopic scale (Fito et al., 1996; Salvatori et al., 1998; Barat et al., 2001; Fito et al., 2001; Chiralt & Fito, 2003; Giraldo et al., 2003; Gras et al., 2003; Mujica-Paz et al., 2003; Zhao & Xie, 2004; Derossi et al., 2010; Derossi et al., 2011) . At first, the porosity fraction of biological tissue is the most important parameter for the application of VI because it presents the void space potentially available for the influx of the external solution. In general, fruits and vegetables show a higher porosity fraction in comparison with meats, fishes and cheeses making them more suitable for the use of VI techniques. In table 1 the porosity fraction values of several foods are reported.

In general, as greater is the porosity fraction as greater is the impregnation level. Fito et al. (1996) showed that for apple, banana, apricot, strawberry and mushroom samples the impregnation level was in the same order of the effective porosity values of fresh vegetables. Gras et al. (2003) studied the calcium fortification of eggplants, carrots and oyster mushroom by VI treatments. The authors reported the maximum X values for eggplants and oyster mushrooms which showed, in comparison with carrots, the greater intercellular spaces respectively of 54%±1 and 41%±2. Fito et al. (1996), studying the HDM and DRP on apple, banana, apricot, strawberry and mushroom samples submitted to VI, showed that the impregnation levels were directly correlated with the porosity values of the vegetables.

However, the porosity value is a not sufficient index to completely characterize and to predict the behavior of food during vacuum impregnation. As reported from Zhao & Xie (2004), during VI three main phenomena are involved: gas outflows, deformation-relaxation of solid matrix and the liquid influx. Since these phenomena simultaneously occur, the result of VI is a consequence of the equilibrium among their kinetics which, in turn, are affected by (Fito et al. 1996):

- Tissue structure (pores and size distribution)
- Relaxation time of the solid matrix, a function of the viscoelastic properties of the material;
- The rate of HDM, a function of porosity, size and shape of capillaries, their connectivity, the viscosity of solution;
- Size and shape of samples.

Food (variety, type)	Porosity fraction (ε, %)	Reference
Apple (Granny Smith)	23.8±1.0	Salvatori et al. (1998)
Apple (Golden Delisious)	27.3±1.1	Mujica-Paz et al. (2003b)
Apple (Gala)	18.3	Silvia Paes et al. (2007)
Mango (Tommy Atkins)	9.9±1.3	Salvatori et al. (1998)
Strawberry (Chandler)	6.3±1.6	Salvatori et al. (1998)
Kiwi fruit (Hayward)	2.3±0.8	Salvatori et al. (1998)
Pear (Passa Crassana)	3.4±0.5	Salvatori et al. (1997)
Plum (President)	2±0.2	Salvatori et al. (1997)
Peach (Miraflores)	2.6±0.5	Salvatori et al. (1998)
Peach (Criollo)	4.6±0.2	Mujica-Paz et al. (2003a)
Apricot (Bulida)	2.2 ± 0.2	Salvatori et al. (1997)
Pienapple (Espanola Roja)	3.7±1.3	Salvatori et al. (1997)
Banana (Giant Cavendish)	~ 9	Fito et al. (1996)
Banana (Macho)	1.6±0.3	Mujica-Paz et al. (2003a)
Orange peel (Valencia Late)	21±0.04	Chafer et al. (2000)
Mandarina peel (Satsuma)	25±0.11	Chafer et al. (2001)
Eggplant (Saroya)	64.1±2	Gras et al. (2001)
Zucchini (Blanco Grise)	4.4±0.9	Gras et al. (2001)
Zucchini	8.39±2.22	Derossi et al. (2011)
Mushroom (Albidus)	35,9±1.9	Gras et al. (2001)
Oyster mushroom	16.1±4	Gras et al. (2001)
Mango (Manila)	15.2±0.1	Mujica-Paz et al. (2003b)
Melon (Reticulado)	13.3±0.6	Mujica-Paz et al. (2003b)
Carrot (Nantesa)	13.7±2	Gras et al. (2001)
Beetroot	4.3±1.3	Gras et al. (2001)
Cherry	~ 30	Vursavus et al (2006)
Persimonn (Rojo Brillante)	4.9±0.8	Igual et al. (2008)
Cheese (Manchego type)	~ 3	Chiralt et al. (2001)

Table 1. Porosity fraction values of several foods

Moreover, further internal variables such as the tortuosity of the internal pathway and the effect of temperature on the mechanical properties of solid matrix greatly affect the phenomena above reported. Furthermore, a synergistic (positive or negative) effects of the internal variables were reported from several authors. Among the above variables the mechanical properties of biological tissues are one of the most important, because HDM is based on the pressure gradient generated by both vacuum pressure externally imposed and the deformation-relaxation of solid matrix. Salvatori et al. (1998) studied the effects of VI on pore volume impregnated from external (osmotic) solution and on deformation phenomena of several fruits. Strawberries, even tough showed a greater porosity fraction (6.3%) in comparison with kiwi fruit and peach, reported a negligible impregnation level (X = 0.2). It was hypothesized that the kinetic of liquid penetration was longer than the relaxation time promoting the deformation (compression) of pores rather than impregnation phenomena. These differences in the rate of HDM and DRP phenomena could be attributed to the microscopic properties of the strawberry tissues such as high tortuosity of the internal pathways and/or size and shape of pores which hindered the influx of the external solution. On the other hand strawberries being a soft material could be characterized from a fast compression rate which reduced the liquid penetration. In agreement with these hypothesis the authors showed a high deformation (compression) index (γ = - 4.0). Also, in the same paper it was reported that mango and peach samples showed positive γ values at the end of relaxation time, stating that an enlargement of the capillaries was still observed when the equilibrium was reached. In this case a high rigidity of the vegetable tissues could have reduced the rate of compression phenomena and increased the liquid penetration. However, among the studied fruits, strawberries and kiwi fruits were those with a lower impregnated pore volume (respectively of 0.2% and 0.89%) and between them kiwi fruits showed the lowest porosity fraction of ~ 2.3%. Mujica-Paz et al. (2003a) studied the effect of vacuum pressure on the pore volume impregnated from an osmotic solution. Figures 11a and 11b respectively show the prediction of X values of apple and others fruit samples as a function of the factor (1 – 1/r). Since r is defined as the ratio between atmospheric pressure and work pressure the x-axis reports an index of pressure gradient intensity.

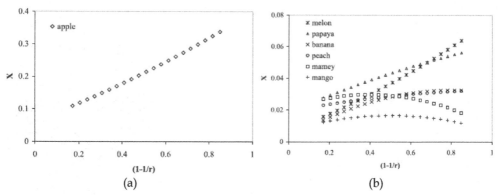

Fig. 11. Prediction of impregnation level value (X) as a function of (1-1/r) for some fruit samples submitted to VI for a vacuum time of 10 minutes and relaxation time of 25 minutes (From Mujica-Paz et al., 2002)

Mango, papaya, namey and peach samples showed a linear increase of X, stating that their vegetable tissues are subject to low deformation phenomena. Instead, banana, mango, namey, and peach showed a linear increase of X values until certain limits (about 400 mbar), after which impregnation level decreased. The authors supposed that these operative conditions could have produced a high vegetable tissue deformation (compression), which reduced the void phase available for liquid penetration. It was concluded that these fruits suffered of a great deformation when submitted to VI treatments in particular operative conditions. In their conclusion the authors highlighted the importance of the internal variables such as the number and the diameter of pores as well as the mechanical properties of solid matrix. Moreover, the spatial distribution of cells and their characteristics as well as the kind of fluid (liquid or gas) present in the intercellular space must be considered for the precise setting of vacuum impregnation treatments. Fito et al. (2001) deeply studied the vacuum impregnation and osmotic dehydration in matrix engineering. Among the obtained results, the authors reported that although orange peel showed a porosity fraction lower than eggplants, the fruit may be considered more suitable for VI treatments. This is because in orange peel the cells in flavedo zone are densely packed but the albedo zone shows large shape cells with a rupture of cell junctions that occurs during fruit ripening. This peculiar cell space distribution gives sponge-like properties with a high impregnation and swelling capacity. Taking into account the effect of the size of samples, it can be state that as thinner the food piece as greater the impregnation level because a greater void space is exposed to the surface. Gras et al. (2003) reported that carrot slices submitted to VI treatments showed an impregnation level of 5.8±0.3, 4.6±0.7 and 3.4±0.3 respectively for samples with a diameter of 10 mm, 15 mm and 25 mm. Derossi et al. (2011), studying the effect of a pulsed vacuum acidification treatment on zucchini slices in different operative conditions, showed that no significant differences were observed when the experiments were performed at 200 and 400 mbar in terms of pH reduction. It was hypothesized that the variability of vegetable microstructure coupled with the small diameter of the samples (~ 1.5 cm) favored the capillary impregnation reducing the effect of an increase of vacuum level. In accordance with this result the authors showed that no differences were observed in terms of porosity value reduction when fresh zucchini slices were submitted to VI at 200 mbar and 400 mbar. However the interaction of all these variables act simultaneously and could interact within them producing unexpected and/or complicated result. For instance, Chiralt et al. (1999) stated that samples with narrow pores treated with viscous liquid tend to show more deformation than impregnation effect.

4. Industrial applications

Back in the past, when HDM was proposed as an additional mass transfer mechanism occurring during osmotic dehydration, the first application of VI techniques had the aim to increase the rate of osmotic dehydration through the impregnation of capillaries with hypertonic solutions. Since then, vacuum osmotic dehydration (VOD) was subject to a great number of scientific papers; in addition to VOD, the application of an initial vacuum pulse followed by the restoration of atmospheric pressure and the maintaining of food immersed into the osmotic solution for a long time was proposed as pulsed vacuum osmotic dehydration (PVOD). In the last 10 years a great number of industrial applications have been studied (Javeri et al., 1991; Ponappa et al., 1993; Rastogi et al., 1996; Fito et al., 2001a; Fito et al., 2001b; Gonzalez-Martinez et al., 2002; Betoret et al., 2003; Giraldo et al., 2003; Gras et al., 2003; Mujica-Paz et al., 2003a ;Mujica-Paz et al., 2003b; Zhao & Xie, 2004; Hofmeister, et al., 2005; Corzo et al., 2007; Silva Paes et al., 2007; Igual et al., 2008; Cruz et al., 2009). In general it is possible to consider two principal aims: 1. The increase of mass transfer; 2. The introduction of chemical and biological compounds in food microstructure with the aim to improve their quality. On these basis, VI may be used as pre-treatment before drying or freezing, as innovative treatment inserted into a more complex processing with several industrial applications. In figure 12 is schematically reported some potential application of VI in food industry (Zhao & Xie, 2004).

4.1 VI as method to increase the rate of mass transfer of food during food processing

As known, several industrial processes, carried out with the aim to extend the shelf life or to improve/change their nutritional, functional or sensorial properties, are based on mass transfer toward foods or *vice versa*. If the purpose is to increase the rate of mass transfer between food and an external solution as in the cases of osmotic dehydration acidification treatments, vacuum impregnation significantly increases the rate of the processes due to HDM and the increase of liquid-product contact area. A large literature concerning the effect of VI on the rate of osmotic dehydration recognized the effectiveness of VI (Shi & Fito, 1994; Fito et al., 1996; Rastogi et al., 1996; Tapia et al., 1999; Cunha et al., 2001;Moreno et al., 2004). Shi & Fito (1994) reported that water loss of apricot samples occurred much faster when vacuum was applied during the experiments. Shi et al. (1995) showed that osmotic dehydration under vacuum promotes a faster water diffusion when apricots, pineapples and strawberries were treated. Rastogi et al. (1996), studying the kinetics of osmotic dehydration of apple and coconut samples showed that the estimated rates were significantly higher in vacuum conditions (235 mbar). Fito et al. (2001b) stated that VI promotes the effective diffusion in the fruit liquid phase when impregnated with low viscosity solution.

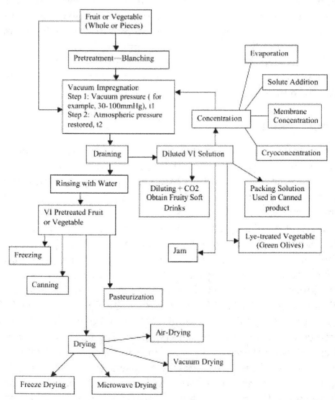

Fig. 12. Potential application of vacuum impregnation techniques in food industry (from Zhao & Xie, 2004)

Figure 13 shows the estimated diffusion coefficient values (De) of different type of fruits submitted to traditional osmotic dehydration and pulsed osmotic dehydration. It is possible to observe higher *De* values in the case of VI for vegetables which have high porosity values stating the effectiveness of the treatment as a consequence of the increase of mass transfer. Fito et al. (1994) found that water loss and solid gain obtained with a PVOD treatment performed at 70 mbar with a t_1 of 5 minutes were greater than those obtained through a traditional OD. Mujica-Paz et al. (2003a) performed osmotic dehydration treatments under vacuum (135-674 mbar, $t_1=t_2=10$ min) and at osmotic pressure measuring water (t = 20 min) activity depression (Da_w, %) for apple, melon and mango samples. The authors showed that in all cases the experiment carried out at sub-atmospheric conditions allowed to obtain the greater Da_w values stating that water loss and solid gain (the two main phenomena involved in the water activity depression) were faster than at atmospheric pressure. Giraldo et al. (2003) studied the kinetic of OD and PVOD treatments of mango samples performed with different osmotic solutions concentrations (35 – 65 °Bx). The authors, using a simplified Fickian equation, estimated De values of $5.9*10^{-10}$ m^2/s, $6.4*10^{-10}$ m^2/s and $9.7*10^{-10}$ m^2/s for PVOD performed with osmotic solution concentration respectively of 65°Bx, 55°Bx and 35°Bx; in the same conditions were estimated De values of $1.8*10^{-10}$ m^2/s, $5.9*10^{-10}$ m^2/s and $5.9*10^{-10}$ m^2/s for OD experiments. These results stated that in PVOD diffusion in the tissues is promoted, due to the removal of gas in the intercellular space and the introduction of hypertonic solution. Moreover, the application of VI as substituted of traditional brining of fishes, meats and cheeses or acidifying-dipping showed to be effective for the increase of mass transfer. Gonzales-Martinez et al. (2002) studied three type of brining methods applied to Machengo type cheese: at atmospheric pressure (BI), at vacuum pressure (VI) and with an initial vacuum pulse of 30 minutes (PVI). The authors estimated diffusion coefficients by using diffusional mass transfer mechanism. They obtained in the upper part of the samples D_e values of $4.4*10^{-10}$ m^2/s, $6.1*10^{-10}$ m^2/s and $9.5*10^{-10}$ m^2/s respectively for BI (traditional), PVI and VI treatments. Also, similar results were obtained for the lower part of the samples. In their conclusion the authors stated that BI in vacuum condition greatly improved the kinetics of salting process as a consequence of hydrodynamic mechanism. Derossi et al. (2010), studying the application of an innovative vacuum acidification treatment (VA) on pepper slices, reported RpH values of 0.968, 0.929 and 0.894 for a traditional acidifying-dipping, a VA performed at 400 mbar (t_1 = 5 min, t_2 = 30 min) and a VA performed at 200 mbar ($t_1=5$ min, $t_2=30$ min) respectively. In agreement with the theoretical principles of VI the authors concluded that VA is able to increase the rate of pH reduction as a consequence of the capillary impregnation which improve the acid solution-product contact area.

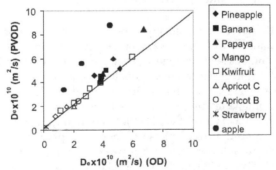

Fig. 13. Comparison of De values obtained from traditional osmotic dehydration (OD) and from pulsed vacuum osmotic dehydration (PVOD). Results obtained from several fruits (From Fito et al., 2001b)

Also Derossi et al. (2001) studied the kinetics of pulsed vacuum acidification treatments (PVA) of zucchini slices. The authors reported that an increase of vacuum did not affect the rate of pH reduction but in all cases the treatments were faster than acidification treatments performed at atmospheric pressure.

4.2 VI as method to improve food quality

As previous reported vacuum impregnation may be considered a technique by which is possible to introduce, in a controlled way, an external liquid in the void phase of foods. The term "external liquid" refers to a liquid obtained by dissolving in water any added chemical components. So, it is clear that VI may be used to introduce several compounds such as anti-browning agents, firming agents, nutritional compounds, functional ingredients, anti-microbial agents, anti-freezing, enzymes, etc. with several aims such as to prolong shelf life, to enrich fresh food with nutritional and/or functional substances, to obtain innovative food formulations as pre-treatments before drying or freezing, etc.

As well known, drying process performed with hot air, microwave, etc., is characterized by high energy consumption hence, high expenses; in this way, the possibility to reduce the initial amount of water in fresh foods (free water) by different techniques would allow to significantly reduce the cost of the process. Several papers showed the effectiveness of a traditional osmotic dehydration before air drying in the reduction of energy consumption. Instead, through a vacuum impregnation treatment applied before air drying two goals may be simultaneously obtained: 1. The reduction of energy costs; 2. The introduction of solutes such as antimicrobial, antibrowning and antioxidant to improve the final quality of dried foods (Sapers et al., 1990; Torreggiani, 1995; Barat et al., 2001b). Fito et al. (2001b) reported an improvement of the drying behavior of several fruits and vegetables submitted to VI pretreatments. Several authors reported that the stability of pigments was enhanced without the use of common compounds for color preservation when VI is applied before drying (Maltini et al., 1991; Torreggiani, 1995). Prothon et al. (2001) reported that water diffusivity of samples was increased when a pre-VI treatment was carried out. On the contrary, the same authors, studying the rehydration capacity of the samples, showed that this property was greater for non-treated samples, probably because both impregnation and DRP reduced the pores in which water flows during rehydration.

Among the stabilization processes, freezing is one of the most important because the use of low temperatures allows to better retain nutritional compounds in comparison with others traditional techniques such as dehydration, pasteurization/sterilization treatments, etc. However the formation of ice crystals leads to several physical damage and drip loss during thawing. Moreover, fluctuation of temperature along the cold chain may promote recrystallization phenomena leading to changes in size and shape of ice as well as their orientation (Zhao & Xie, 2004; Cruz et al., 2009). Vacuum impregnation was shown to be a useful method to incorporate in void phase of foods cryoprotectants and cryostabilizers, such as hypertonic sugar solution, antifreeze protein (AFP), high methoxyl pectin, etc. Martinez-Monzo et al. (1998) showed that as higher the concentration of sugar solution used during VI treatments as lower the freezable water content in food which reduces the drip loss during thawing. The same authors studied the potential application of VI treatment to introduce concentrate grape musts and pectin solution with the aim to decrease the damages of freezing on apple samples. It was observed that the both aqueous solutions were effective in the reduction of drip loss during thawing. In particular, grape must was able to reduce the freezable water content; instead pectin solution increased glassy transition temperature of liquid phase. Cruz et al. (2009) carried out several experiments to evaluate the potential application of vacuum impregnation with antifreeze protein (AFP) with the aim to preserve the quality of watercress leaves. Results stated that samples treated with AFP showed small ice crystals in comparison with water impregnated samples, due to the ability of the proteins to bound water molecules modifying the natural ice crystal formation. Figures 14a,14b and 14c show the results of wilting test on watercress leaves. It can be observed that samples impregnated with AFP showed a similar turgidity to the raw samples and higher than the control (water vacuum impregnated samples). This is in agreement with the formation of small ice crystals that during the thawing at room temperature led a reduction of cellular damage. Moreover, it is worth noting that this behavior could produce a better retention of nutrients traditionally lost during thawing. Ralfs et al. (2003) showed that carrot tissues vacuum impregnated with AFP had a higher mechanical strength in comparison with samples submitted to VI with ultrapure water.

(a)   (b)   (c)

Fig. 14. Wilting test of watercress leaves. (a) raw samples; (b) vacuum impregnated with water; (c) vacuum impregnated with AFP.

Xie and Zhao (2003) showed that vacuum impregnation of strawberries and marionberries with cryoprotectan solutions (HCFS and high methyl pectin) enriched with 7.5% of calcium gluconal highly enhanced the texture and the reduction of drip loss on frozen-thawed samples. In particular the authors reported an increase of compression force in a range of 50-100% and a reduction of drip loss of 20-50% in comparison with untreated samples. According to literature, the authors attributed these results to the reduction of freezable water content. Furthermore, it is worth noting that the reduction of water content as a consequence of VI treatment reduces energy consumption during freezing.

In the last years, on the basis of the consumer interest on the relation between the assumption of correct diet and its health benefits, the possibility to obtain foods whit high nutritional and functional properties exponentially increased the efforts of food scientists and industries in this research field. In this way, vacuum impregnation is a useful techniques to fill the void phase of foods with nutritional and functional ingredients. However, the scientific results concerning this application of VI are still few. Fito et al. (2001a) were the first scientists that evaluated the feasibility of VI to obtain innovative fresh functional foods (FFF). During their experiments the authors studied the impregnation of several vegetables with calcium and iron salt solution taking into account the solubility of these salts in water. In particular, the experiments had the aim to estimate the possibility to obtain fortified samples containing the 25% of the recommended daily intake (RDI) for each specific salts. As example, Figure 15 reports the weight of fruits that contain the 25% of RDI as a function of calcium lactate concentration.

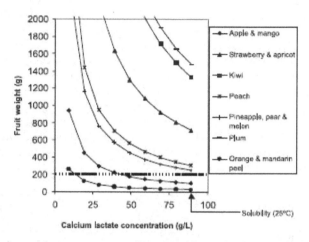

Fig. 15. Weight of some fruits containing the 25% of RDI of calcium salts.

The trends shown in the figure were estimated taking into account the porosity fraction of each type of fruit and by assuming the void phase completely impregnated with the external solution. For instance, by using a calcium lactate solution with a concentration of 50 g/L, less than 100g of orange peel should be consumed to assume the 25% of RDI. In the same way, about 600g and 1200g of peach and strawberry or apricot should be, respectively, assumed. Gras et al. (2003) studied the application of VI with the purpose to obtain eggplant, carrot and oyster mushroom fortified with calcium salts. The authors reported that eggplants and oyster mushroom may be considered as more appropriate to obtain FFF due to their high porosity in comparison with carrots. Hirinoka et al. (2011) studied the application of VI for the enrichment of whole potato with ascorbic acid. Also the authors compared the AA content of samples submitted to VI with the untreated one after a steam cooking treatment over boiling water for 25 minutes and during storage at 5°C for two weeks. Figures 16a, 16b, 16c and 16c show the visual aspect of potato samples submitted to VI and immersed at atmospheric pressure in red ink solution. The figures clearly show that the immersion of potato in solution at atmospheric pressure did not allow to impregnate capillaries of potatoes and the same results were obtained with a VI without apply any restoration time ($t_2 = 0$). Instead after a restoration time of 3 h a significant impregnation was observed.

As expected ascorbic acid content significantly decrease (about 42%) after steam cooking. Nevertheless, VI-cooked samples had an acid ascorbic concentration 22 time higher than samples only submitted to steam cooking (raw-cooking). Also, the VI-cooking samples showed an AA content of ~ 100 mg/100g which is twice of the FAO value (45mg/100g) and close to the RDA value. Furthermore, in another series of experiments Xie & Zhao (2003) used VI to enrich apple, strawberry and marionberry with calcium and zinc. The experiments performed with high corn syrup solution enriched with calcium and zinc showed that a 15-20% of RDI of calcium more than 40% RDI of zinc could be obtained in 200g of impregnated apple fresh-cut samples.

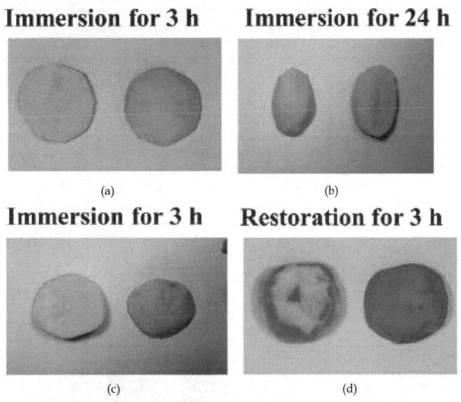

Immersion for 3 h   Immersion for 24 h

(a)   (b)

Immersion for 3 h   Restoration for 3 h

(c)   (d)

Fig. 16. Potato samples immersed in red ink solution without vacuum (a,b), after a vacuum time of 3 h (without restoration time) and after a restoration time of 3 h (From Hironaka et al., 2011).

Figure 17 reports the ascorbic acid content of whole potato submitted to VI and cooked over boiling water for 25 minutes and the controls (un-VI samples cooked).

Vacuum impregnation could be a method to produce a numerous series of innovative probiotic foods. For instance, Betoret et al. (2003) studied the use of VI to obtain probiotic enriched dried fruits. The authors performed VI treatments on apple samples by using apple juice and whole milk containing respectively *Saccharomyces cerevisiae* and *Lactobacillus casei* (*spp. Rhamnosus*) with a concentration of 10^7–10^8 cfu/ml. Results allowed to state that, combining VI and low temperature air dehydration, it was possible to obtain dried apples with a microbial content of 10^6–10^7 cfu/g. However, despite the wide number of the potential industrial application, shelf life extension is one of the most important. So, due to its unique advantage vacuum impregnation may be considered a useful methods to introduce inhibitors for microbial growth and/or chemical degradation reactions; nevertheless, the scientific literature concerning the application of VI in this field of research is still poor. Tapia et al. (1999) used a complex solution containing sucrose (40°Bx), phosphoric acid (0.6% w/w), potassium sorbate (100 ppm) and calcium lactate (0.2%) to increase the shelf life of melon samples. Results showed that foods packed in glass jars and covered with syrup maintained a good acceptance for 15 days at 25°C. Welty-Chanes et al. (1998), studying the feasibility of VI for the production of minimally processed oranges reported that the samples were microbiologically stable and showed good sensorial properties for 50 days when stored at temperature lower than 25°C. Derossi et al. (2010) and Derossi et al. (2011) proposed an innovative vacuum acidification (VA) and pulsed vacuum acidification (PVA) to improve the pH reduction of vegetable, with the aim to assure the inhibition of the out-grow of *Clostridium botulinum* spores in the production of canned food. The results stated the possibility to obtain a fast reduction of pH without the use of high temperature of acid solution as in the case of acidifying-blanching. However, the authors reported the effect of VI on visual aspect of vegetable that need to be considered for the industrial application, because the compression-deformation phenomena could reduce the consumer acceptability. Guillemin et al. (2008) showed the effectiveness of VI for the introduction of pectinmethylesterase which enhances fruit firmness.

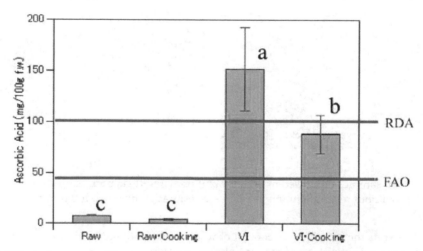

Fig. 17. Effect of steam cooking on ascorbic acid content of whole potato submitted to vacuum impregnation. VI solution: 10% AA, p = 70 cm Hg, t1=1h, t2= 3 h)

5. Conclusion

Although vacuum impregnation was for the first time proposed at least 20 years ago, it may be still considered an emerging technology with high potential applications. Due to its unique characteristics, VI is the first food processing based on the exploitation of three dimensional food microstructure. It is performed by immerging food in an external solution and applying a vacuum pressure (p) for a time (t₁). Then, the restoration of atmospheric pressure maintaining the foods into the solution for a relaxation time (t_2) allows to complete the process. During these steps three main phenomena occurs: the out-flow of native liquid and gases from the pores; the influx of external solution inside capillaries; deformation-relaxation of solid matrix. The influx of external liquid occurs under the action of a pressure gradient between the pores and the pressure externally imposed; this is known as hydrodynamic mechanisms (HDM). However, on the basis of its nature, VI is a very complex treatment and its results are affected from several external and internal variables. The former are the operative conditions above reported coupled with the temperature and viscosity of external solution. The latter are characterized from the microscopic and mesoscopic properties of food architecture such as length and diameter of pores, their shapes, the tortuosity of internal pathways, the mechanical (viscoelastic) properties of biological tissues, the high or low presence of gas and/or liquid inside capillaries, etc. VI has shown to be very effective in a wide number of industrial applications. The impregnation, causing a significant increase of the external solution/product contact area, is an important method to increase the mass transfer of several solid-liquid operation such as osmotic dehydration, acidification, brining of fish and meat products, etc. VI may be used as pretreatment before drying or freezing, improving the quality of final product and reducing cost operations due to the removal of native liquid (water) from the pores. Furthermore, the possibility to introduce, in a controlled way, an external solution enriched with any type of components catch light on a high number of pubblications. Indeed, VI has been used to extend shelf life, to produce fresh fortified food (FFF), to enrich food with nutritional/functional ingredients, to reduce the freezing damage, to obtain foods with innovative sensorial properties, to reduce oxidative reaction, to reduce browning, etc. Furthermore, from an engineering point of view some advantages may be considered: 1. it is a fast process (usually it is completed in few minutes); it needs low energy costs; it is performed at room temperature; the external solution may be reused many times. Nevertheless, the applications of VI at industrial scale are still poor. This problem may be attributed to the lack of industrial plants in which it is possible to precise control the operative conditions during the two steps of the process. Also, some technical problems need to be solved. For instance, as reported from Zhao & Xie (2004), the complete immersion of foods into the external solution is a challenge for the correct application of VI. Often, fruits and vegetables tend to float due to their low density in comparison with external solution as in the case of osmotic solution. The current VI is applied by stirring solution with the aim to keep food pieces inside solution with the drawback of an increase of energy costs and possible damages of foods. Furthermore, the lack of information for industries on the advantage of these techniques reduces its application at industrial scale.

6. References

Aguilera, J.M. (2005). Why food microstructure?. *Journal of Food Engineering*, Vol. 67, pp. 3-11.

Andres, I., Salvatori, D., Chiralt, A., & Fito, P. (2001). Vacuum impregnation viability of some fruits and vegetables. In: Osmotic dehydration and vacuum impregnation,

P. Fito, A. Chiralt, Barat, J.M., Spiess, W.E.L., Behsnilian, D., Barat, J.M. (eds.). Behsnilian.

Atares, L., Chiralt, A., & Gonzales-Martinez C. (2008). Effect of solute on osmotic dehydration and rehydration of vacuum impregnated apple cylinders (cv. Granny Smith). *Journal of Food Engineering*, Vol. 89, pp. 49-56.

Barat, J.M., Chiralt, A., & Fito, P. (1998). Equilibrium in cellular food osmotic solution systems as related to structure. *Journal of Food Science*, Vol. 63, pp. 836-840.

Barat, J.M., Chiralt, A., & Fito, P. (2000). Structural changes kinetics in osmotic dehydration of apple tissue. In *Proceeding of the 21th International Drying Simposium* IDS2000, Paper No. 416, Amsterdam, Elsevier.

Barat, J.M., Chiralt, A., Fito, P. (2001b). Effect of osmotic solution concentration, temperature and vacuum impregnation pretreatment on omostic dehydration kinetics of apple slices. Food Science and Technology International, Vol. 7, pp. 451-456.

Barat, J.M., Fito, P., & Chiralt, A. (2001a). Modeling of simultaneous mass transfer and structural changes in fruit tissues. *Journal of Food Engineering*, Vol. 49, pp. 77-85.

Betoret, N., Puente, I., Diaz, M.J., Pagan, M.J., Garcia, M.J., Gras, M.L., Marto, J. & Fito, P. (2003). Development of probiotic-enriched dried fruits by vacuum impregnation. *Journal of Food Engineering*, Vol. , No. (2-3), pp. 273-277.

Chafer, M., Gonzales-Martinez, C., Ortola, M.D., Chiralt, A., & Fito, P. (2000). Osmotic dehydration of mandarin and oragn peel by using rectified grape must. Proceedings of the 12th international drying symposium. IDS 2000, Elsevier Science, Amsterdam, Paper n. 103.

Chafer, M., Ortola, M.D., Martinez-Monzo, J., Navarro, E., Chiralt, A., & Fito, P. (2001). Vacuum impregnation and osmotic dehydration on mandarin peel. Proceedings of the ICEF8. Lancaster: Technomic Publishing Company.

Chiralt, A., & Fito, P. (2003). Transport mechanisms in osmotic dehydration: the role of the structure. *Food Science Technology International*, 9 (3), 179-186.

Chiralt, A., Fito, P., Andres, A., Barat, J.M., Martinez-Monzo, J, & Martinez-Navarrete, N. (1999). Vacuum impregnation: a tool in minimally processing of foods. In: *Processing of Foods: Quality Optimization and Process Assessment.* Oliveira F.A.R., & Oliveira, J.C. (eds), Boca Raton: CRC press, pp. 341-356.

Chiralt, A., Fito, P., Barat, J.M., Andres, A., Gonzales-Martinez, C., Escrichr, I., & Camacho, M.M. (2001). Use of vacuum impregnation in food salting process. *Journal of Food Engineering*, Vol. 49, pp. 141-151.

Corzo, O., Brancho, N., Rodriguez, J., & Gonzales, M. (2007). Predicting the moisture and salt contents of sardine sheets during vacuum pulse osmotic dehydration. *Journal of Food Engineering*, Vol. 80, pp. 781-780.

Cruz, R.M.S., Vieira, M.C., Silva, C.L.M. (2009). The response of watercress (*Nasturtium officinale*) to vacuum impregnation: effect of and antifreeze protein type I. *Journal of Food Engineering*, Vol. 95, pp. 339-345.

Cunha, L.M., Oliveira, F.A.R., Aboim, A.P., frias, J.M., & Pinheiro-Torres, A. (2001). Stochastic approach to the modelling of water losses during osmotic dehydration and improved parameter estimation. *International Journal of Food Science and Technology*, Vol. 36, pp. 253-262.

Datta, A.K. (2007a). Porous media approaches to studying simultaneous heat and mass transfer in food processes. I: Problem formulations. *Journal of Food Engineering*, pp. 80-95.

Datta, A.K. (2007b). Porous media approaches to studying simultaneous heat and mass transfer in food processes. II: Property data and representative results. *Journal of Food Engineering*, Vol. 80, pp. 96-110.

Derossi, A., De Pilli, T., La Penna, M.P., & Severini, C. (2011). pH reduction and vegetable tissue structure changes of zucchini slices during pulsed vacuum acidification. *LWT- Food Science and Technology*, Vol. 44, pp. 1901-1907.

Derossi, A., De Pilli, T., Severini, C. (2010). Reduction in the pH of vegetables by vacuum impregnation: A study on pepper. *Journal of Food Engineering*, Vol. 99, pp. 9-15.

Fito, P. (1994). Modelling of vacuum osmotic dehydration of foods. Journal of Food Engineering, 22, 313-318.

Fito, P., & Chiralt, A. (1994). An Update on vacuum osmotic dehydration. In *Food Preservation by Moisture Control: Fundamentals and Application*, G.V. Barbosa-Canovas and J. Welti-Chanes, eds., pp. 351-372, Technomic Pub. Co., Lancaster, PA.

Fito, P., & Chiralt, A. (2003). Food Matrix Engineering: The Use of the water-structure-functionality ensemble in dried food Product development. *Food Science Technology International*, Vol. 9(3), pp. 151-156.

Fito, P., & Pastor, R. (1994). *Non-diffusional mechanism occurring during vacuum osmotic dehydration (VOD)*. Journal of Food Engineering, 21, 513-519.

Fito, P., Andres, A., Chiralt, A., & Pardo, P. (1996). *Coupling of Hydrodynamic mechanisms and Deformation-Relaxation Phenomena During Vacuum treatments in Solid Porous Food-Liquid Systems*. Journal of Food Engineering, 27, 229-241.

Fito, P., Andres, A., Pastor, R. & Chiralt, A. (1994). Modelling of vacuum osmotic dehydration of foods. In: *Process optimization and minimal processing of foods*, Singh, P., & Oliveira, F. (eds.), pp. 107-121, Boca Raton: CRC Press.

Fito, P., Chiralt, A., Barat, J.M., Andres, A., Martinez-Monzo, J., & Martinez-Navarrete, N. (2001b). Vacuum impregnation for development of new dehydrated products. *Journal of Food Engineering*, Vol. 49, pp. 297-302.

Fito, P., Chiralt, A., Betoret, M., Gras, M.C., Martinez-Monzo, J., Andres, A., & Vidal, D. (2001a). Vacuum impregnation and osmotic dehydration in matrix engineering. Application in functional fresh food development. *Journal of Food Engineering*, Vol. 49, pp. 175-183.

Giraldo, G., Talens, P., Fito, P., & Chiralt, A. (2003). Influence of sucrose solution concentration on kinetics and yield during osmotic dehydration of mango. *Journal of Food Engineering*, Vol. 58, pp. 33-43.

Gonzalez-Martinez, C., Chafer, M., Fito, P., Chiralt, A. (2002). Development of salt profiles on Machengo type cheese during brining. Influence of vacuum pressure. *Journal of Food Engineering*, Vol. 53. Pp. 67-73.

Gras, M., Vidal-Brotons, D., Betoret, N., Chiralt, A., & Fito, P. (2002). The response of some vegetables to vacuum impregnation. *Innovative Food Science and Emerging Technologies*, Vol. 3, pp. 263-269.

Gras, M.L., Fito, P., Vidal, D., Albors, A., Chiralt, A., & Andres, A. (2001). The effect of vacuum impregnation upon some properties of vegetables. *Proceedings of the ICEF8*. Technomic Publishing Company. Lancanster

Gras, M.L:, Vidal, D., Betoret, N., Chiralt, A., & Fito, P. (2003). Calcium fortification of vegetables by vacuum impregnation. Intercations with cellular matrix. *Journal of Food Engineering*, Vol. 56, pp. 279-284.

Guillemin, A., Degraeve, P, Noel, C., & Saurel, R. (2008). Influence on impregnation solution viscosity and osmolarity on solute uptake during vacuum impregnation of apple cubes (var. Granny Smith). *Journal of Food Engineering*, Vol. 86, pp. 475-483.

Halder, A., Dhall, A., & Datta, A.K. (2007). An improved, easily implementable porous media based model for deep-fat frying Part I: Model development and input parameters. *Food and Bioproducts Processing*.

Hironaka, K., Kikuchi, M., Koaze, H., Sato, T., Kojima, M., Yaamamoto, K., Yasuda, K., Mori, M., & Tsuda, M. (2011). Ascorbic acid enrichment of whole potato tuber by vacuum-impregnation. *Food Chemistry*, Vol 127, pp. 1114-1118.

Hofmeister, L.C., Souza, J.A.R., & Laurindo, J.B. (2005). Use of dye solutions to visualize different aspect of vacuum impregnation of Minas Cheese. *LWT – Food Science and Technology*, Vol. 38, pp. 379-386.

Igual, M., Castello, M.L., Ortola, M.D., & Andres, A. (2008). Influence of vacuum impregnation on respiration rate, mechanical and optical properties of cut persimmon. *Journal of Food Engineering*, Vol. 86, pp. 315-323.

Javeri, R.H., Toledo, R., & Wicker, L. (1991). Vacuum infusion of citrus pectinmethylesterase and calcium effects on firmness of peaches. *Journal of Food Science*, Vol. 56, pp. 739-742.

Lewis, M.J. (1993). *Propiedades fisicas de los alimentos y de los sistemas de procesado*. Ed. Acribia, Zaragoza, Espana.

Maltini, E., Pizzocaro, F., Torreggiani, D., & Bertolo, G. (1991). Effectiveness of antioxidant treatment in the preparation of sulfur free dehydrated apple cubes. In 8th World Congress: Food Science and Technology, Toronto, Canada, pp. 87-91.

Martinez-Monzo, J., Martinez Navarrete, N., Chiralt, A., & Fito, P. (1998). Mechanical and structural change in apple (var. Granny Smith) due to vacuum impregnation with cryoprotectans. *Journal of Food Science*, Vol. 63 (3), pp. 499-503.

Mebatsion, H.K., Verboven, P., Ho, Q.T., Verlinden, B.E., & Nicolai, B.M. (2008). Modelling fruit (micro)structures, why and how?. Trends in Food Science & Technolgy, 19, 59-66.

Moreno, J., Bugueno, G., Velasco, V., Petzold, G., & Tabilo-Munizaga, G. (2004). Osmotic dehydration and vacuum imprengation on physicochemical properties of Chilean Papaya (Carica candamarcensis). *Journal of Food Science*, Vol. 69, pp. 102-106.

Mujica-Paz, H., Valdez-Fragoso, A., Lopez- Malom A., Palou, E., & Welti-Chanes, J. (2002). Impregnation properties of some frutis at vacuum pressure. Journal of Food Engineering, Vol.56, pp. 307-314.

Mujica-Paz, H., Valdez-Fragoso, A., Lopez-Malo, A., palou, E., & Welti-Chanes (2003). Impregnation and osmotic dehydration of some fruits: effect of the vacuum pressure and syrup concentration. *Journal of Food Engineering*, Vol. 57, pp. 305-314.

Ponappa, T., Scheerens, J.C., & Miller, A.R. (1993). Vacuum infiltration of polyamines increases firmness of strawberry slices under various storage conditions. *Journal of Food Science*, Vol. 58, pp. 361-364.

Prothon, F., Ahrme, L.M., Funebo, T., Kidman, S., Langton, M., & Sjoholm, I. (2001). Effects of combined osmotic and microwave dehydration of apple on texture, microstructure and rehydration characteristics. *Lebensmittel-Wissenschaft und technologie*, Vol. 34, pp. 95-101.

Ralfs, J.D., Sidebottom, C.M., Ormerod, A.P. (2003). Antifreeze proteins in 444 vegetables. World intellectual property organization, patent WO 03/055320 A1, pp. 1-8.

Rastogi, N.K., & Raghavarao, K.S.M.S. (1996). Kinetics of Osmotic dehydration under vacuum. *Lebensm.-Wiss. U.-Technol.*, Vol. 29, pp. 669-672.

Salvatori, D. (1997). Osmotic dehydration of fruits: Compositional and structural changes at moderate temperatures. Ph.D. Thesis.

Salvatori, D., Andres, A., Chiralt, A., & Fito, P. (1998). The response of some properties of fruits to vacuum impregnation. *Journal of Food Process Engineering*, Vol. 21, pp. 59-73.

Sapers, G.M., Garzarella, L., & Pilizota, V. (1990). Application of browning inhibitors to cut apple and potato by vacuum and pressure infiltration. Journal of Food Science, Vol. 55, pp. 1049-1053.

Shi, X.Q., & Fito, M.P. (1994). Mass transfer in vacuum osmotic dehydration of fruits: A mathematical model approach. *Lebensm.-Wiss.-u.-Technol*, Vol. 27, pp. 67-72.

Shi, Z.Q., Fito, P., Chiralt, A. (1995). Influence of vacuum treatments on mass transfer during osmotic dehydration of fruits. *Food Research International*, Vol. 21, pp. 59-73.

Tapia, M.S., Ranirez, M.R., Castanon, X., & Lopez-Malo, A. (1999) Stability of minimally treated melon (*Cucumis melon, L.*) during storage and effect of the water activity depression treatment. No. 22D-13. *Presented at 1999 IFT annual meeting*, Chicago, IL.

Torquato, S. (2000). Modeling of physical properties of composite materials. *International Journal of Solids and Structures*, Vol. 37, pp. 411-422.

Torregiani, D. (1995). Technological aspect of osmotic dehydration in foods. In: Food Preservation by moisture control. Fundamentals and Applications, Barbosa-Canovas, G.V., & Welti-Chanes, J. (eds.), Lancaster: Technomic Publisher Co. Inc., pp 281-304.

Vursavus, K., Kelebek, H., & Selli, S. (2006). A study on some chemical and physico-mechanic properties of three sweet cherry varieties (Prunus avium L.). Journal of Food Engineering, Vol. 74, pp. 568-575.

Welti-Chanes, J., santacruz, C., Lopez-Malo, A., & Wesche-Ebeling, P. (1998). Stability of minimally processed orange segments obtained by vacuum dehydration techniques. *No. 34B-8. Presented at 1998 IFT annual meeting*, Atlanta, GA.

Xie, J., & Zhao, Y. (2003). Improvement of physicochemical and nutritional qualities of frozen Marionberry and by vacuum impregnation pretreatment with cryoprotectants and minerals. *Journal of Horticultural Science and Biotechnology*, Vol. 78, pp. 248-253.

Zao, Y., & Xie, J. (2004). Practical applications of vacuum impregnation in fruit and vegetable processing. *Trend in Food Science & Technology*, Vol. 15, pp. 434-451.

A Discussion Paper on Challenges and Proposals for Advanced Treatments for Potabilization of Wastewater in the Food Industry

D. B. Luiz[1,2], H. J. José[1] and R. F. P. M. Moreira[1]
[1]*Federal University of Santa Catarina*
[2]*Embrapa Fisheries and Aquaculture*
Brazil

1. Introduction

The World Commission on Environment and Development (WCED, 1987, apud Burkhard et al., 2000) defines sustainable development as "development that meets the needs of the present without compromising the ability of future generations to meet their own needs". Sustainable production is a realistic dilemma in food industries, especially in slaughterhouses and the meat processing industry due to the many types of processes involved (Oliver et al., 2008; Casani et al., 2005).

Unfortunately, in many cases the implementation of environmental actions for industrial purposes depends on large-scale economic incentives (Cornel et al., 2011). Concomitantly, the increased costs of fresh water, wastewater treatment and waste disposal are important economic incentives for wastewater reuse and biomass-to-energy actions in industries. Hence, environmental, financial and social sustainability must be achieved together. Since pollution is predominantly due to human behavior, social sustainability requires the commitment of managers, workers and consumers (Burkhard et al., 2000).

Water scarcity is a reality in many world regions. Water pollution and overexploitation, climate change, urbanization, industrialization and increases in the world population and, consequently, food consumption and production are the main factors aggravating the global fresh water crisis (EPA, 2004; Clevelario et al., 2005; Khan et al., 2009; Luiz et al., 2009, 2011). Hence, sustainability in water and wastewater management is a current requirement of industries in order to promote the minimization of fresh water consumption and reduction of wastewater production preserving high-quality groundwaters (Cornel et al., 2011).

Concerning the water consumption in food industries, this input is mostly used for cleaning and disinfection, cooling and heating (Oliver et al., 2008), and also for processing food for human consumption and for sanitary uses. International organizations (e.g. Codex Alimentarius, 2001, 2007) have recognized and stimulated the implementation of direct and indirect wastewater reuse and indirect potable reuse with techniques that take into account hygienic concerns (Casani et al., 2005) to avoid risking adverse effects on the product integrity, the environment and the health of workers and customers. However, these procedures will be universal in the near future.

Currently, conventional wastewater techniques are not satisfactory to produce reuse wastewater with drinking water quality, further procedures being necessary. Advanced oxidation processes (AOPs) are the most suitable tertiary treatment. AOPs are based on chemical oxidation and can degraded any kind of organic matter with highly reactive and non-selective radicals, mainly hydroxyl radicals ($\bullet OH$) (Oller et al., 2010; Luiz et al., 2010). AOPs can involve a combination of strong oxidants (e.g. O_3 and H_2O_2), ultraviolet (UV), semiconductor catalysts (e.g. TiO_2 and ZnO) and ultrasound (15 kHz to 1 MHz), and the most common examples are: H_2O_2/UV, H_2O_2/O_3, $H_2O_2/O_3/UV$, O_3/UV, TiO_2/UV, $H_2O_2/catalyst$, Fenton (Matilainen and Sillanpää, 2010).

There are many studies and reviews on advanced processes to treat secondary wastewater from industries which also address the challenges and limitations to water reuse and the hygienic concerns especially in food industries. However, the challenges associated with treating a secondary wastewater to provide drinking water quality become clear when real wastewater is treated in batch and in pilot scale. The limitations are several, particularly in food industries. Hence, this review aims to present an overview and a discussion regarding the challenges and proposals related to the advanced treatment of food industry wastewater to provide drinking water quality, elucidating some experimental and theoretical questions surrounding such processes. Valuable experimental advice which is not usually found in research papers will be provided, for instance, which oxidation treatment should be chosen, which initial experiments should be carried, and which methodology should be followed to evaluate the kinetics constants in different situations will be provided, along with methodology issues. Some previous papers by the authors will be discussed and also some unpublished data on experiences in slaughterhouses and the meat processing industry. The oxidation treatments discussed are ozonation and many AOPs; ultraviolet treatment will also be addressed.

Due to the wide variety of processes and food products, food-processing wastewater can be a complex mixture, and this must be taken into account when considering the recycling, reuse, reconditioning for recycling or reuse, treatment or disposal of water. It is known that the segregation of wastewater streams in at an industrial plant – separating the effluent of each process – or at least the combining of the most similar streams in terms of physical-chemical and microbiological characteristics, enables an optimal treatment for each type of wastewater, energy savings, greater efficiency and lower cost of disposal and reuse. Nevertheless, in most plants the wastewater from all process – including from the toilets – are collected together, making the treatment difficult and costly. Hence, ascertaining the basic nature of food industry wastewater and its variability is the first challenge. This information is important to selecting the best combination of processes through which to treat the effluent.

Besides the need to remove minerals (by filtration) and organic matter, it is also necessary eliminate nitrogen compounds, especially in the dairy and meat industries. Advanced oxidation processes can remove simultaneously organic matter and nitrate; however, there are many parameters that should be taken into account, e.g., concentration of organic compounds and oxygen-free media. Some industrial plants also need to remove a specific recalcitrant organic compound which can be oxidized by advanced treatments; however, the effectiveness of the treatment is mainly dependent on whether the oxidant is selective or non-selective, the presence of oxidant scavengers and the dosage of oxidant.

Given that the growing demand for water, limited access to water in some regions and increased concern regarding the environmental impact of industrial activities on the environment are aspects driving the research on and implementation of water reuse in industrial plants, the theme discussed in this review will be of interest to several fields of chemistry, as well as food, chemical and environmental engineering.

2. UV radiation

UV radiation can degrade organic compounds via two routes: direct photolysis and photo-oxidation via radical generation. In the first route, direct photolysis or photodegradation, the efficiency is directly related to the ability of the target organic compound to absorb UV radiation at the wavelength used (λ) (Beltran et al., 1993). The UV absorption leads to direct excitation and the breakdown of organic pollutants (Rincón; Pulgarin, 2006). Lau et al. (2007) leading to the formation of excited radicals (R•) by UV excitation of organic compounds (RH). These radicals can be converted into stable molecules (dimers) via a dimerization process (Equations 1-6) which is favored by the presence of O_2 in water. If the free radical chain is interfered with (termination reactions, Equations 4-6), UV treatment may be less effective (Lau et al., 2007).

Initiation:

$$RH \xrightarrow{h\upsilon} R^{\bullet} + H^{\bullet} \tag{1}$$

Propagation:

$$R^{\bullet} + O_2 \longrightarrow ROO^{\bullet} \tag{2}$$

$$RH + ROO^{\bullet} \longrightarrow ROOH + R^{\bullet} \tag{3}$$

Termination:

$$ROO^{\bullet} + ROO^{\bullet} \longrightarrow ROOR + O_2 \tag{4}$$

$$ROO^{\bullet} + R^{\bullet} \longrightarrow ROOR \tag{5}$$

$$R^{\bullet} + R^{\bullet} \xrightarrow{h\upsilon} RR \tag{6}$$

According to Beltran-Heredia et al. (2001), for pH <7.0, the rate of photodecomposition increases with increasing pH, which may be due to an increase in the generation of free radicals (R•) (Equations 1-3).

The second route is photo-oxidation via oxidative processes by radical generation. When natural water matrices are used, the presence of nitrate, iron (III) and/or organic matter can provide •OH due to photo-oxidation of these compounds by UV or other AOP in combination with UV, such as UV/H_2O_2, UV/O_3, UV/H_2O_2/O_3.

Nitrate ions absorb UV radiation by acting as an internal filter to UV light and, in parallel, can form •OH through the mechanism detailed below (Equations 7-11) (Neamtu and Frimmel, 2006).

$$NO_3^- \xrightarrow{h\nu} NO_2^- + O \tag{7}$$

$$NO_3^- \xrightarrow{h\nu} NO_2^\bullet + O^- \bullet \tag{8}$$

$$2NO_2^\bullet + H_2O \longrightarrow NO_2^- + NO_3^- + 2H^+ \tag{9}$$

$$O + H_2O \longrightarrow 2 \bullet OH \tag{10}$$

$$O^- \bullet + H_2O \longrightarrow \bullet OH + HO^- \tag{11}$$

For wavelengths longer than 300 nm, complex aqueous ferric iron $Fe(OH)_2^+$, $Fe(OH)_4^-$, $Fe(OH)^{2+}$, Fe^{3+}, $Fe_2(OH)_2^{4+}$) can generate ferrous ions and •OH radicals by photolysis through the internal transfer of electrons (Equation 12) (Espinoza et al., 2007).

$$Fe(OH)^{2+} \xrightarrow{h\nu} Fe^{2+} + HO^\bullet \tag{12}$$

Dissolved organic matter absorbs UV light and can generate reactive radicals such as singlet oxygen, superoxide anions ($\bullet O_2^-$), •OH radicals and peroxyl radicals (ROO•). These reactive transients can degrade organic pollutants following different paths, and consequently they are not degraded only by direct photolysis (Chin et al. 2004; Neamtu and Frimmel, 2006).

The UV radiation is also used for disinfection, especially of reuse water. Ultraviolet radiation is bactericidal because it causes damage to the nucleic acids (DNA and RNA) of microorganisms (bacteria and viruses), inactivating them. Consequently, UV radiation prevents micro-organisms from multiplying because the photoproducts formed from nucleic acids (e.g., pyrimidine dimers) inhibit replication and transcription (Achilleos et al., 2005). The peak absorbance of nucleic acid is around 260 nm, but below 230 nm absorbance is also high. UV lamps, most of which emitted light at 254 nm (low pressure mercury lamps) are commonly used in studies on the disinfection of aqueous matrices, but polychromatic UV lamps can also be effective in inactivating certain microorganisms (Hijnen et al., 2006).

The use of UV disinfection is increasing mainly due to the fact that it is less expensive than chlorine disinfection, safer than chlorine gas, does not form organochlorine and is effective against *Cryptosporidium* and *Giardia*, while chlorine is not. Additionly, UV disinfection does not form by-products or residual toxicity, and has a low regrowth of bacteria. However, UV radiation requires water with low turbidity to allow the radiation to be effective (Achilleos et al., 2005).

As solar radiation is composed of approximately 4% of UVA/UVB (295-400 nm), besides visible light (400-800 nm) and infrared radiation, this natural radiation has been used for disinfection, color reduction and dissolved organic matter removal in surface waters (Kulovaara et al., 1995, Martín-Domínguez et al., 2005, Davies et al., 2009).

A Discussion Paper on Challenges and Proposals for Advanced Treatments for Potabilization of
Wastewater in the Food Industry

39

3. Ozonation

Ozone decomposes to oxygen after generation, so it cannot be stored and must be generated on-site. As this oxidant is a gas with low solubility in water (1.05 g L^{-1}), the efficiency of the gas–liquid transfer of ozone is the main variable affecting the effectiveness of organic compound oxidation by ozone in water matrices (Sievers, 2011). The main decomposition reactions of ozone in pure water at ambient temperature (around 20°C) are Equations 13-23.

Initiation:

$$O_3 + OH^- \longrightarrow HO_2^- + O_2$$

$k = 70$ M^{-1}s^{-1} (Staehling; Hoigne, 1982)   (13)
$k = 48$ M^{-1}s^{-1} (Forni et al., 1982)

Propagation:

$$O_3 + HO_2^- \longrightarrow \cdot OH + O_2^- \cdot + O_2$$

$k = 2{,}8 \times 10^6$ M^{-1} s^{-1} (Staehling; Hoigne, 1982)   (14)

$$O_3 + O_2^- \cdot \longrightarrow O_3^- \cdot + O_2$$

$k = 1{,}6 \times 10^9$ M^{-1} s^{-1} (Bühler et al., 1984)   (15)

$$(pH \geq 8)\ O_3 + O_2^- \cdot \rightleftharpoons O_3^- \cdot + O_2$$

$k_+ = 1{,}9 \times 10^3$ s^{-1} (Elliot; McCracken, 1989)   (16)
$k_- = 3{,}5 \times 10^9$ M^{-1} s^{-1} (Elliot; McCracken, 1989)

$$O^- \cdot + H_2O \longrightarrow \cdot OH + OH^-$$

$k = 10^8$ s^{-1} (von Gunten, 2003a)   (17)

$$(pH \leq 8)\ \ O_3^- \cdot + H^+ \rightleftharpoons HO_3 \cdot$$

$k_+ = 5{,}2 \times 10^{10}$ M^{-1} s^{-1} (Bühler et al., 1984)   (18)
$k_- = 3{,}3 \times 10^2$ s^{-1} (Bühler et al., 1984)

$$HO_3 \cdot \longrightarrow \cdot OH + O_2$$

$k = 1{,}1 \times 10^5$ s^{-1} (Bühler et al., 1984)   (19)

$$O_3 + \cdot OH \longrightarrow \cdot HO_2 \cdot + O_2$$

$k = 3{,}0 \times 10^9$ M^{-1} s^{-1} (Bahnemann; Hart, 1982)   (20)

$$HO_2^- + H^+ \rightleftharpoons H_2O_2$$

$pK_a = 11{,}7 \pm 0{,}2$ (Behar et al., 1970)   (21)

$$HO_2 \cdot \rightleftharpoons H^+ + O_2^- \cdot$$

$pK_a = 4{,}8$ (Behar et al., 1970)   (22)
$K_{equilíbrio} = 1{,}3 \times 10^{-5}$ M^{-1} (Behar et al., 1970)

$$\cdot OH \rightleftharpoons O^- \cdot + H^+$$

$pK_a = 11{,}84$ (Elliot; McCracken, 1989)   (23)

Where pK$_a$ is the acid dissociation constant. The termination reactions are any reactions between •O$_2^-$,•O$_3^-$, HO$_2^-$, HO$_2$• and •OH (Bühler et al. 1984; Nadezhdin, 1988).

Analyzing the mechanism described above (Equations 13-23), it appears that the decomposition of ozone can be accelerated by increasing the pH, since hydroxyl ions (OH-) (Equation 13) start this process and the HO$_2^-$ concentration is pH-dependent (Equation 21) (Andreozzi et al., 1999). As the decomposition of ozone leads to the formation of H$_2$O$_2$ (Equation 21), the addition of hydrogen peroxide will promote the decomposition of ozone and increase the formation of •OH. Therefore, the combination O$_3$+H$_2$O$_2$ provides the advanced oxidation process (AOP) O$_3$/H$_2$O$_2$.

In the decomposition reactions (organic matter oxidation reactions), there are two mechanisms: direct attack of ozone and free radical attack. The direct attack of ozone is selective and usually occurs in the case of unsaturated compounds (carbon-carbon double or triple bonds - π bonds), aromatic rings, amines and sulfides (von Gunten, 2003a). The primary reaction of ozone with a certain compound (C) occurs through the generation of an electrophilic additional intermediate compound (C-O_3) which decomposes via consecutive reactions until the formation of stable compounds (von Gunten, 2003a).

The direct attack of ozone on tertiary amine occurs by hydrolysis and demethylation, generating the corresponding aldehyde and secondary amine, which is the most stable compound containing the nitrogen atom. Similarly, the same reactions occur with the secondary and primary amine until the formation of the most stable organic compound (Muñoz and von Sonntag, 2000, Lange et al., 2006, Luiz et al., 2010). The oxidation of ammonia through direct attack by ozone is very slow, but it generates the same product as the radical attack: NO_3^-. The replacement of hydrogen atoms by alkyl groups (C_nH_{2n+1}) in the amine nitrogen increases the rate constant of oxidation via ozone (k). The k value for oxidation via ozonation of ammonia (NH_3) is 20 M^{-1} s^{-1}, of diethylamine is 9.1 x 10^5 M^{-1} s^{-1} and of triethylamine is 4.1 x 10^6 M^{-1} s^{-1} (von Gunten, 2003a).

In the radical attack, radical ions are formed during the oxidative degradation of ozone in water, the main ones being: oxygen free radical (O-•), superoxide radical (•O_2-) and hydroxyl radical (•OH) (Hoigné; Bader, 1983, Luiz et al., 2010).

The free radical attack is usually faster than the direct attack of ozone. However, in an ozonation process, the concentration of ozone is higher than that of its degradation radicals and hence the oxidation is mainly due to molecular attack. There are molecules that are slowly degraded by ozone treatments (because the molecular attack is selective) and rapidly degraded in advanced oxidation processes in which the attack by •OH radicals predominates (non-selective, higher removal efficiency and the mineralization of organic compounds) (Lau et al., 2007). Therefore, in tertiary wastewater treatment, it is necessary to know which organic contaminants are present in the wastewater in order to verify whether they can be oxidized by simple ozonation or if an AOP would be more effective.

Ozone is a powerful disinfectant, causing cell inactivation through direct damage to the membrane and cell wall, disruption of enzymatic reactions and DNA damage, including in the case of highly resistant pathogens such as protozoan *Cryptosporidium parvum* (EPA, 2004; von Gunten, 2003b). Compared with chlorination, disinfection with ozone is a better option for water bodies which are not organochlorine receivers and its excess can be easily dissipated (Achilleos et al. 2005; EPA, 2004). However, since the concentration required to inactivate the most resistant organisms is very high, there may be the formation of non-desirable disinfection by-products, especially bromates which have carcinogenic potential (von Gunten, 2003b).

4. Advanced oxidation processes (AOP)

Advanced oxidation processes (AOPs) degrade organic contaminants into carbon dioxide, water and inorganic anions through the action of transient oxidizing species, especially hydroxyl radicals (•OH). AOPs are clean processes because they do not generally require post-treatment or the final disposal of potential waste.

A Discussion Paper on Challenges and Proposals for Advanced Treatments for Potabilization of
Wastewater in the Food Industry

41

The •OH radicals are not selective, thus they have the ability to degrade all organic substances present in a liquid or gas to be treated. The abstraction of H for the reaction of •OH and organic compounds in the presence of saturated oxygen is the main route for the shortening of the chains of ketones, aldehydes and carboxylic acids (Mellouk et al., 2003).

4.1 UV/H₂O₂

Ultraviolet (UV) treatment promotes photolysis (or photo-oxidation) of organic compounds. To achieve a greater degradation, there are processes that combine UV radiation with other chemical agents such as O_3 and H_2O_2 (Parkinson et al., 2001). Among the variety of AOPs available and their combinations, the reagent H_2O_2 is frequently used due to its stability during transport and storage, almost infinite solubility in water, and low installation and operation costs compared to other processes, such as the application of O_3 (Alfano et al., 2001). In UV/H_2O_2 treatment, radiation with a wavelength shorter than 300 nm (UV C irradiation) breaks H_2O_2 into •OH radicals (Equation 24) which oxide the pollutant compounds. However, for the organic compound degradation to be effective, and mineralization of the compounds to occur (into CO_2 and H_2O), the H_2O_2 concentration must be above the stoichiometric demand and the reaction time under UV radiation must be sufficient, as previously determined experimentally (Alfano et al., 2001).

The photolysis of water yields dissolved oxygen (Equation 24) which can favour the photodegradation of organic compounds (Equation 2). Nevertheless the photolysis of hydrogen peroxide by UV radiation promotes the formation of stronger oxidants: hydroxyl radicals (•OH) (Equation 25). The mechanism of AOP UV/H_2O_2 treatment includes initiation, propagation and termination reactions, and degradation reactions of the organic compounds (RH) (Equations 25-32) (Alfano et al., 2001).

$$2H_2O \xrightarrow{h\nu} O_2 + 4H^+ + 4e^- \tag{24}$$

Initiation:

$$H_2O_2 \xrightarrow{h\nu} 2 \, ^\bullet OH \tag{25}$$

Propagation:

$$H_2O_2 + \, ^\bullet OH \xrightarrow{k_2} HO_2^\bullet + H_2O \tag{26}$$

$$H_2O_2 + HO_2^\bullet \xrightarrow{k_3} \, ^\bullet OH + H_2O + O_2 \tag{27}$$

Termination:

$$2 \, ^\bullet OH \xrightarrow{k_4} H_2O_2 \tag{28}$$

$$2HO_2^\bullet \xrightarrow{k_5} H_2O_2 + O_2 \tag{29}$$

$$\bullet OH + HO_2\bullet \xrightarrow{k_6} H_2O + O_2 \tag{30}$$

Decomposition:

$$RH + \bullet OH \xrightarrow{k_7} products \tag{31}$$

$$RH + HO_2\bullet \xrightarrow{k_8} products \tag{32}$$

The propagation (Equations 26-27) and termination (Equations 28-30) reactions indicate that when there is H_2O_2 in excess and, subsequently, an excess of $\bullet OH$ radicals, the recombination of these radicals is favored (Equation 28) reversing the initiation reaction (Equation 25) and inhibiting the oxidation reactions of organic compounds (Equations 31-32).

Yang et al. (2005) found that there is no significant effect of pH on the degradation efficiency if the treatment is only via photodegradation by UV radiation. However, in a UV/H_2O_2 reaction the generation of hydroxyl radicals ($\bullet OH$) may be affected by the presence of high concentrations of H^+ and OH^-. Many authors have observed higher rates of decomposition of organic contaminants at low pH (Song et al., 2008). However, Yang et al. (2005) found the efficiency of mineralization of the complex sodium ethylenediamine tetraacetic acid (Na-EDTA) decreased from 98% (in ultra-pure water and without pH correction) to around 70% with initial pH values of 2.3 and 11.6, respectively.

The commercial and industrial applications of UV/H_2O_2 processes require the determination of optimal oxidant and irradiation dosages. Therefore, the development of a mathematical model for these processes can help the determination of such variables (Crittenden et al., 1999). Usually, the kinetic model used for common pollutants in aqueous solutions is a pseudo-first-order model by photodegradation (UV radiation) and also by the AOP H_2O_2/UV (Chen et al. 2007; Song et al. 2008; Luiz et al., 2009).

Dosages below 15 mg L^{-1} of hydrogen peroxide do not have a significant toxic effect on bacteria when the system is not exposed to light (in darkness) (Rincón; Pulgarin, 2006; Sciacca et al., 2010). However, hydrogen peroxide in the extracellular region weakens the cell walls, making the bacteria more sensitive to oxidative stress and this may explain the observed inactivation at lower concentrations. Therefore, although H_2O_2 is not directly toxic to the bacteria, the cleavage products, for instance, the hydroxyl ($\bullet OH$) and superoxide ($HO_2\bullet$ and $\bullet O_2^-$) radicals generated by the addition of another agent in the system, such as UV, O_3, ferric or ferrous ions (Fenton reaction) (Sciacca et al., 2010) can be toxic. Such species can induce reactions in lipids, proteins and DNA, inactivating the cells.

4.2 O_3/H_2O_2, O_3/UV and O_3/H_2O_2/UV

The AOPs O_3/H_2O_2, O_3/UV and O_3/H_2O_2/UV have been developed to treat wastewater highly contaminated by both organic matter and microorganisms. The effectiveness of these treatments is due to synergistic interaction of the positive effects of three reactions: direct ozonation (molecular attack), photolysis and oxidation by hydroxyl radicals ($\bullet OH$) (Lau et al., 2007, Luiz et al., 2010). Some important factors, such as quantifying the speed of photons

A Discussion Paper on Challenges and Proposals for Advanced Treatments for Potabilization of
Wastewater in the Food Industry

43

absorbed in the heterogeneous medium (ozone gas in water matrices), and assessing the degree of ozone reactivity with unsaturated organic compounds, still need to be better understood (Legrini et al., 1993).

The mechanism of the AOPs O_3/H_2O_2, O_3/UV and $O_3/H_2O_2/UV$ are very similar to each other, and all have the hydroxyl radical ($\bullet OH$) as the most important intermediate compound (Beltran et al. 1995; Legrini et al., 1993). In the AOP O_3/H_2O_2, besides the initiation reaction of ozone decomposition, another initiation reaction is the decomposition of hydrogen peroxide by ozone into hydroxyl radicals (Equation 33). The decomposition of hydrogen peroxide by ozone (Equation 33) is very slow (Legrini et al., 1993), but it enhances the other ozone decomposition reactions which generate more $\bullet OH$ radicals (Equations 13-23). Therefore, the reaction rate can be increased with increasing the pH, because the ionic form of H_2O_2 (HO_2^-) reacts with O_3 (Staehelin; Hoigné, 1982) (Equation 14) followed by other reactions of propagation, termination and decomposition.

$$O_3 + H_2O_2 \rightarrow O_2 + HO_2^{\bullet} + {}^{\bullet}OH \tag{33}$$

The absorbance spectrum of ozone indicates that it has a greater absorption at 254 nm than H_2O_2, 3300 and 18.6 M^{-1} cm^{-1}, respectively, and the influence of the internal effects of the system is lower in an O_3/UV system, due to the effect of suspended solids and aromatic compounds which can act as a UV filters. Nevertheless, the quantum yield of $\bullet OH$ radicals in the AOP O_3/UV (around 0.1) is lower than in H_2O_2/UV (Sievers, 2011). The quantum yield of a photochemical reaction is the number of events divided by the number of absorbed photons of a specific wavelength during the same period time (Khun et al., 2004). The absorption of UV light by O_3 in water leads, firstly, to the generation of H_2O_2 (Equation 34) (Legrini et al., 1993). Therefore, the other reactions of propagation, termination and decomposition are all cited for O_3 treatments, and for O_3/H_2O_2 H_2O_2/UV (Equations 14-33).

$$O_3 + H_2O \xrightarrow{h\nu} H_2O_2 + O_2 \tag{34}$$

The AOP $O_3/H_2O_2/UV$ should be used to enhance the generation of hydroxyl radicals, as its initiation reactions are the same as those of ozonation, H_2O_2/UV, O_3/H_2O_2 and O_3/UV (Equations 13, 25, 33 and 34). All other reactions are of the types propagation, termination and decomposition (Equations 14-32).

AOPs promote a higher degree of mineralization compared to other tertiary oxidative treatments (such as simple ozonation), ensuring decreased levels or the absence (if the mineralization is total) of oxidized intermediate products that, usually, reduce the initial toxicity (Rosal et al., 2008).

4.3 Homogeneous fenton and fenton-like pocess

A widespread AOP is the homogeneous Fenton process involving the reaction of hydrogen peroxide (H_2O_2) with the Fenton catalyst (ferrous ion - Fe(II) - dissolved in acid aqueous media) generating hydroxyl radicals ($\bullet OH$) and ferric ions (Fe(III)). Fe(III) ions undergo reduction to Fe(II) mainly through the action of the oxidant H_2O_2 ensuring the continuation of the catalytic reaction. The main reactions of the Fenton process are represented by Equations 35 to 41 (Garrido-Ramirez et al. 2010; Herney-Ramirez et al. 2010; Navalon et al., 2010, Umar et al., 2010).

$$Fe^{2+} + H_2O_2 \rightarrow Fe^{3+} + OH + OH^- \tag{35}$$

$$Fe^{3+} + H_2O_2 \rightarrow Fe^{2+} + HO_2 + H^+ \tag{36}$$

$$OH + H_2O_2 \rightarrow HO_2 + H_2O \tag{37}$$

$$OH + Fe^{2+} \rightarrow Fe^{3+} + OH^- \tag{38}$$

$$Fe^{3+} + HO_2 \rightarrow Fe^{2+} + O_2 + H^+ \tag{39}$$

$$Fe^{2+} + HO_2 + H^+ \rightarrow Fe^{3+} + H_2O_2 \tag{40}$$

$$2HO_2 \rightarrow H_2O_2 + O_2 \tag{41}$$

There are many variations of the Fenton process and the main ones are:
- Modified Fenton or Fenton-like reaction: ferric ion or another transition metal (such as copper) is used instead of ferrous ion (Herney-Ramirez et al., 2010);
- Heterogeneous Fenton: supported catalyst is used in a solid matrix (heterogeneous system);
- Photo-Fenton: the regeneration of Fe(II) is accelerated due to photoreduction of Fe(III), and the generation of hydroxyl radicals is increased due to the photolysis of hydrogen peroxide (Equation 25) (Umar et al., 2010);
- Electro-Fenton: continuous generation of H_2O_2 in acidic solution by reducing gaseous O_2 in carbon cathodes (Sirés et al., 2007);

Other variations: heterogeneous Fenton-like, heterogeneous photo-Fenton, Photo-Fenton-like, and heterogeneous Photo-Fenton-like processes.

Equation 42 describes the Fenton reaction and its variations should occur at acid pH through the occurrence of hydrogen peroxide decomposition in the presence of a ferrous ion catalyst (Umar et al., 2010).

$$2Fe^{2+} + H_2O_2 + 2H^+ \rightarrow 2Fe^{3+} + 2H_2O \tag{42}$$

The homogeneous Fenton process is carried out in four stages: (1) pH adjustment (optimal pH is around 2.8), (2) oxidation reaction through the action of oxidizing radicals (especially •OH), (3) neutralization followed by coagulation (formation of ferric hydroxide complexes), and (4) precipitation (Umar et al., 2010). Therefore, the organic substances are removed by oxidation and by coagulation. Thus, the homogeneous Fenton process has the great disadvantage of producing, at the end of the process, solid waste (sludge) with a high metal concentration, generating environmental risks and extra costs due to the post-treatment of sludge to recover

A Discussion Paper on Challenges and Proposals for Advanced Treatments for Potabilization of
Wastewater in the Food Industry

45

part of the catalyst (Herney-Ramirez et al., 2010, Umar et al., 2010). Hence, the heterogeneous
Fenton process eliminates the step of catalyst precipitation and removal at the end of the
process (Garrido-Ramirez et al. 2010; Herney-Ramirez et al. 2010 ; Navalon et al., 2010).
Among the various factors that affect the efficiency of the Fenton process, pH and the
$H_2O_2/Fe(II)$ ratio are the most important ones. In reactions with a low $H_2O_2/Fe(II)$ ratio,
chemical coagulation is predominant in the removal of organic matter, whereas with high
ratios the chemical oxidation is dominant (Umar et al., 2010).

4.4 Heterogeneous AOPs

The heterogeneous AOPs using a semiconductor as the catalyst and ultraviolet radiation
have the advantage of not requiring the addition of a strong oxidant to produce hydroxyl
radicals (photocatalysis process, e.g., the AOP TiO_2/UV). However, the addition of an
oxidant can accelerate the production of oxidizing radicals and increase the speed of organic
matter removal (e.g. the AOP $TiO_2/UV/H_2O_2$).
Photocatalysis is the catalysis of a photochemical reaction on the surface of a solid, usually a
semiconductor, TiO_2 being the most used (Fujishima et al., 2008). In this reaction the
oxidation of organic pollutants occurs in the presence of an oxidizing agent, followed by
activation of a semiconductor with light energy (Ahmed et al., 2010) at wavelengths ≤ 400
nm (Malato et al., 2009).
The catalyst is activated when it absorbs photons with energies greater than the value of
its "band gap" energy resulting in the excitation of electrons from the valence band to the
conduction band forming a hole (h^+) in the valence band. This hole promotes the
oxidation of pollutants (in the case of photo-catalytic oxidation of organic pollutants) or
water, in the latter case producing hydroxyl radicals (OH •). The electron (e^-) of the
conduction band reduces the oxygen or other oxidative agent in the absence of O_2
(Ahmed et al., 2010). The oxidizing agent binds to the conduction band, where the
reduction occurs through the gaining of electrons (e^-), while the organic compound to be
degraded binds to the valence band, where oxidation occurs, donating e^- and receiving h^+.
Thus the organic compound is the reducing agent or electron donor (hole scavenger)
(Malato et al., 2009).
Photocatalysis is an AOP because, during the photochemical reactions and the activation of
the catalyst, hydroxyl radicals are formed (Malato et al., 2009, Hermann, 2010). Hence, these
treatments may reduce the toxicity of wastewaters by total or partial mineralization of
synthetic organic compounds transforming them into more biodegradable substances.
There are photocatalytic processes using solar energy, representing a Green Chemistry
technology approach (Hermann, 2010), with the following benefits: 1) process carried out at
ambient temperature and pressure, 2) the oxidizing agent can be oxygen obtained from the
atmosphere, 3) many catalysts are low-cost, safe and reusable, and 4) use of solar energy as
an energy source (Malato et al., 2009).

5. Water and wastewater management

Prior to the implementation of a tertiary process for wastewater treatment aiming at water
reuse to reduce the consumption of fresh drinking water, it is essential to minimize and to
optimize the water consumption in all processes of an industrial plant. Thus, it is essential to

understand the industrial process in detail, and to develop a methodology for water management, presenting alternatives for minimizing water consumption and effluent generation, obeying the specific laws of the industry sector.

The authors of this review carried out a water and wastewater management (W2M) study in a slaughterhouse based on water consumption and effluent generation in the various sectors. The W2M proposed was aimed at minimizing water consumption and the evaluation of possibilities for water and indirect potable reuse in food industries.

5.1 Case study in a meat processing plant
The case study was carried out in a meat processing plant located in the west of Santa Catarina State (southern Brazil) where water pollution and overexploitation, the uneven distribution of rainfall throughout the seasons and long periods of drought especially in summer, have become a problem. The industrial activities include the slaughter and processing of poultry and swine. The industry has its own drinking water treatment plant (DWTP) and wastewater treatment plant (WWTP). The major water resource of this industry is a river. Its DWTP produces around 8,600 m^3/day of drinking water and the WWTP treats around 7,900 m^3/day of wastewater. The first step of the wastewater treatment is equalization, followed by coagulation, flocculation, and flotation. The last step is the secondary treatment: activated sludge.

5.1.1 Water minimization
To evaluate the possibilities for water minimization, it was first necessary to analyze the current Brazilian laws concerning slaughterhouses for poultry and swine. These guidelines provide the technical standards required for the facilities and equipment used in the slaughter of animals and meat industrialization, including water quality and minimum water consumption in each process. As a further step, a water balance was carried out for all processes. The processes or equipment which have the greatest water demand are usually the most important and the easiest points for which to verify the possibilities to avoid the misuse and overexploitation of fresh water (drinking water from plant's DWTP). Hence, four high consumption points were identified with the potential for reducing the fresh water consumption in-line with the current Brazilian legislation: (1) pre-cooling of giblets, (2) washing of poultry carcasses before pre-chiller, (3) transportation of giblets, poultry necks and feet, and (4) washing of swine carcasses after buckling. It was found that in the first two processes the water consumption could be reduced; in the third process the devices could be substituted by others which do not need water to function properly; and in the last process there was an unnecessary step, according to the legislation. The minimization of the water demand achieved in these four process steps alone was approximately 806 m^3/d.

5.1.2 Direct wastewater reuse
After the minimization of water use, the most important action is to prioritize direct recycling and reuse of wastewater, without the need for advanced reconditioning or treatment. Hence, following the water balance, the wastewater with the possibility for direct or indirect recycling or reuse was evaluated physically, chemically and microbiologically to verify if and where it could be recycled and reused. Four wastewaters were identified as having a real possibility of reuse after carrying out all of the characterization analysis: (1)

A Discussion Paper on Challenges and Proposals for Advanced Treatments for Potabilization of
Wastewater in the Food Industry

47

defrosting of refrigerating and freezing chambers, (2) purging of condensers, (3) cooling of smoke fumigator chimneys, and (4) sealing and cooling of vacuum pumps. These four wastewaters totalized approximately 1,383 m³/day of wastewater with the possibility for reuse in processes without direct contact with food products, that is, in non-potable uses (e.g., as cooling water, for flushing toilets or as irrigation around the plant), thus saving fresh potable water. These wastewaters had similar water quality parameters indicating that they could be mixed before undergoing the same reconditioning treatment. Hence, the mixed wastewater was also characterized and it was shown that it could be simply treated to remove total suspended solids and chlorinated/disinfected before reuse as, for example, cooling water.

The successful results achieved from the proposed W2M (*i.e.*, financial saving of 28.2% and 25.4% in the fresh water requirement - Table 4) prompted the managers of the pilot industrial plant to suggest and to develop other research projects related to water and wastewater reuse in partnership with Brazilian government funded research centers, to develop indirect wastewater technologies. These results are consistent with the work of Tokos et al. (2009) who presented several mixed-integer nonlinear programming models for the optimization of a water network in a brewery plant. The authors presented a theoretical reduction of around 27% in the fresh water demand and associated costs.

Condition	Water flow [m³/dia]	Water saving [%]	Annual Costs[1] [$]
Production in 2007	8,616.0	-	1,539,353
Theoretical production after water minimization	7,810.0	9.4	1,366,000
Theoretical production after wastewater reuse	7,216.8	16.0	1,256,996
Theoretical production after water minimization and wastewater reuse	6,410.0	25.4	1,104,731

[1]Considering costs in 2007: $0.10 and $0.42 per m³ to treat water (DWTP) and wastewater (WWTP) respectively, in pilot industrial plant.

Table 1. Water and financial savings

5.1.3 Indirect wastewater reuse: indirect potable reuse

Additionally, to reduce even further the percentage of fresh water consumption, indirect wastewater reuse can be carried out after reconditioning of the secondary effluent (*i.e.*, after secondary activated-sludge treatment) by applying tertiary treatments such as advanced oxidation processes (AOPs). This tertiary treated water could be used in other processes without contact with food products, that is, non-potable uses (*e.g.* as cooling water, boiler feed water, toilet flushing water or for irrigation around the plant) (Cornel et al., 2011; Luiz et al., 2011).

Luiz and co-workers (2009, 2011) evaluated some AOPs in bench- and pilot-scale tests with the aim of providing reclaimed water with drinking water quality (according to Brazilian legislation) for industrial water reuse in processes without direct contact with food products. In the first study (Luiz et al., 2009) evaluated the kinetics of the photo-induced degradation of color and UV_{254} under UV radiation with and without the addition of H_2O_2 to treat secondary wastewater after ferric sulfate coagulation. The H_2O_2/UV treatment was 5.2 times faster than simple UV application in removing aromatic compounds.

In the second study (Luiz et al., 2011) four tertiary hybrid treatments using a pilot plant with a capacity of 500 L/h were evaluated. This pilot plant consisted of a pre-filtration system, an oxidation (H_2O_2) or second filtration system and a UV radiation device. The best combination was pre-filtration followed by H_2O_2 addition and UV radiation (AOP H_2O_2/UV). In addition to H_2O_2/UV treatment, the authors also carried out experiments in pilot scale with O_3, O_3/UV and O_3/H_2O_2. However, H_2O_2/UV was found to be a faster and more efficient option to treat slaughterhouse wastewater to the drinking water quality standard, with the exception of one quality parameter, that is, nitrate.

Hence, another treatment was evaluated by the authors to remove the nitrate content, allowing the reuse of the wastewater with drinking water quality in cases without direct contact with food products, reducing the fresh water consumption and conserving natural water resources. The p-n junction photocatalyst $p-ZnO/n-TiO_2$ was prepared by decomposition of zinc nitrate and photodeposition on TiO_2 and it was used to reduce nitrate ions in aqueous solution and in synthetic and real slaugherthouse wastewater.

Heterogeneous photocatalytic reduction of nitrate over semiconductor materials have also been developed as a promising method for controlling the concentration of nitrate in water. Metal doping (Pt, Pd, Rh, Pt-Cu, Cu, Fe, Bi) and addition of hole scavengers are essential for the reductive removal of nitrate (Rengaraj and Li, 2007; Sá et al., 2009; García-Serrano et al., 2009; Kim et al., 2008; Li et al., 2010). Hole scavengers are electron donors, such as methanol, benzene, oxalic acid, formic acid and humic acids which have been used to improve photocatalytic efficiency (Xu et al., 2010; Li and Wasgestian, 1998; Rengaraj and Li; 2007).

There were a number of studies related to the photocatalytic activity of TiO_2 or ZnO coupled with metals, and the results showed that these catalysts presented more efficient charge separation, and increased lifetime of the charge carriers, and an enhanced interfacial charge transfer to adsorb substrates (Shifu et al., 2008). Coupled photocatalyst ZnO/TiO_2 has showed higher photocatalytic activity than that of single one, since the p-n junction that is formed by the integration of p-type ZnO and n-type TiO_2 contributes to enhance the electrons/holes separation (Shifu et al., 2008).

The selectivity, conversion and activity data indicated $ZnO-TiO_2$ was a better catalyst than TiO_2 (Table 2). This indicates that Zn ion exerted its role as a doping ion, promoting the charge separation of the pairs of vacancy-electron photoproduced by acting as an electron sink, and consequently increase of electrons reaching the surface for reaction to take place (Sá et al., 2009; Ranjit and Viswanathan, 1997).

Catalyst	Selectivity [%]	Conversion [%]	Activity [$\mu mol_{NO_3^-}$ (min $g_{catalisador}$)$^{-1}$]
TiO_2	70.71	87.46	0.92
$ZnO-TiO_2$	95.45	91.67	14.24

Table 2. Selectivity (S_N) and conversion ($C_\%$) within reaction time = 120 min and activity within reaction time = 20 min.

Prior to the photocatalytic reactions, the real wastewater (RSW) was filtered 50 micron cartridge filter, activated carbon filter and 10 micron cartridge filter to remove suspended

A Discussion Paper on Challenges and Proposals for Advanced Treatments for Potabilization of
Wastewater in the Food Industry

49

particles (RSW-F) simulating the microfiltration of the previous studies (Luiz et al., 2009, 2011). Due to low TOC concentration in RSW-F, formic acid was added as hole scavenger (Zhang et al., 2005; Rengaraj; Li, 2007; Sá et al., 2009; Malato et al., 2009; Wehbe et al., 2009). Formic acid was added in enough amount to satisfy the stoichiometric relationship nitrate:formic acid (Equation 43) without formation of intermediates (including nitrite and ammonia) (Zhang et al., 2005). Hence, the evaluated molar ratios were CHOOH:nitrate = 2.7, 1.6 and 1.0, respectively [HCOOH] = 1636, 1000 and 500 mg L^{-1} or 427, 261 and 130 mg C L^{-1}, and catalyst 1 g L^{-1}.

$$2NO_3^- + 5HCOO^- + 7H^+ \rightarrow N_2 + 5CO_2 + 6H_2O \tag{43}$$

Therefore, after the application of microfiltration for the removal of mainly suspended solids and the catalytic photoreduction of nitrate with the addition of a hole scavenger (carbon source), it was necessary to remove the excess of this carbon source. Hence, after nitrate was below the Brazilian legislation limit, (CHOOH$_{initial\ (0\ min)}$:H$_2$O$_{2,\ initial\ (120\ min)}$ = 1:1 (M:M)) was added into the reaction to promote another advanced oxidative process (AOP): H$_2$O$_2$/TiO$_2$/UV (Figure 1). Hence, after all these steps, the wastewater achieved drinking water quality, being the main critical parameters shown in Table 3.

	RSW[2]	RSW-F[2]	A[1]	B[1]	C[1]	BRL[3]
pH	6.8 ± 0.6	6.2 – 6.5	6.20	6.15	6.00	6.0 - 9.50
Turbidity, FTU	40 ± 33	22 – 25	2.00	2.00	1.00	5
Apparent color, Pt-Co unit	217 ± 177	11 – 13	3.00	3.00	10.00	15
Total coliforms, CFU/100 mL	> 10[5]	> 10[5]	0	0	0	0
Nitrate, mg N L^{-1}	45.9 ± 17.7	39.8 – 41.9	1.20	4.50	4.00	10
Nitrite, mg N L^{-1}	-	3.74 – 3.77	0.06	0.08	0.21	1
Ammonia, mg N L^{-1}	-	0.4 – 1.0	0.30	0.90	0.30	1.24

[1] Composite samples (RSW). Composite samples after filtration (RSW-F)
[2] A: [CHOOH]$_i$ = 1636 mg L^{-1} / H$_2$O$_2$ = 1209 mg L^{-1}; B: [CHOOH]$_i$ = 1000 mg L^{-1} / H$_2$O$_2$ = 739 mg L^{-1}; C: [CHOOH]$_i$ = 500 mg L^{-1} / H$_2$O$_2$ = 370 mg L^{-1}. Zn-TiO$_2$ catalyst = 1 g L^{-1}. Reaction time: 300 min.
[3] BRL: Brazilian regulatory legislation (Brazilian drinking water standards: Brazilian Ministry of Health Administrative Ruling 518/2004).

Table 3. Average quality of the secondary effluent of the slaughterhouse and quality parameters of the wastewater after the treatments: filtration (RSW-F, Table 1) + TiO$_2$/UV/Argon (reaction time 120 min) + H$_2$O$_2$/TiO$_2$/UV (reaction time 300 min).

Table 3 shows the average characterization of three composite samples collected to characterize the secondary effluent from the slaughterhouse (RSW). The real effluent slaughterhouse wastewater used in those studies was collected after treatment (coagulation+flocculation followed by aerobic biological treatment) from a Brazilian poultry slaughterhouse. The point to collect real treated effluent was the one located immediately outside the secondary sedimentation tank. A composite sample of 24h (1 L of sample collected per hour for 24 hours) was collected and characterized. All organic, inorganic and microbiological parameters of Brazilian drinking water standards (Brazilian Ministry of

Health Administrative Ruling 518/2004) were analyzed, and only nitrate, nitrite, color, turbidity and total coliform were above the limit. The proposal treatments were able to reduce those parameters and also remove micropollutants that are not removed by secondary treatment (e.g. activated sludge).

The influent of the slaughterhouse RSW contained the mixture of sewage from meat processing and sanitary sewage, the treated wastewater still contained persistent pollutants arising from animal and human use, e.g., drugs and personal care products. The target compounds nonyl- (NP) and octylphenol (OP), linear alkylbenzene sulfonates (LAS), triclosan and ibuprofen could be detected and confirmed applying a LTQ Orbitrap mass spectrometer (Thermo Electron, Bremen, Germany) to the SPE-extracts after HPLC-(HR)MS. Besides these pollutants octylphenol- (OPEO) and nonylphenol ethoxylates (NPEO), polyethylene glycols (PEG), erythromycin, sulfamethoxazole and sulfadimidine were also observed.

The AOPs evaluated were also very effective to remove the micropollutants present in the already treated slaughterhouse wastewater. Among all the compounds found by LC-MS in the samples only NP, NPEO and PEG were still observable in the samples after the tertiary treatments proposed, however, in lower amounts.

Figure 1 shows the suggested flowchart of the process suggested by the authors to treat secondary wastewater with high concentrations of nitrogen compounds and recalcitrant organic compounds as antibiotics, medicament and personal care products which are commonly found in sanitary, domestic and industrial wastewater.

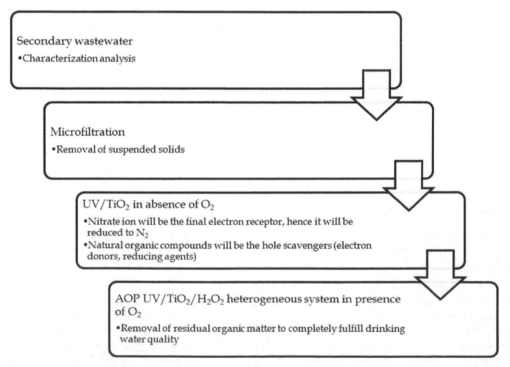

Fig. 1. Suggested process to treat secondary wastewaters with high concentration of nitrate and residual organic matter.

A Discussion Paper on Challenges and Proposals for Advanced Treatments for Potabilization of
Wastewater in the Food Industry

51

6. Selection of an AOP and alternatives for water reuse

As proposed in section 5.1 (Case Study in a Meat Processing Plant), experiments in bench and pilot scale should be carried out before the selection of an AOP for the tertiary treatment of the wastewater intended for reused (Sievers, 2011). This initial research would define: (i) the best treatment for each kind of industrial wastewater considering the reused water quality required and hence which improvements should be sought (*e.g.* color or turbidity removal, target recalcitrant organic compound removal, disinfection); (ii) the kind and quantity of radical scavengers present in the wastewater (*e.g.* humic substances); (iii) the dosage of oxidant (if used) and the impact of its residual concentration on the subsequent use or disposal of the tertiary treated wastewater, and (iv) the energy and investment costs (Sievers, 2011).

The tertiary treated water could be used in other processes without contact with food products (non-potable uses) where public health and product integrity would not be compromised. Therefore, "this major health concern makes it imperative for governments and the global community to implement proper reuse planning and practices, emphasizing public health and environmental protection" (EPA, 2004). Boiler feed water, cooling water, toilet flushing water or irrigation around the plant are some alternatives for the use of reclaimed water (Cornel et al., 2011; Luiz et al., 2011). Cooling water may be the most attractive option because it does not need to be of high quality, and usually a large amount is required (Cornel et al., 2011).

A sentence at EPA (2004) resume the past and the future of wastewater reuse (indirect potable reuse): "Notwithstanding the fact that some proposed, high profile, indirect potable reuse projects have been defeated in recent years due to public or political opposition to perceived health concerns, indirect potable reuse will likely increase in the future" (EPA, 2004).

7. References

Achilleos, A., Kythreotou, N. & Fatta, D. (2005). Development of tools and guidelines for the promotion of sustainable urban wastewater treatment and reuse in agricultural production in the Mediterranean countries: Task 5 - Technical Guidelines on Wastewater Utilisation. European Commission - Euro Mediterranean Partnership.

Ahmed, S., Rasul, M.G., Martens, W. N., Brown, R. & Hashib, M.A. (2010). Heterogeneous photocatalytic degradation of phenols in wastewater: A review on current status and developments, *Desalination*, 261 (1-2), 3-18.

Alfano, O. M., Brandi, R. J. & Cassano, A. E. (2001). Degradation kinetics of 2,4-D in water employing hydrogen peroxide and UV radiation, *Chem. Eng. J.*, 82, 209-218.

Andreozzi, R., Caprio, V., Insola, A. & Marotta, R. (1999) Advanced oxidation processes (AOP) for water purification and recovery, *Catal. Today*, 53, 51-59.

Bahnnemann, D. & Hart, E. J. (1982). Rate Constants of the Reaction of the Hydrated Electron and Hydroxyl Radical with Ozone In Aqueous Solution, *J. Phys. Chem.*, 86, 252-255.

Behar, C., Czapski, G. & Duchovny, I. (1970). Carbonate radical in flash photolysis and pulse radiolysis of aqueous carbonate solutions, *J. Phys. Chem.*, 74 (10), 2206-2210.

Beltrán, F. J., Ovejero, G., Acedo, B. (1993). Oxidation of atrazine in water by ultraviolet radiation combined with hydrogen peroxide, *Water Res.*, 27 (6), 1013-1021.

Beltrán, F. J., Ovejero, G., Garcia-Araya, J. F. & Rivas, J. (1995). Oxidation of Polynuclear Aromatic Hydrocarbons in Water. 2. UV Radiation and Ozonation in the Presence of UV Radiation, *Ind. Eng. Chem. Res.*, 34, 1607-1615.

Beltran-Heredia, J., Torregrosa, J., Dominguez, J. R. & Peres, J. A. (2001). Kinetics of the Oxidation of p-Hydroxybenzoic Acid by the H_2O_2/UV System, *Ind. Eng. Chem. Res.*, 40, 3104-3108.

Bühler, R. E., Staehlin, J. & Hoigné, J. (1984). Ozone decomposition in water studied by pulse radiolysis. 1. HO_2/O_2^- and HO_3/O_3^- as intermediates, *J. Phys. Chem.*, 88, 2560-2564.

Burkhard, R., Deletic, A. & Craig, A. (2000) Techniques for water and wastewater management: a review of techniques and their integration in planning, *Urban Water*, 2, 197-221.

Casani, S., Rouhany, M. & Knøchel, S. (2005). A discussion paper on challenges and limitations to water reuse and hygiene in the food industry. *Water Res.* 39, 1134-1146.

Chen, L., Zhou, H.-Y. & Deng, Q.-Y. (2007). Photolysis of nonylphenol ethoxylates: the determination of the degradation kinetics and the intermediate products. *Chemosphere*, 68 (2), 354-359.

Chin, Y.-P., Miller, P. L., Zeng, L., Cawley, K. & Weavers, L. K. (2004). Photosensitized degradation of bisphenol A by dissolved organic matter. *Envir. Sci. Tech.*, 38 (22), 5888-5894.

Clevelario, J., Neves, V., Oliveira, P. T. T. M., Costa, V. G., Amendola, P., Rocha, R. M. & Costa, J. J. G. (2005). Water Statistics in Brazil: An Overview. IWG-Env, International Work Session on Water Statistics, Vienna.

Codex Alimentarius (2001) Proposed Draft Guidelines for the Hygienic Reuse of Processing Water in Food Plants, CX/FH 01/9. Joint FAO/WHO Food Standards Programme, 34th Session, Bangkok, Thailand.

Codex Alimentarius (2007) Report of the Thirty-Eighth Session of the Codex Committee on Food Hygiene, Alinorm 07/30/13. Joint FAO/WHO Food Standards Programme, 38th Session, Rome, Italy.

Cornel P, Meda A and Bieker S (2011) Wastewater as a Source of Energy, Nutrients, and Service Water. In: Peter Wilderer (ed.) *Treatise on Water Science*, Oxford: Academic Press, vol. 4, 337-375.

Crittenden, J. C., Hu, S., Hand, D. W. & Green, S. A. (1999). A kinetic model for H_2O_2/UV process in a completely mixed batch reactor. *Water Res.*, 33 (10), 2315-2328.

Davies, C. M., Roser, D. J., Feitz, A. J. & Ashbolt, N. J. (2009). Solar radiation disinfection of drinking water at temperate latitudes: Inactivation rates for an optimised reactor configuration. *Water Res.*, 43 (3), 643-652.

Elliot, A. J. & McCracken, D. R. (1989). Effect of temperature on O⁻ reactions and equilibria: a pulse radiolysis study. *Radiat. Phys. Chem.*, 33 (1), 69-74, 1989.

EPA. *Guidelines for Water Reuse*. Washington, DC: Environmental Protection Agency, 2004.

Espinoza, L. A., Neamtu, M. & Frimmel, F. H. The effect of nitrate, Fe(III) and bicarbonate on the degradation of bisphenol A by simulated solar UV-irradiation, *Water Res.*, 41 (19), 4479-4487, 2007.

Forni, L., Bahnemann, D. & Hart, E.J. (1982). Mechanism of the Hydroxide Ion Initiated Decomposition of Ozone in Aqueous Solution, *J. Physical Chem.*, 86 (2), 255-259.

Fujishima, A., Zhang, X. & Tryk, D.A. (2008). TiO_2 photocatalysis and related surface phenomena, *Surf. Sci. Rep.*, 63, 515–582.

Garrido-Ramírez, E.G., Theng, B.K.G. & Mora, M.L. (2010). Clays and oxide minerals as catalysts and nanocatalysts in Fenton-like reactions — A review, *Appl. Clay Sci.*, 47, 182-192.

Herney-Ramírez, J., Vicente, M. A. & Madeira, L.M. (2010). Heterogeneous photoFenton oxidation with pillared claybased catalysts for wastewater treatment: A review, *Appl. Catal. B Environ*, 98, 10–26.

Herrmann, J-M. (2010). Fundamentals and misconceptions in photocatalysis, *J. Photoch. Photobio. A.*, 216(2-3), 85-93.

Hijnen, W. A., Beerendonk, E. F. & Medema, G. J.(2006). Inactivation credit of UV radiation for viruses, bacteria and protozoan (oo)cysts in water: A review, *Water. Res.*, 40 (1), 3-22.

Hoigné, J. & Bader, H. (1983) Rate constants of reactions of ozone with organic and inorganic compounds in water – I: Non-dissociating organic compounds. *Water Res.*, 17, 173-183.

J. García-Serrano, E. Gómez-Hernández, M. Ocampo-Fernández & U. Pal. (2009). Effect of Ag doping on the crystallization and phase transition of TiO2 nanoparticles, Current Applied Physics, 9, 1097–1105.

Khan, S., Khan, M.A., Hanjra, M.A. & Mu, J. (2009). Pathways to reduce the environmental footprints of water and energy inputs in food production, *Food Policy*, 34, 141–149,

Khun, H. J., Braslavsky, S. E. & Schmidt, R. (2004). Chemical Actinometry (IUPAC Technical Report), *Pure Appl. Chem.*, 76 (12), 2105-2146.

Kim, Y., Lee, J., Jeong, H., Lee, Y., Um, M.-H., Jeong, K. M., Yeo, M.-K. & Kang, M. (2008). Methyl orange removal over Zn-incorporated TiO_2 photo-catalyst, *Journal of Industrial and Engineering Chemistry*, 14 (3), 396-400.

Kulovaara, M., Backlund, P. & Corin, N. (1995). Light-induced degradation of DDT in humic water, *Sci. Total Environ.*, 170, 185-191.

Lange, F., Cornelissen, S., Kubac, D., Sein, M. M., von Sonntag, J. & Hannich, C. B., et al. (2006). Degradation of macrolide antibiotics by ozone: A mechanistic case study with clarithromycin, *Chemosphere*, 65, 17-23.

Lau, T. K., Chu, W. & Graham, N. Reaction pathways and kinetics of butylated hydroxyanisole with UV, ozonation and UV/O_3 processes. Water Res., 41, 765-774, (2007).

Legrini, O., Oliveros, E. & Braun, A. M. (1993). Photochemical Processes for Water Treatment. *Chem. Rev.*, 93, 671-678.

Li, L., Xu, Z., Liu, F., Shao, Y., Wang, J., Wan, H. & Zheng, S. (2010). Photocatalytic nitrate reduction over $Pt-Cu/TiO_2$ catalysts with benzene as hole scavenger, *Journal of Photochemistry and Photobiology A: Chemistry*, 212, 113-121.

Luiz, D. B., Genena, A. K., José, H. J., Moreira, R. F. & Schröder, H. F. (2009). Tertiary treatment of slaughterhouse effluent: degradation kinetics applying UV radiation or H_2O_2/UV, *Water Sci. Technol.*, 60 (7), 1869-1874.

Luiz, D. B., Genena, A. K., Virmond, E., José, H. J., Moreira, R. F., Gebhardt, W. & Schröder, H. F. (2010). Identification of degradation products of erythromycin A arising from ozone and AOP treatment, *Water Environ. Res.*, 82, 797-805.

Luiz, D.B., Silva, G. S., Vaz, E.A.C., José H.J. & Moreira, R.F.P.M. (2011). Evaluation of Hybrid Treatments to Produce High Quality Reuse Water, *Water Sci. Technol.*, 63(9), 46-51.

Malato, S., Fernández-Ibáñez, P., Maldonado, M.I., Blanco, J. & Gernjak, W. (2009). Decontamination and disinfection of water by solar photocatalysis: Recent overview and trends, *Catalysis Today*, 147, 1–59.

Malato, S., Fernández-Ibáñez, P., Maldonado, M.I., Blanco, J. & Gernjak, W. (2009). Decontamination and disinfection of water by solar photocatalysis: Recent overview and trends, *Catalysis Today*, 147, 1–59.

Martín-Domínguez, A., Alarcón-Herrera, M. T., Martín-Domínguez, I. R. & González-Herrera, A. (2005). Efficiency in the disinfection of water for human consumption in rural communities using solar radiation, *Sol. Energy*, 78 (1), 31-40.

Matilainen, A. & Sillanpää, M. (2010). Removal of natural organic matter from drinking water by advanced oxidation processes, *Chemosphere*, 80, 351–365

Mellouki, A., Le Bras, G. & Sidebottom, H. (2003). Kinetics and Mechanisms of the Oxidation of Oxygenated Organic Compounds in the Gas Phase, *Chem. Rev.*, 103, 5077–5096.

Muñoz, F. & von Sonntag, C. (2000). The reactions of ozone with tertiary amines including the complexing agents nitrilotriacetic acid (NTA) and ethylenediaminetetraacetic acid (EDTA) in aqueous solution. *J. Chem. Soc., Perkin Trans.*, 2, 2029-2033.

Nadezhdin, A. D. (1988). Mechanism of Ozone Decomposition in Water. The Role of Termination. *Ind. Eng. Chem. Res.*, 27 (4), 548-550.

Navalon, S., Alvaro, M. & Garcia, H. (2010). Heterogeneous Fenton catalysts based on clays, silicas and zeolites. *Applied Catalysis B: Environmental*, 99, 1–26.

Neamtu, M. & Frimmel, F. H. (2006). Photodegradation of endocrine disrupting chemical nonylphenol by simulated solar UV-irradiation. *Sci. Total Environ.* , 369 (1-3), 295-306.

Oliver, P., Rodríguez, R. & Udaquiola, S. (2008). Water use optimization in batch process industries. Part 1: design of the water network. *J. Cleaner Product.*, 16, 1275-1286.

Oller, I., Malato, S. & Sánchez-Pérez, J.A. (2010). Combination of Advanced Oxidation Processes and biological treatments for wastewater decontamination - A review. *Science of the Total Environmen*, 409 (20), 4141-4166.

Parkinson, A., Barry, M. J., Roddick, F. A. & Hobday, M. D. (2001). Preliminary toxicity assessment of water after treatment with UV-irradiation and UVC/H_2O_2. *Water Res.* , 35 (15), 3656-3664.

Ranjit, K. T.; Viswanathan, B. (1997). Photocatalytic reduction of nitrate and nitrite ions over doped TiO_2 catalysts. *Journal of Photochemistry and Photobiology A: Chemistry*, 107, 215-220.

Rengaraj, S. & Li, X.Z. Enhanced photocatalytic reduction reaction over Bi^{3+}-TiO_2 nanoparticles in presence of formic acid as a hole scavenger. *Chemosphere*, 66, 930–938, 2007.

Rincón, A. G. & Pulgarin, C. (2006). Comparative evaluation of Fe^{3+} and TiO_2 photoassisted processes in solar photocatalytic of water. *Appl. Catal. B-Environ.*, 63, 222-231.

Rosal, R., Rodríguez, A., Perdigón-Melón, J. A., Mezcua, M., Hernando, M. D. & Léton, P., et al. (2008). Removal of pharmaceuticals and kinetics of mineralization by O_3/H_2O_2 in a biotreated municipal wastewater. *Water Res.*, 42, 3719-3728.

Sá, J., Agüera, C. A., Gross, S. & Anderson, J. A. (2009). Photocatalytic nitrate reduction over metal modified TiO_2. *Applied Catalysis B: Environmental*, 85, 192–200.

Sciacca, F., Rengifo-Herrera, J. A., Wéthé, J. & Pulgarin, C. (2010). Dramatic enhancement of solar disinfection (SODIS) of wild Salmonella sp. in PET bottles by H_2O_2 addition on natural water of Burkina Faso containing dissolved iron. *Chemosphere*, 78 (9), 1186-1191.

Shifu, C., Wei, Z. & Sujuan, Z. (2008). Preparation, characterization and activity evaluation of p-n junction photocatalyst p-ZnO/n-TiO_2. *Applied Surface Science*, 255, 2478-2484.

Sievers, M. (2011). Advanced Oxidation Processes. In: Peter Wilderer (ed.) *Treatise on Water Science*, Oxford: Academic Press, vol. 4, pp. 377-408.

Sirés, I., Oturan, N., Oturan, M.A., Rodríguez, R.M. & Garrido, J.A., Brillas, E. (2007). Electro-Fenton degradation of antimicrobials triclosan and triclocarban. *Electrochimica Acta*, 52, 5493-5503.

Song, W., Ravindran, V. & Pirbazari, M. (2008). Process optimization using a kinetics model for the ultraviolet radiation-hydrogen peroxide decomposition of natural and synthetic organic compounds in groundwater. *Chem. Eng. Sci.*, 63, 3249-3270.

Staehelin, J. & Hoigne, J. Decomposition of Ozone in Water: Rate of Initiation by Hydroxide Ions and Hydrogen Peroxide. *Environ. Sci. Technol.*, 16, 676–681, 1982.

Tokos,H. & Zorka, N.P. (2009). Synthesis of batch water network for a brewery plant. *Journal of Cleaner Production*, 17, 1465–1479

Umar, M., Aziz, H. A. & Yusoff, M. S. (2010). Trends in the use of Fenton, electro-Fenton and photo-Fenton for the treatment of landfill leachate. *Waste Management*, 30 (11), 2113-2121.

von Gunten, U. (2003a). Ozonation of drinking water: Part I. Oxidation kinetics and product formation. *Water Res.*, 37, 1443-1467.

von Gunten, U. (2003b). Ozonation of drinking water: Part II. Disinfection and by-product formation in presence of bromide, iodide or chlorine. *Water Res.*, 37, 1469-1487.

Wehbe, N., Jaafar, M., Guillard, C., Herrmann, J.-M., Miachon, S., Puzenat, E. & Guilhaume, N. (2009). Comparative study of photocatalytic and nonphotocatalytic reduction of nitrates in water. *Applied Catalysis A: General*, 368, 1-8.

Y. Li & F. Wasgestian. (1998). Photocatalytic reduction of nitrate ions on TiO2 by oxalic acid. *Journal of Photochemistry and Photobiology A: Chemistry*, 112, 255-259.

Yang, C., Xu, Y. R., Teo, K. C., Goh, N. K., Chia, L. S. & Xie, R. J. (2005). Destruction of organic pollutants in reusable wastewater using advanced oxidation technology. *Chemosphere*, 59 (3), 441-445.

Zhang, F., Jin, R., Chen, J., Shao, C., Gao, W., Li, L. & Guan, N. (2005). High photocatalytic activity and selectivity for nitrogen in nitrate reduction on Ag/TiO_2 catalyst with fine silver clusters. *Journal of Catalysis*, 232, 424–431.

3

Microorganism-Produced Enzymes in the Food Industry

Izabel Soares, Zacarias Távora,
Rodrigo Patera Barcelos and Suzymeire Baroni
Federal University of the Bahia Reconcavo / Center for Health Sciences
Brazil

1. Introduction

The application of microorganisms, such as bacteria, yeasts and principally fungi, by the food industry has led to a highly diversified food industry with relevant economical assets. Fermentation, with special reference to the production of alcoholic beverages, ethyl alcohol, dairy products, organic acids and drugs which also comprise antibiotics are the most important examples of microbiological processes.

The enzyme industry, as it is currently known, is the result of a rapid development of biotechnology, especially during the past four decades. Since ancient times, enzymes found in nature have been used in the production of food products such as cheese, beer, wine and vinegar (Kirk et al., 2002).

Enzymes which decompose complex molecules into smaller units, such as carbohydrates into sugars, are natural substances involved in all biochemical processes. Due to the enzymes' specificities, each substratum has a corresponding enzyme.

Although plants, fungi, bacteria and yeasts produce most enzymes, microbial sources-produced enzymes are more advantageous than their equivalents from animal or vegetable sources. The advantages assets comprise lower production costs, possibility of large-scale production in industrial fermentors, wide range of physical and chemical characteristics, possibility of genetic manipulation, absence of effects brought about by seasonality, rapid culture development and the use of non-burdensome methods. The above characteristics make microbial enzymes suitable biocatalysts for various industrial applications (Hasan et al., 2006).

Therefore, the identification and the dissemination of new microbial sources, mainly those which are non-toxic to humans, are of high strategic interest. Besides guaranteeing enzyme supply to different industrial processes, the development of new enzymatic systems which cannot be obtained from plants or animals is made possible and important progress in the food industry may be achieved.

2. Fungus of industrial interest

Owing to progress in the knowledge of enzymes, fungi acquired great importance in several industries since they may improve various aspects of the final product.

In fact, the fungi kingdom has approximately 200 species of *Aspergillus* which produce enzymes. They are isolated from soil, decomposing plants and air. *Aspergillus* actually produces a great number of extracellular enzymes, many of which are applied in biotechnology. *Aspergillus flavus*, *A. niger*, *A. oryzae*, *A. nidulans*, *A. fumigatus*, *A. clavatus*, *A. glaucus*, *A. ustus* and *A. versicolor* are the best known.

The remarkable interest in *Aspergillus niger*, a species of great commercial interest with a highly promising future and already widely applied in modern biotechnology, is due to its several and diverse reactions (Andersen et al, 2008).

Moreover, *A. niger* not only produces various enzymes but it is one of the few species of the fungus kingdom classified as GRAS (Generally Recognized as Safe) by the Food and Drug Administration (FDA). The species is used in the production of enzymes, its cell mass is used as a component in animal feed and its fermentation produces organic acids and other compounds of high economic value (Couto and Sanroman, 2006; Mulimania and Shankar, 2007).

2.1 Microbial enzymes for industries
2.1.1 Pectinase enzyme

Plants, filamentous fungi, bacteria and yeasts produce the pectinase enzymes group with wide use in the food and beverages industries. The enzyme is employed in the food industries for fruit ripening, viscosity clarification and reduction of fruit juices, preliminary treatment of grape juice for wine industries, extraction of tomato pulp (Adams et al., 2005), tea and chocolate fermentation (Almeida et al. 2005; da Silva et al., 2005), vegetal wastes treatment, fiber degumming in the textile and paper industries (Sorensen, et al. 2004; Kaur, et al. 2004, Taragano, et al., 1999, Lima, et al., 2000), animal nutrition, protein enrichment of baby food and oil extraction (Da Silva et al., 2005, Lima, et al., 2000).

The main application of the above mentioned enzyme group lies within the juice processing industry during the extraction, clarification and concentration stages (Martin, 2007). The enzymes are also used to reduce excessive bitterness in citrus peel, restore flavor lost during drying and improve the stableness of processed peaches and pickles. Pectinase and β-glucosidase infusion enhances the scent and volatile substances of fruits and vegetables, increases the amount of antioxidants in extra virgin olive oil and reduces rancidity.

The advantages of pectinase in juices include, for example, the clarification of juices, concentrated products, pulps and purees; a decrease in total time in their extraction; improvement in the production of juices and stable concentrated products and reduction in waste pulp; decrease of production costs; and the possibility of processing different types of fruit (Uenojo and Pastore 2007). For instance, in the production of passion fruit juice, the enzymes are added prior to filtration when the plant structure's enzymatic hydrolysis occurs. This results in the degradation of suspended solids and in viscosity decrease, speeding up the entire process (Paula, et al., 2004).

Several species of microorganisms such as *Bacillus*, *Erwinia*, *Kluyveromyces*, *Aspergillus*, *Rhizopus*, *Trichoderma*, *Pseudomonas*, *Penicillium* and *Fusarium* are good producers of pectinases (De Gregorio, et al., 2002). Among the microorganisms which synthesize pectinolytic enzymes, fungi, especially filamentous fungi, such as *Aspergillus niger* and *Aspergillus carbonarius* and *Lentinus edodes*, are preferred in industries since approximately 90% of produced enzymes may be secreted into the culture medium (Blandino et al., 2001).

In fact, several studies have been undertaken to isolate, select, produce and characterize these specific enzymes so that pectinolytic enzymes could be employed not only in food processing but also in industrial ones. High resolution techniques such as crystallography and nuclear resonance have been used for a better understanding of regulatory secretion mechanisms of these enzymes and their catalytic activity. The biotechnological importance of microorganisms and their enzymes triggers a great interest toward the understanding of gene regulation and expression of extracellular enzymes.

2.1.2 Lipases

Lipolytic enzymes such as lipases and esterases are an important group of enzymes associated with the metabolism of lipid degradation. Lipase-producing microorganisms such as *Penicillium restrictum* may be found in soil and various oil residues. The industries Novozymes, Amano and Gist Brocades already employ microbial lipases.

Several microorganisms, such as *Candida rugosa, Candida antarctica, Pseudomonas alcaligenes, Pseudomonas mendocina* and *Burkholderia cepacia,* are lipase producers (Jaeger and Reetz, 1998). Other research works have also included *Geotrichum* sp. (Burkert et al., 2004), *Geotrichum candidum* DBM 4013 (Zarevúcka et al., 2005), *Pseudomonas cepacia, Bacillus stearothermophilus, Burkholderia cepacia* (Bradoo et al., 2002), *Candida lipolytica* (Tan et al., 2003) *Bacillus coagulans* (Alkan et al., 2007), *Bacillus coagulans* BTS-3 (Kumar et al., 2005), *Pseudomonas aeruginosa* PseA (Mahanta et al., 2008), *Clostridium thermocellum* 27405 (Chinn et al., 2008), *Yarrowia lipolytica* (Dominguez et al., 2003) and *Yarrowia lipolytica* CL180 (Kim et al., 2007).

The fungi of the genera *Rhizopus, Geotrichum, Rhizomucor, Aspergillus, Candida* and *Penicillium* have been reported to be producers of several commercially used lipases.

The industrial demand for new lipase sources with different enzymatic characteristics and produced at low costs has motivated the isolation and selection of new lipolytic microorganisms. However, the production process may modify their gene expression and change their phenotypes, including growth, production of secondary metabolites and enzymes. Posterior to primary selection, the production of the enzyme should be evaluated during the growth of the promising strain in fermentation, in liquid medium and / or in the solid state (Colen et al., 2006). However, it is evident that each system will result in different proteins featuring specific characteristics with regard to reactions' catalysis and, consequently, to the products produced (Asther et al., 2002).

2.1.3 Lactase

Popularly known as lactase, beta-galactosidases are enzymes classified as hydrolases. They catalyze the terminal residue of b-lactose galactopiranosil (Galb1 - 4Glc) and produce glucose and galactose (Carminatti, 2001). Lactase's production sources are peaches, almonds and certain species of wild roses; animal organisms, such as the intestine, the brain and skin tissues; yeasts, such as *Kluyveromyces lactis, K. fragilis* and *Candida pseudotropicalis*; bacteria, such as *Escherichia coli, Lactobacillus bulgaricus, Streptococcus lactis* and *Bacillus* sp; and fungi, such as *Aspergillus foetidus, A. niger, A. oryzae* and *A. Phoenecia.*

The b-galactosidase may be found in nature, or rather, in plants, particularly almonds, peaches, apricots, apples, animal organs such as the intestine, the brain, placenta and the testis.

Lactase is produced by a widely diverse fungus population and by a large amount of microorganisms such as filamentous fungus, bacteria and yeast (Holsinger, 1997; Almeida and Pastore, 2001).

Beta-galactosidase is highly important in the dairy industry, in the hydrolysis of lactose into glucose and galactose with an improvement in the solubility and digestibility of milk and dairy products. Food with low lactose contents, ideal for lactose-intolerant consumers, is thus obtained (Mahoney, 1997; Kardel et al. 1995; Pivarnik et al., 1995). It also favors consumers who are less tolerant to dairy products' crystallization, such as milk candy, condensed milk, frozen concentrated milk, yoghurt and ice cream mixtures, (Mahoney, 1998; Kardel et al., 1995). It also produces oligosaccharides (Almeida and Pastore, 2001), the best biodegradability of whey second to lactose hydrolysis (Mlichová; Rosenberg, 2006).

2.1.4 Cellulases

Cellulases are enzymes that break the glucosidic bonds of cellulose microfibrils, releasing oligosaccharides, cellobiose and glucose (Dillon, 2004). These hydrolytic enzymes are not only used in food, drug, cosmetics, detergent and textile industries, but also in wood pulp and paper industry, in waste management and in the medical-pharmaceutical industry (Bhat and Bhat, 1997).

In the food industry, cellulases are employed in the extraction of components from green tea, soy protein, essential oils, aromatic products and sweet potato starch. Coupled to hemicellulases and pectinases they are used in the extraction and clarification of fruit juices. After fruit crushing, the enzymes are used to increase liquefaction through the degradation of the solid phase.

The above enzymes are also employed in the production process of orange vinegar and agar and in the extraction and clarification of citrus fruit juices (Orberg 1981). Cellulases supplement pectinases in juice and wine industries as extraction, clarification and filtration aids, with an increase in yield, flavor and the durability of filters and finishers (Pretel, 1997).

Cellulase is produced by a vast and diverse fungus population, such as the genera *Trichoderma, Chaetomium, Penicillium, Aspergillus, Fusarium* and *Phoma*; aerobic bacteria, such as *Acidothermus, Bacillus, Celvibrio, pseudonoma, Staphylococcus, Streptomyces* and *Xanthomonas*; and anaerobic bacteria, such as *Acetovibrio, Bacteroides, Butyrivibrio, Caldocellum, Clostridium, Erwinia, Eubacterium, Pseudonocardia, Ruminococcus* and *Thermoanaerobacter* (Moreira & Siqueira, 2006; Zhang et al., 2006). *Aspergillus* filamentous fungi stand out as major producers of cellulolytic enzymes. It is worth underscoring the filamentous fungus *Aspergillus niger*, a fermenting microorganism, which has been to produce of cellulolytic enzymes, organic acids and other products with high added value by solid-state fermentation processes. (Castro, 2006, Chandra et. al., 2007, Castro & Pereira Jr. 2010)

2.1.5 Amylases

Amylases started to be produced during the last century due to their great industrial importance. In fact, they are the most important industrial enzymes with high biotechnological relevance. Their use ranges from textiles, beer, liquor, bakery, infant feeding cereals, starch liquefaction-saccharification and animal feed industries to the chemical and pharmaceutical ones.

Currently, large quantities of microbial amylases are commercially available and are almost entirely applied in starch hydrolysis in the starch-processing industries.

The species *Aspergillus* and *Rhizopus* are highly important among the filamentous fungus for the production of amylases (Pandey et al., 1999, 2005). In the production of amyloglucosidase, the species *Aspergillus niger, A. oryzae, A. awamori, Fusarum oxysporum, Humicola insolens, Mucor pusillus, Trichoderma viride* . Species Are producing α-amylase. *Aspergillus niger, A. fumigatus, A. saitri, A. terreus, A. foetidus foetidus, Rhizopus, R. delemar* (Pandey et al. 2005), with special emphasis on the species of the genera *Aspergillus* sp., *Rhizopus* sp. and *Endomyces* sp (Soccol et al. 2003).

In fact, filamentous fungi and the enzymes produced thereby have been used in food and in the food-processing industries for decades. In fact, their GRAS (Generally Recognized as Safe) status is acknowledged by the U.S. Food and Drug Administration in the case of some species such as *Aspergillus niger* and *Aspergillus oryzae*.

The food industry use amylases for the conversion of starch into dextrins. The latter are employed in clinical formulas as stabilizers and thickeners; in the conversion of starch into maltose, in confectioneries and in the manufacture of soft drinks, beer, jellies and ice cream; in the conversion of starch into glucose with applications in the soft-drinks industry, bakery, brewery and as a subsidy for ethanol production and other bioproducts; in the conversion of glucose into fructose, used in soft drinks, jams and yoghurts (Aquino et al., 2003, Nguyen *et al.*, 2002).

Amylases provide better bread color, volume and texture in the baking industry. The use of these enzymes in bread production retards its aging process and maintains fresh bread for a longer period. Whereas fungal α-amylase provides greater fermentation potential, amyloglucosidase improves flavor and taste and a better bread crust color (Novozymes, 2005). Amylases are the most used enzymes in bread baking (Giménez et al. 2007; Haros; Rosell, Leon; Durán).

Amylases have an important role in carbon cycling contained in starch by hydrolyzing the starch molecule in several products such as dextrins and glucose. Dextrins are mainly applied in clinical formulas and in material for enzymatic saccharification. Whereas maltose is used in confectioneries and in soft drinks, beer, jam and ice cream industries, glucose is employed as a sweetener in fermentations for the production of ethanol and other bioproducts.

The above amylases break the glycosidic bonds in the amylose and amylopectin chains. Thus, amylases have an important role in commercial enzymes. They are mainly applied in food, drugs, textiles and paper industries and in detergent formulas (Peixoto et al. 2003; Najafpour, Gupta et al., 2002; Asghar et al. 2006; Mitidieri et al., 2006).

Results from strains tested for the potential production of amylases, kept at 4°C during 10 days, indicated that the wild and mutant strains still removed the nutrients required from the medium by using the available substrate. This fact showed that cooling maintained intact the amylase's activities or that a stressful condition for the fungus caused its degradation and thus consumed more compounds than normal (Smith, et al., 2010).

The best enzyme activity of microbial enzymes occurs in the same conditions that produce the microorganisms' maximum growth. Most studies on the production of amylases were undertaken from mesophilic fungi between 25 and 37°C. Best yields for α-amylase were achieved between 30 and 37°C for *Aspergillus* sp.; 30°C for *A. niger* in the production of amyloglucosidase 30°C in the production of α-amylase by *A. oryzae* (Tunga, R.; Tunga B.S, 2003), 55°C by thermophile fungus *Thermomonospora*, and 50°C by *T. lanuginosus* in the production of α-amylase (Gupta et al., 2003). However, no reports exist whether increase in enzyme activity after growth of fungus in ideal conditions and kept refrigerated at 4°C for 10 days has ever been tested.

2.1.6 Proteases

Proteases are enzymes produced by several microorganisms, namely, *Aspergillus niger, A. oryzae, Bacillus amyloliquefaciens, B. stearotermophilus, Mucor miehei, M. pusillus*. Proteases have important roles in baking, brewing and in the production of various oriental foods such as soy sauce, miso, meat tenderization and cheese manufacture.

Man's first contact with proteases activities occurred when he started producing milk curd. Desert nomads from the East used to carry milk in bags made of the goat's stomach. After long journeys, they realized that the milk became denser and sour, without understanding the process's cause. Curds became thus a food source and a delicacy. Renin, an animal-produced enzyme, is the protease which caused the hydrolysis of milk protein.

Proteases, enzymes that catalyze the cleavage of peptide bonds in proteins, are Class 3 enzymes, hydrolases, and sub-class 3.4, peptide-hydrolases. Proteases may be classified as exopeptidases and endopeptidases, according to the peptide bond to be chain-cleaved.

Recently proteases represent 60% of industrial enzymes on the market, whereas microbial proteases, particularly fungal infections, are advantageous because they are easy to obtain and to recover (Smith et al, 2009).

An enzyme extract (Neves-Souza, 2005), which coagulates milk and which is derived from the fungus *Aspergillus niger* var. *awamori*, is already produced industrially.

Although the bovine-derived protease called renin has been widely used in the manufacture of different types of cheese, the microbial-originated proteases are better for coagulant (CA) and proteolytic (PA) activities (PA). The relationship AC / AP has been a parameter to select potentially renin-producing microbial samples. The higher the ratio AC / AP, the most promising is the strain. It features high coagulation activity, with fewer risks in providing undesirable characteristics from enhanced proteolysis (Melo et al, 2002).

The microbial proteases have also been important in brewery. Beer contains poorly soluble protein complexes at lower temperature, causing turbidity when cold. The use of proteolytic enzymes to hydrolyze proteins involved in turbidity is an alternative for solving this problem.

Most commercial serine proteinases (Rawling et al, 1994), mainly neutral and alkaline, are produced by organisms belonging to the genus *Bacillus*. Whereas subtilisin enzymes are representatives of this group, similar enzymes are also produced by other bacteria such as *Thermus caldophilus, Desulfurococcus mucosus* and *Streptomyces* and by the genera *Aeromonas* and *Escherichia coli*.

In their studies and observations on the activities of proteases from *Bacillus*, Singh and Patel (2005), Silva, and Martin Delaney (2007); Sheri and Al-Mostafa (2004) and others evaluated their properties for a better performance in pH and temperature ranges.

2.1.7 Glucose oxidase

Glucose oxidase [E.C. 1.1.3.4] (GOx) is an enzyme that catalyzes the oxidation of beta-D-glucose with the formation of D-gluconolactone. The enzyme contains the prosthetic group flavin adenine dinucleotide (FAD) which enables the protein to catalyze oxidation-reduction reactions.

Guimarães et al. (2006) performed a screening of filamentous fungi which could potentially produce glucose-oxidase. Their results showed high levels of GOx in *Aspergillus versicolor* and *Rhizopus stolonifer*. The literature already suggests that the genus *Aspergillus* is a major GOx producer.

The enzyme is used in the food industry for the removing of harmful oxygen. Packaging materials and storage conditions are vital for the quality of products containing probiotic microorganisms since the microbial group's metabolism is essentially anaerobic or microaerophilic (MattilaSandholm et al., 2002). Oxygen level during storage should be consequently minimal to avoid toxicity, the organism's death and the consequent loss of the product's functionality.

Glucose oxidase may be a biotechnological asset to increase stability of probiotic bacteria in yoghurt without chemical additives. It may thus be a biotechnology alternative.

2.1.8 Glucose isomerase

Glucose isomerase (GI) (D-xylose ketol isomerase; EC 5.3.1.5) catalyzes the reversible isomerase from D-glucose and D-xylose into D-fructose and D-xylulose, respectively. The enzyme is highly important in the food industry due to its application in the production of fructose-rich corn syrup.

Interconversion of xylose into xylulose by GI is a nutritional requirement of saprophytic bacteria and has a potential application in the bioconversion of hemicellulose into ethanol. The enzyme is widely distributed among prokaryotes and several studies have been undertaken to enhance its industrial application (Bhosale et al, 1996).

The isolation of GI in *Arthrobacter* strains was performed by Smith et al. (1991), whereas Walfridsson et al. (1996) cloned gene xylA of *Thermus themophilus* and introduced it into *Saccharomyces cerevisiae* to be expressed under the control of the yeast PGK1 promoter. The search for GI thermostable enzymes has been the target of protein engineering (Hartley et al., 2000).

In fact, biotechnology has an important role in obtaining mutants with promising prospects for the commercialization of glucose isomerase enzyme.

The development of microbial strains which use xylan with prime matters for the growth or selection of GI-constituted mutants should lead towards the discontinuation of the use of xylose as an enzyme production inducer.

2.1.9 Invertase

Invertase is an S-bD-fructofuranosidase obtained from *Saccharomyces cerevisiae* and other microorganisms. The enzyme catalyzes the hydrolysis from sucrose to fructose and glucose. The manufacture of inverted sugar is one of invertase's several applications. Owing to its sweetening effects which are higher than sucrose's, it has high industrial importance and there are good prospects for its use in biotechnology.

Invertase is more active at temperatures and pH ranging respectively between 40° and 60° C and between 3.0 and 5.0. When invertase-S is applied at 0.6% rate in a solution of sucrose 40% w / w at 40°C, it inverts 80% of sucrose after 4h. 20min.

When Cardoso et al. (1998) added invertases to banana juice to assess its sweetness potential, they reported an increase in juice viscosity besides an increase in sweetness.

Alternaria sp isolates from soybean seed were inoculated in a semi-solid culture and the microorganism accumulated large amounts of extracellular invertase, which was produced constitutively without the need for an inductor.

Microorganisms, such as filamentous fungi, are good producers of invertase with potential application in various industrial sectors.

Gould et al. (2003) cultivated the filamentous fungus *Rhizopus* sp in wheat bran medium, and obtained invertase identified as polyacrylamide gel. Another potentially producing fungus invertase is *Aspergillus casiellus*. It was inoculated in soybean meal medium and after 72 hours its crude extract was isolated (Novak et al., 2010). Since most invertases used in industry are produced by yeasts, underscoring the search for fungi that produce it in great amounts is a must.

3. Final considerations

Perspectives for biotechnological production of enzymes by microorganisms.

Biotechnology is an important tool for a more refined search for microorganisms with commercial assets. Microorganisms have existed on the planet Earth during millions of years and are a source of biotechnological possibilities due to their genetic plasticity and adaptation.

The isolation of new species from several and different habitats, such as saltwater and freshwater, soils, hot springs, contaminated soils, caves and hostile environments is required. Microorganisms adapted to these conditions may have great biotechnological potential.

Methods such as the selection of mutants are simple ways to obtain strains or strains with enzymatic possibilities and these methods are widely used by researchers in academic pure science laboratories.

Geneticists also employ the recombination and selection of mutants, which feature promising characteristics, in new strains. This method consists of transferring genetic material among contrasting genotypes, obtain recombinants and use the selection for the desired need.

The recombinant DNA technology (TDR) is a very useful method under three aspects: it increases the production of a microbial enzyme during the fermentation process; it provides enzymes with new properties suitable for industrial processes, such as thermostability and ability to function outside the normal pH range; it produces enzymes from animal- and vegetable-derived microorganisms.

Extremophile microorganisms are potentially producing enzymes with useful characteristics for high temperature industrial processes.

Microorganisms that grow at low temperatures have important biotechnological assets since their enzymes are more effective at low temperatures and enables contamination risks in continuous fermentation processes. This will shorten fermentation time and enhance energy saving.

The DNA sequencing technologies have advanced greatly in recent years and important progress on genes that synthesize proteins and thus determine their function in organisms has been achieved.

Genomes of several microorganisms have been sequenced, including those which are important for the food industry, such as *Saccharomyces cerevisiae, Bacillus subtilis, Lactococcus Latis, Lactobacillus acidophilus, Lactobacillus* sp, and *Streptococcus thermophilus*. These genomes have revealed several new genes, most of which codify enzymes.

Microorganisms are potential producers of enzymes useful for the food industry. Biotechnological tools are available for the selection and obtaining of strains and for strains which increase enzymes' production on a large scale.

Progress and achievements in this area will bring improvements in the food industry and, consequently, a better health quality for mankind.

4. References

Alkan, H., Baisal, Z., Uyar, F., Dogru, M. (2007). Production of Lipase by a Newly Isolated *Bacillus coagulans* Under Solid-State Fermentation Using Melon Wastes. *Applied Biochemistry and Biotechnology*, 136, 183-192.

Almeida, C.; Brányik, T.; Moradas-Ferreira, P.; Teixeira, J. (2005). *Process Biochem.*, 40, 1937.

Almeida, M.M. de; PASTORE G.M.(2001). Galactooligossacarídeos-Produção e efeitos benéficos, Ciência e tecnologia de Alimentos, Campinas, SBCTA, Vol. 35, No. 1/2, p.12-19.

Al-Sheri, M. A.; Mostafa, S.Y. (2004). Production and some properties of protease produced by *Bacillus licheniformis* isolated from Tihamet Aseer, Saudi Arabia. *Payuistan Journal of Biological Sciences*, Vol.7, p.1631-1635.

Andersen,M R; Nielsen, M L; Nielsen, J. (2008). Metabolic model integration of the bibliome, genome, metabolome and reactome of *Aspergillus niger, Molecular, Systems Biology*, Vol. 4, No.178, 1-13.

Aquino, A.C.M.M.; jorge, J.A.; terenzi, H.F.; polizeli, M.L.T.M. (2003). Studies on athermostable a-amylase from thermophilic fungus Scytalidium *thermophilum*. *Appl.Microbiol. Biotechnol.*, 61: 323-328, 2003.

Asghar, M.; Asad M. J.; Rehman, S.; Legge, R. L. A. (2006). Thermostable α-amylase from a Moderately Thermophilic *Bacillus subtilis* Strain for StarchProcessing. *Journal of Food Engineering*, Vol.38, p. 1599-1616.

Asther, M., Haon, M., Roussos, S., Record, E., Delattre, M., Meessen-Lesage, L., Labat, M., Asther, M. (2002). Feruloyl esterase from *Aspergillus niger* a comparison of the production in solid state and submerged fermentation. *Process Biochemistry*, Vol. 38, 685-691.

Bhat, M. K. (2000). *Biotechnol. Adv.*, 18, 355.*Biochem. 37*, 497.

Bhosale, S.H.; Rao, M.B.; Deshpande,V.V.(1996). Molecular and industrial aspects of glucose isomerase. *Microbiol. Rev*, Vol.60, No.2, p.280-300.

Blandino, A.; Dravillas, K.; Cantero, D.; Pandiella, S. S.; Webb, C.(2001). *Process*

Bradoo, S., Rathi, P., Saxena, R.K., Gupta, R. (2002). Microwave-assisted rapid characterization of lipase selectivities. *Journal of Biochemistry*, 51, 115-120.

Cardoso, M.H.; Jackix, M.N.H.; Menezes, H.C.; Gonçalves, E. B.; Marques, S.V.B.(1998). Efeito da associação de pectinase, invertase e glicose isomerase na qualidade do suco de banana. Ciênc. Tecnol. Aliment. Vol. 18,No.3, ISSN 0101-2061.

Castro. A. M.; Pereira Jr, N.(2010). Produção, propriedades e aplicação de celulases na hidrólise de resíduos agroindustriais. *Quimica. Nova*, Vol.33, No.1, PP.181-188.

Chandra, M. S.; Viswanath, B.; Rajaseklar Reddy, B. (2007). Cellulolytic enzymes on lignocellulosic substrates in solid state fermentation by *Aspergillus niger*. *Indian Journal of Microbiology*, Vol.47, pp.323-328.

Chinn, M.s., Nokes, S.E., Strobel, H.J. (2008). Influence of moisture content and cultivation duration on *Clostridium thermocellum* 27405 endproduct formation in solid substrate cultivation on Avicel. *Bioresource Technology*, Vol. 99, 2664-2671.

Colen, G., Junqueira, R.G., Moraes-Santos, T. (2006). Isolation and screening of alkaline lipase-producing fungi from Brazilian savanna soil.*World Journal of Microbiology & Biotechnology*, Vol. 22, 881-885.

Couto, S.R.; Sanroman M A. (2006). Application of solid state fermentation to food industry – A review, *Journal of Food Engineering*, Vol.76, No.3, pp.291-302.

Da Silva, E. G.; Borges, M. F.; Medina, C.; Piccoli, R. H.; Schwan, R. F.; (2005) *FEMS Yeast Res. 5*, 859.

De Gregorio, A.; Mandalani, G.; Arena, N.; Nucita, F.; Tripodo, M. M.; Lo Curto, R. B. (2002). SCP and crude pectinase production by slurry-state fermentation of lemon pulps. *Bioresour. Technol.*, Vol.83, No.2, p. 8994.

Dillon, Aldo. Celulases. *In:* SAID, S.; PIETRO, R. C. L. (2004). *Enzimas como agentes biotecnológicos.* Ribeirão Preto: *Legis Summa*, p. 243-270.

Dominguez, A.; Costas, M.; Longo, M.A.; Sanromán, A. (2003). "A novel application of solid culture: production of lipases by *Yarrowia lipolytica*", *Biotechnology Letters* , Vol.25, p. 1225-1229.

Giménez, A.; Varela, P. Salvador, A.; Ares, G; Fiszman, S.; and Garitta, L. (2007). Shelf life estimation of brown pan bread: A consumer approach. *Food Quality and Preference*, Barking, Vol.18, No.2, pp. 196-204.

Glazer, A. N.; Nikaido, H. *Microbiol biotechnology.* (1995). New York: W.H.Freeman, 662 p.

Goesaert, H; Brijs, K.; Veraverbeke, W.S. Courtin, C.M. Gebruers, K. and Delcour, J.A. (2005). Wheat flour constituents: how they impact bread quality, and how to impact their functionality. *Trends in Food* Science & *Technology*, Cambridge, Vol.16, No.1-3, pp. 12-30.

Gomes, E. Guez, M. A. U, Martin, N, Silva, R. (2007). Enzimas termoestáveis: fontes, produção e aplicação industrial, *Química Nova*, Vol.30, No.1,pp 136-145.

Goto, C. E; Barbosa, E. P.; Kistner, L. C. L.; GANDRA, R. F.; ARRIAS,V. L.; , R. M. (1998). Production of amylases by *Aspergillus fumigatus*. *Revista de Microbiologia*, v.29, p.99-103.

Goulart, A. J.; Adalberto, P.R.; Monti, R. (2003). Purificação parcial de invertase a partir de *Rhizopus sp* em fermentação semi-sólida. *Alim. Nutr.*, Vol.14, No.2, pp. 199-203.

Gupta, R. et al. Microbial α-Amylases. (2003). Biotechnological Perspective. *Process Biochemistry*, Vol.38, No.11, p. 1-18.

Haros, M.; ROSELL, C. M.; BENEDITO, C. (2002). Effect of different carbohydrases on fresh bread texture and bread staling. *Eur. Food Res. Technol.*, Berlin,Vol.215, No. 5, pp. 425-430.

Hartley, B.S.; Hanlon, N. Robin, J.; Rangarajan, J.; Ragaranjan M. (2000). Glucose isomerase: insights into protein engineering for increased thermostability. *Biochimica et Biophysica Acta (BBA) - Protein Structure and Molecular Enzymology*, Vol. 1543, Issue 2, p.294-335.

Hasan, F.; Shah, A. A.; Hameed, A.; (2006), "Industrial application of microbial lipase." *Enzyme and Microbial technology*, Vol.39, No.2, pp. 235-251.

Holsinger, V. H.; kilgerman, K. H. (1991). Application of lactose in dairy foods and other foods containing lactose. *Food Technology*, Vol.45, No.1, pp. 94-95.

Houde, A.; Kademi, A.; Leblanc, D. (2004).Lipases and their industrial applications: an overview. *Appl. Biochem. Biotechnol.*, Clifton, Vol.118, No.1-3, pp. 155–170.

Jaeger, K.E & Reetz, M.T., (1998). Microbial lipases form versatile tools for biotechnology. *Tibtech*, Vol.16, pp 396-403.

Kashyap, D. R.; Chandra, S.; Kaul, A.; Tewari, R. (2000). *World J. Microbiol. Biotechnol. 16*, 277.

Kardel, G.; Furtado, M.M.; Neto, J.P.M.L. (1995). Lactase na Indústria de Laticínios (Parte 1). *Revista do Instituto de Laticínios "Cândido Tostes".* Juiz de Fora, Vol.50, No.294, pp.15-17.

Kaur, G.; Kumar, S.; Satyanarayama, T. (2004). *Bioresour. Technol. 94*, 239.

Kim, J.-T., Kang, S. G., Woo, J.-H., Lee, J.-H., Jeong, B.C., Kim, S.-J., (2007). Screening and its potential application of lipolytic activity from a marine environment:characterization of a novel esterase from *Yarrowia lipolytica* CL180. *Applied Microbiology and Biotechnology*, Vol.74, pp 820-828.

Kirk, O.; Borchert, T. V.; Fuglsang, C. C. (2002). Industrial enzyme applications. *Current Opinion Biotechnology*, Vol. 13, pp. 345 - 351.

Kumar, S., Kikon, K., Upadhyay, A., Kanwar, S.S, Gupta, R., (2005). Production, purification, and characterization of lipase from thermophilic and alkaliphilic *Bacillus coaguluns* BTS-3. *Protein Expression Purificati*, Vol. 41, 38-44.

León, A. E.; Durán, E.; Barber, C. B. (2002).Utilization of enzyme mixtures to retard bread crumb firming. *Journal of Agricultural and Food Chemistry*, Easton, Vol..50, No.6, p. 1416-1419.

Lima, A. S.; Alegre, R. M.; Meirelles, A. J. A. (2000). *Carbohydr. Polym. 50*, 63.

Mattila-Sandholm, T.; Crittenden, R.; Mogensen,G.; Fondén, R.; Saarela, M. Technological challenges for future probiotic foods. *Int. Dairy J*. Vol:12, pp. 173-182.

Mahoney, R.R. (1997), Lactose: Enzymatic Modification. In: *Lactose, water, salts and vitamins*, London, *Advanced Dairy Chemistry*, Vol.3, p.77-125.

Melo, I.S.; Valadares-Inglis, M.C.; Nass, L.L.; Valois, A.C.C. (2002). *Recursos Genéticos & Melhoramento- Microrganismos*. (1 ed), Embrapa, ISBN 85-85771-21-6- Jaguariauna-São Paulo- SP.

Milichová, Z.; Rosenberg, M. (2006). Current trends of β-galactosidase application in food techonology. *Journal of Food an Nutrition Research*, Vol.45, No.2, p. 47-54.

Mitidieri, S.; Martinelli, A. H. S.; SCHRANK, A.; VAINSTEIN, M. H. (2006). Enzymatic detergent formulation containing amylase from *Aspergillus niger*: A comparative study with commercial detergent formulations. *Bioresource Technology*, Vol.97, p. 1217-1224.

Najafpour, G. D.; Shan, C. P. (2003). Enzymatic hydrolysis of molasses. *Bioresource Technology*, v. Vol.86, p. 91-94.

Nguyen, Q.D.; Rezessy SZABO, J.M.; Claeyssens, M.; STALS, I.; Hoschke, A. (2002). Purification and characterisation of amylolytic enzymes from thermophilic fungus*Thermomyces lanuginosus* strain ATCC 34626. *Enzimes. Microbial Technol.*, Vol.31, pp.345-352.

Novaki, L.; Hasan, S.D.M.; Kadowaku, M.K. Andrade, D. (2010). Produção de invertase por fermentação em estado sólido a partir de farelo de soja. Engevista, Vol.12, No.2. pp. 131-140.

Pandey, A., Benjamin, S., Soccol, C.R., Nigam, P., Kriger, N., Soccol, V.T. (1999). The realm of microbial lipases in biotechnology.*Biotechnology Applied Biochemist*, Vol. 29, 119-131.

Patel, R. M.; Singh, S. P. (2005). Extracellular akaline protease from a newly isolated haloalkaliphilic *Bacillus* sp.: Production and Optimization. *Process Biochemistry*, Vol.40, pp.3569-3575.

Paula, B.; Moraes, I. V. M.; Castilho, C.C.; Gomes, F. S.; Matta, V. M.; Cabral, L. M. C. (2004). Melhoria na eficiência da clarificação de suco de maracujá pela combinação dos processos de microfiltração e enzimático. Boletim CEPPA, Vol.22, No.2, pp. 311-324.

Peixoto, S.C.; jorge, J.A.; Terenzi, H.F.; Polizeli, M.L.T.M. (2003). Rhizopus microsporus var. rhizopodiformis: a thermotolerant fungus with potential for production of thermostable amylases. *Int. Microbiol.*, Vol.6, pp.269-273.

Pretel, M.T.(1997). Pectic enzymes in fresh fruit processing: optimization of enzymic peeling of oranges. *Process Biochemistry.* Vol.32, No.1, pp. 43-49.

Pivarnik, L. F.; Senegal, A.G.; Rand, A.G. (1995). Hydrolytic and transgalactosil activities of commercial β-galactosidase (lactase) in food processing. *Advances in Food and Nutrition Research*, New York, Vol.38, p. 33.

Rawling, N.D., Barret, A. (1994). Families of serine peptidases. *Meth. Enzymol.*, Vol.244, pp.18-61.

Shankar, S.K.; Mulimania, V.H. (2007). β-Galactosidase production by *Aspergillus* oryzae *Bioresource Technology*, Vol.98, No.4, pp. 958-961.

Silva, C.R.; Delatorre, A. B.; Martins, M. L. L. (2007). Effect of the culture conditions on the production of an extracellular protease by thermophilic *Bacillus* sp. and some properties of the enzymatic activity. *Brazilian Journal of Microbiology*, Vol.38, pp.253-258.

Silva, G.A.B; Almeida, W.E.S; Cortes, M.S; Martins, E.S. (2009). Produção e caracterização de protease obtida por *gliocladium verticilloides* através da fermentação em estado sólido de subprodutos agroindustriais. *Revista Brasileira de Tecnologia Agroindustrial*, Vol.03, No.01, pp. 28-41. ISSN: 1981-3686.

Smith, C. A.; Rangarajan, M.; Hartley, B. S. (1991). D-Xylose (D-glucose) isomerase from Arthrobacter strain N.R.R.L. B3728. Purification and properties. *Biochemestry Journal*, Vol.1; No.277, pp. 255–261.

Soccol, C.R. & Vandenberghe, L.P.S., (2003). Overview of applied solidstate fermentation in Brazil. *Biochemical Engineering Journal*, Vol. 13, pp. 205-218.

Sorensen, J. F.; Krag, K. M.; Sibbesen, O.; Delcur, J.; Goesaert, H.; Svensson, B.; Tahir, T. A.; Brufau, J.; Perez-Vendrell, A. M.; Bellincamp, D.; D'Ovidio, R.; Camardella, L.; Giovane, A.; Bonnin, E.; Juge, N. (2004). *Biochim. Biophys. Acta, 1696*, 275.

Systems Biology, Vol.4, No.178, 1-13.

Tan, T., Zhang, M., Wang, B., Ying, C., Deng, Li. (2003). Screening of high lipase producing *Candida* sp. And production of lipase by fermentation. *Process Biochemistry*, Vol.39, pp.459-465.

Taragano, V. M.; Pilosof, A. M. R. (1999). *Enzyme Microb. Technol. 25*, 411.

Tunga, R.; Tunga, B. S. (2003).Extra-cellular Amylase Production by *Aspergillus oryzae* Under Solid State Fermentation. Japan: *International Center for Biotechnology*, Osaka University, 12 p.

Uenojo, M., Pastore, G. M. (2007). Pectinases: Aplicações Industriais e Perspectivas. *Química Nova*, Vol.30, No. 2, pp. 388-394.

Walfridsson, M.; Bao, X.; Anderlund, M.; Lilius, G. Bulow, L.; Hahn-Hagerdal, B. (1996). Ethanolic fermentation of xylose with Saccharomyces cerevisiae harboring the *Thermus thermophilus* xylA gene, which expresses an active xylose (glucose) isomerase. *Appl. Environ. Microbiol.*, Vol.62, No.12, pp. 4648-4651.

Zarevúcka, M., Kejík, Z., Saman, D., Wimmer, Z., Demnerová, K. (2005). Enantioselective properties of induced lipases from *Geotrichum*. *Enzyme and Microbial Technology*, Vol.37, pp.481-486.

4

Freezing / Thawing and Cooking of Fish

Ebrahim Alizadeh Doughikollaee
University Of Zabol
Iran

1. Introduction

One of the greatest challenges for food technologists is to maintain the quality of food products for an extended period. Fish and shellfish are perishable and, as a result of a complex series of chemical, physical, bacteriological, and histological changes occurring in muscle, easily spoiled after harvesting. These interrelated processes are usually accompanied by the gradual loss or development of different compounds that affect fish quality. Fresh seafood has a high commercial value for preservation, and the sensory and nutritional loss in conventionally frozen/thawed fish is a big concern for producers and consumers. This chapter present the effect of Freezing/Thawing and Cooking on the quality of fish.

2. Freezing

Freezing is a much preferred technique to preserve food for long period of time. It permits to preserve the flavour and the nutritional properties of foods better than storage above the initial freezing temperature. It also has the advantage of minimizing microbial or enzymatic activity. The freezing process is governed by heat and mass transfers. The concentration of the aqueous phase present in the cell will increase when extra ice crystal will appear. This phenomenon induces water diffusion from surrounding locations. Of course, intra cellular ice induces also an increase of the concentration of the intra cellular aqueous phase. The size and location of ice crystals are considered most important factors affecting the textural quality of frozen food (Martino et al., 1998). It has been recognized that the freezing rate is critical to the nucleation and growth of ice crystals. Nucleation is an activated process driven by the degree of supercooling (the difference between the ambient temperature and that of the solid-liquid equilibrium). In traditional freezing methods, ice crystals are formed by a stress-inducing ice front moving from surface to centre of food samples. Due to the limited conductive heat transfer in foods, the driving force of supercooling for nucleation is small and hence the associated low freezing rates. Thus, the traditional freezing process is generally slow, resulting in large extracellular ice crystal formations (Fennema et al., 1973; Bello et al., 1982; Alizadeh et al., 2007a), which cause texture damage, accelerate enzyme activity and increase oxidation rates during storage and after thawing.

Pressure shift freezing (PSF) has been investigated as an alternative method to the existing freezing processes. The PSF process is based on the principle of water-ice phase transition under pressure: Elevated pressure depresses the freezing point of water from 0°C to -21°C at about 210 MPa (Bridgman, 1912). The sample is cooled under pressure to a temperature just above the melting temperature of ice at this pressure. Pressure is then fast released resulting in supercooling, which enhanced instantaneous and homogeneous nucleation throughout the cooled sample (Kalichevsky et al., 1995). Ice crystal growth is then achieved at atmospheric pressure in a conventional freezer. Pressure shift freezing (PSF), as a new technique, is increasingly receiving attention in recent years because of its potential benefits for improving the quality of frozen food (Cheftel et al., 2002; Le Bail et al., 2002). PSF process has been demonstrated to produce fine and uniform ice crystals thus reducing ice-crystal related textural damage to frozen products (Chevalier et al., 2001; Zhu et al., 2003; Otero et al., 2000; Alizadeh et al., 2007a). From a point of view of the tissue damage, pressure shift freezing seemed to be beneficial, causing a very smaller cell deformation than the classic freezing process.

2.1 Freezing process

Freezing is the process of removing sensible and latent heat in order to lower product temperature generally to -18 °C or below (Delgado & Sun, 2001; Li & Sun, 2002). Figure 1 shows a typical freezing curve for the air blast freezing (ABF). The initial freezing point was about -1.5 °C and was observable at the beginning of the freezing plateau (Alizadeh et al., 2007a). The temperature dropped slowly at follow because of the water to ice transition. This freezing point depression has been classically observed in several freezing trials (not always) and has been recognized to be due to the presence of solutes and microscopic cavities in the food matrix (Pham, 1987). The nominal freezing time was used to evaluate the freezing time. The nominal freezing time is defined by the International Institute of Refrigeration as the time needed to decrease the temperature of the thermal centre to 10 °C below the initial freezing point (Institut International du Froid, 1986).

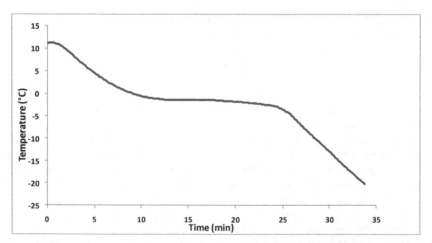

Fig. 1. A typical freezing curve of Atlantic salmon fillets obtained in air-blast freezing (Alizadeh, 2007).

Figure 2 shows a typical Pressure shift freezing curve. The process began when the unfrozen fish sample was placed in the high-pressure vessel. The temperature appeared to drop a little bit and a slight initiation of freezing can be detected at the surface of the sample after the sample was immersed into the ethanol/water medium (-18 °C) of the refrigerated bath (Alizadeh et al., 2007a). Pressurization (200 MPa) induced a temperature increase due to the adiabatic heat generated. It took about 57 min for the sample to be cooled to -18 °C without freezing which is close to the liquid-ice I equilibrium temperature (Bridgman, 1912).

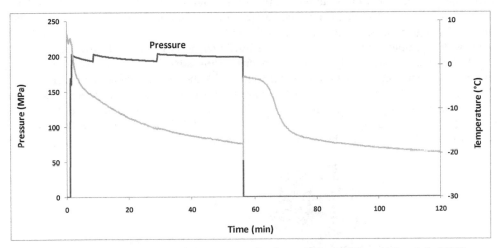

Fig. 2. A typical Pressure shift freezing curve of Atlantic salmon fillets (Alizadeh, 2007).

Then, the quick release of pressure created a large supercooling, causing a rapid and uniform nucleation, due to the shift in the freezing point back to the normal (-1.5°C) and the rapid conversion of the sensible heat (from -18 to -1.5 °C) to the latent heat. After depressurization, the temperature reached a stable temperature (-1.5 °C) for freezing at atmospheric pressure because of the latent heat release. The final step of the PSF process was similar to conventional freezing at atmospheric pressure.

2.2 Fish microstructure during freezing

Ice crystallization strongly affects the structure of tissue foods, which in turn damages the palatable attributes and consumer acceptance of the frozen products. The extent of these damages is a function of the size and location of the crystals formed and therefore depends on freezing rate. It is mentioned that slow freezing treatments usually cause texture damage to real foods due to the large and extracellular ice crystals formed (Fennema et al., 1973). Clearly, most area was occupied with the cross-section of the ice crystals larger than the muscle fibers. This means that the muscle tissue was seriously deformed after the air blast freezing at low freezing rate (1, 62 cm/h) which may cause an important shrinkage of the cells and formation of large extracellular ice crystals but it was very difficult to determine if these ice crystals were intra or extra-cellular (Figure 3). On the other hand, the intra and extracellular ice crystal have been seen during air blast freezing at high freezing rate (2, 51 cm/h). It is possible to observe the muscle fibers and analyse the size of intracellular ice crystal (Alizadeh, 2007).

The pressure shift freezing (PSF) process created smaller and more uniform ice crystals. A higher degree of supercooling should be expected during the pressure shift freezing experiments because of the rapid depressurization and the smaller ice crystals observed in the samples frozen by PSF at higher pressure. Burke et al. (1975) reported that there was a 10-fold increase in the rate of ice nucleation for each °C of supercooling. Thus, a higher pressure and lower temperature resulted in more intensive nucleation and formation of a larger number of small ice crystals. Moreover, PSF at a higher pressure is carried out at lower temperature, creating a larger temperature difference between the sample and the surrounding for final freezing completion after depressurization. This could also be a major factor affecting the final ice-crystal size in the PSF samples. Micrographs in Figure 3 also show well isotropic spread of ice crystals in the fish tissues, especially for the 200 MPa treatments. This is because the isostatic property of pressure allows isotropic supercooling and homogeneous ice nucleation. It is quite clear that the muscle fibers in the PSF treated samples (Figure 3) were well kept as compared with their original structures. Therefore, conventional freezing problems like tissue deformation and cell shrinkage could be much reduced or avoided using PSF process (Martino et al., 1998; Chevalier et al., 2000; Zhu et al., 2003; Sequeira Munoz et al., 2005; Alizadeh et al., 2007a).

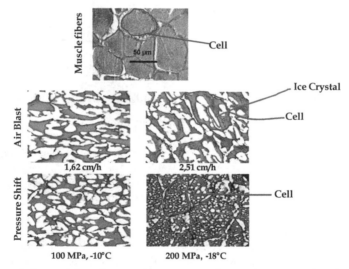

Fig. 3. Ice crystals formed in Atlantic salmon tissues during freezing (Alizadeh, 2007).

2.3 Ice crystal evolution during frozen storage

The evolution of the size of the ice crystal is important during frozen storage. It is difficult to evaluate the extracellular ice crystal for air blast freezing. But the size of high freezing rate extracellular ice crystals is smaller than low freezing rate ones. Alizadeh et al. (2007a) reported that the evolution of the intracellular ice crystal is not significant ($P<0.05$) during 6 months of storage for the air-blast (-30 °C, 4 m/s) and pressure (100 MPa) shift freezing. But for pressure shift freezing (200 MPa), the ice crystal size is changed after 6 months storage. Theoretically during frozen storage, small ice crystals have a tendency to melt and to aggregate to larger ones. It is known that the smallest ice crystals are the most unstable during storage. Indeed, the theory of ice nucleation permits to calculate the free energy of ice crystals as the sum of a surface free energy and of a volume free energy. The volume free energy increases faster than the surface free energy with increasing radius, explaining why the smaller ice crystals are more unstable. Thus the size of the ice crystals for pressure shift freezing (200 MPa) was stable for the first 3 months and then the size of the ice crystals tended to coarsen for longer storage (up to 6 months). In comparison, the size of the ice crystals obtained by pressure (100 MPa) shift freezing were much stable in size, demonstrating that a high pressure level is not necessarily required when prolonged frozen storage duration is envisaged (Alizadeh et al., 2007a).

3. Thawing process

The methodology and technique used for freezing and thawing processes play an important role in the preservation of the quality of frozen foods. Conventional thawing generally occurs more slowly than freezing, potentially causing further damages to frozen food texture. The thawing rate during conventional thawing processes is controlled by two main parameters outside the product: the surface heat transfer coefficient and the surrounding medium temperature. This medium temperature is supposed to remain below 15 °C during thawing, to prevent development of a microbial flora. The heat transfer coefficient then stays as the only parameter affecting the thawing rate at atmospheric pressure. Hence, the small temperature difference between the initial freezing point and room temperature does not allow high thawing rates (Chourot et al., 1996). Figure 4 shows a typical air blast thawing (ABT) curve. The temperature augmented to reach the melting point and temperature plateau appeared during this process.

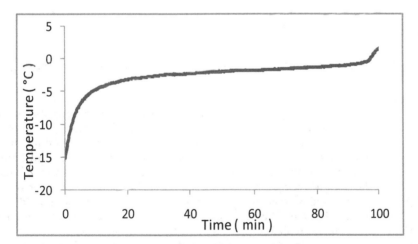

Fig. 4. A typical thawing curve of Atlantic salmon fillets obtained in air-blast thawing (Alizadeh, 2007).

Rapid thawing at low temperatures can help to prevent the loss of food quality during thawing process (Okamoto and Suzuki, 2002). This is obviously a challenge for traditional thawing processes, because the use of lower temperatures reduces the temperature difference between the frozen sample and the ambient, which is the principal driving force for the thawing process.

Pressure assisted thawing (PAT) may be attractive in comparison to conventional thawing when the quality and freshness are of primary importance. Figure 5 shows a typical pressure assisted thawing curve. Temperature increased slightly during the period of sample preparation (about 4 min) before pressurization due to the temperature difference between the sample and the medium in pressure chamber. During pressurisation the temperature decreases according to the depression of the ice-water transition under pressure (Bridgman, 1912). Then there was a temperature plateau due to the large amount latent heat needed for melting. The temperature rose quickly when thawing was completed. During the depressurization, the sample and the pressure medium were instantaneously cooled because of the positive coefficient of thermal expansion of water. To avoid ice crystal formation due to adiabatic cooling, sample temperature must be brought to a minimum level above 0 °C before releasing pressure (Cheftel et al., 2000).

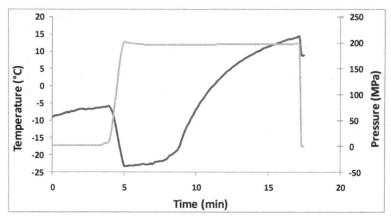

Fig. 5. A typical Pressure assisted thawing curve of Atlantic salmon fillets (Alizadeh, 2007).

3.1 Texture quality

Texture is an important quality parameter of the fish flesh. It is an important characteristic for consumer and also an important attribute for the mechanical processing of fillets by the fish food Industries. One critical quality factor influenced by freezing is food texture. Many foods are thawed from the frozen state and eaten directly, or cooked before consumption. In some cases, the texture of the thawed material is close to that of the fresh and unfrozen food. In other cases, the texture may be changed by the freezing process and yet result in a thawed product that is still acceptable to consumers. The texture of fish is modified after freezing and thawing (Figure 6). Pressure generally caused an increase in the toughness in comparison to conventional freezing and thawing (Chevalier et al., 2000; Zhu et al., 2004; Alizadeh et al., 2007b). This increase was attributed to the denaturation of proteins caused by high pressure processing. On the other hand, high pressure process was deleterious in some other aspects, mainly related to the effect of pressure on protein structures: high-pressure treatment (200 MPa) of Atlantic salmon muscle produced a partial denaturation with aggregation and insolubilization of the myosin (Alizadeh et al., 2007b). Freezing process is an important factor affecting textural quality of the fish. It is interesting to note that pressure shift freezing (200 MPa, -18 °C) induced formation of smaller and more regular ice crystals compared with air blast freezing (Chevalier et al., 2000; Alizadeh et al., 2007a; Martino et al., 1998). A tentative explanation could be that pressure shift freezing were less subjected to ice crystals injuries. Injuries involve a release of proteases (calpains and cathepsins) which are able to hydrolyse myofibrillar proteins and then to lead to quick textural changes (Jiang, 2000).

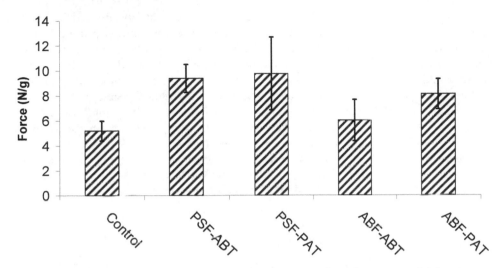

Fig. 6. Effect of Freezing (PSF, ABF) and thawing (PAT, ABT) on the texture of Atlantic salmon fillets (Alizadeh, 2007).

3.2 Colour changes

The first quality judgement made by a consumer on a food at the point of sale is its visual appearance. Appearance analyses of foods (colour and texture) are used in maintenance of food quality throughout and at the end of processing. Colour is one of the most important appearance attribute of food materials, since it influences consumer acceptability (Saenz et al., 1993).Various factors are responsible for the loss of colour during processing of food products. These include non-enzymatic and enzymatic browning and process conditions such as pH, acidity, packaging material and duration and temperature of storage.

The colour of fish is changed after freezing and thawing processes. This changes (assessed by very high colour differences ΔE) can be seen mainly caused by a strong increase in lightness (L*) and decrease for both redness (a*) and yellowness (b*) after pressure shift freezing. But this is opposite of those obtained for air blast freezing after thawing (Alizadeh et al., 2007b). Colour modifications and particularly modifications of lightness could be consequences of protein modifications. Changes in myofibrillar and sarcoplasmic proteins due to pressure could induce meat surface changes and consequently colour modifications (Ledward, 1998). The thawing process had little impact on overall colour change in fish after pressure shift freezing. But the discolouration of the flesh was visible with naked eyes after pressure assisted thawing (Alizadeh et al., 2007b). Murakami et al. (1992) also reported that an increase in all colour values (L*, a*, b*) of tuna when thawed by high pressure (50-150 MPa). This increase was stronger with increasing pressure. Furthermore, colour changes seem to be influenced by temperature, as lower temperatures caused stronger changes under the same pressure.

3.3 Drip loss

Drip loss is not only disadvantageous economically but can give rise to an unpleasant appearance and also involves loss of soluble nutrients. Drip loss during thawing is caused by irreversible damage during the freezing, storage (recrystallization), and thawing process (Pham & Mawson, 1997). Freezing can be considered as a dehydration process in which frozen water is removed from the original location in the product to form ice crystals. During thawing, the tissue may not reabsorb the melted ice crystals fully to the water content it had before freezing. This leads to undesirable release of exudate (drip loss) and toughness of texture in the thawed muscle (Mackie, 1993). Slow freezing produces larger extracellular ice crystals and resulting in more tissue damage and thawing loss. Thus, low freezing rate (air blast freezing) resulted in more drip than high freezing rate (pressure shift freezing) (Alizadeh et al., 2007b; Chevalier et al., 1999; Ngapo et al., 1999).

As shown in Figure 7, the freezing process was generally much more important than thawing for drip loss. Drip loss was reduced for all pressure shift freezing process irrespective to the thawing process but the air blast freezing resulted in a higher drip loss.

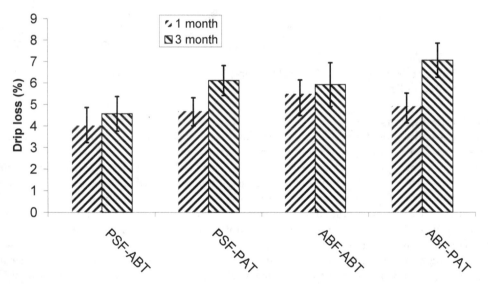

Fig. 7. Effect of Freezing (PSF, ABF) and thawing (PAT, ABT) on the drip loss of Atlantic salmon fillets (Alizadeh, 2007).

The pressure assisted thawing reduced the drip volume after conventional freezing. It can be assumed that during a slow thawing process, (corresponding to atmospheric pressure thawing), crystal accretion might occur leading to a mechanical damage of the cell membrane while thawing, and consequently in an increase of the drip volume. Pressure assisted thawing (PAT) reduced the thawing time and thus might have minimized the phenomenon of crystal accretion (Alizadeh et al., 2007b).

Few studies have reported the application of high pressure technology process for fish. Murakami et al. (1992) observed drip loss reduction in high pressure technology treated tuna meat. Chevalier et al. (1999) found that high freezing rate or high pressurization rate reduced thawing drip loss of whiting fillets (*Gadus merlangus*), but drip loss was reduced only by prolonging holding time of pressure as compared to atmospheric pressure thawing. Rouillé et al. (2002) found that high pressure technology processing (100, 150 and 200 MPa) of Spiny dogfish (*Squalus acanthias*) significantly reduced drip loss when compared with atmospheric thawing.

Crystal growth might enhance shrinkage of muscle fibers and even disrupt the cellular structure, resulting in a greater drip loss during frozen storage. Storage temperatures should be below -18 °C for optimum quality. Some studies suggest that drip loss may increase during extended frozen storage (Alizadeh et al., 2007b; Awonorin & Ayoade, 1992). Finally, drip loss seems to be a complicated process, and more studies are necessary for better understanding the phenomenon related to drip formation during pressure assisted thawing process.

4. Cooking process

Thermal processing techniques are widely used to improve eating quality and safety of food products, and to extend the shelf life of the products. Cooking is a heating process to alter the eating quality of foods and to destroy microorganisms and enzymes for food safety. Sous vide is a French term that means under vacuum. Sous vide involves the cooking of fish inside a hermetically sealed vacuum package. Cooking under vacuum (sous vide technology) defines those foods that are cooked in stable containers and stored in refrigeration. Because these products are processed at low temperatures (65 to 95 °C), the sensorial and nutritional characteristics are maximized in comparison with the sterilized products. The final product is not sterile and its shelf life depends on the applied thermal treatment and storage temperature. Figure 8 shows a typical cooking (Water bath) curve. The cooking was finished when the temperature reached at +80°C, then put in the ice water at 0°C for cooling.

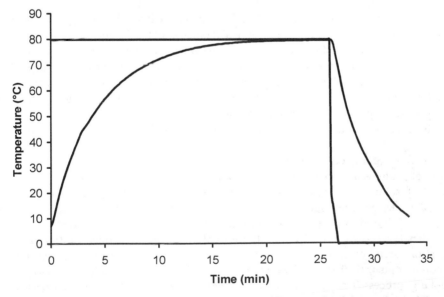

Fig. 8. Curve time/temperature during cooking under vacuum of Atlantic salmon fillets (Alizadeh, 2007).

Shaevel (1993) reported that the sous-vide process can also be used for cooking meat: this entails vacuum sealing the meat portions in plastic pouches, cooking in hot water vats for up to 4 h followed by rapid cooling at 1°C. Cooking time varied from food to food due to variation in heat transfer rates and the size of the food pieces.

4.1 Texture quality

Change in food texture was associated with heat treatment of the food such as cooking. It has been shown that thermal conditions (internal temperature) during meat cooking have a significant effect on all the meat texture profile parameters (cohesiveness, springiness, chewiness, hardness, elasticity). These reach their optimum level in the 70–80 °C range. As observed by Palka and Daun (1999), increasing the temperature to 100 °C causes the meat structure to become more compact due to a significant decrease in fiber diameter. During heating, at varying temperatures (37–75 °C), meat proteins denature and cause structural changes such as transversal and longitudinal shrinkage of muscle fibers and connective tissue shrinkage. Another effect is the destruction of cell membranes and the aggregation of sarcoplasmic proteins (Offer, 1984).

Pressure shift freezing (PSF) and cooking have an important effect on the quality of texture. Cooking process has more effect on texture than pressure shift freezing (Alizadeh et al., 2009). Meanwhile, the pressure shift freezing minimized the drip loss after cooking process. A partial cooking is a favorable fact for the pressure shift freezing, taking into account that a high proportion of fish is exposure to a cooking process before consuming. High cooking temperature can shorten cooking time and hence processing period, but it also causes high cooking loss and lower texture quality.

4.2 Protein denaturation

Denaturation can be defined as a loss of functionality caused by changes in the protein structure due to the disruption of chemical bonds and by secondary interactions with other constituents (Sikorski et al., 1976). Structural and spatial alterations can cause a range of textural and functional changes, such as the development of toughness, loss of protein solubility, loss of emulsifying capacity, and loss of water holding capacity (Miller et al., 1980; Awad et al., 1969; Dyer, 1951). In general during fish heating, sarcoplasmic and myofibrillar proteins are coagulated and denatured. The extent of these changes depends on the temperature and time and affects the yields and final quality of the fishery product.

Differential scanning calorimetry (DSC) can also be used to investigate the thermal stability of proteins and to estimate the cooking temperature of the seafood products. The proteins of salmon are denatured after freezing and cooking processes (Figure 9). Principal peaks are corresponding to myosin (42,5°C), sarcoplasmic proteins (55,5°C), collagen (64°C) and actin (73°C) (Schubring, 1999). Alizadeh et al. (2009) found that a partial denaturation of proteins, mainly to myofibrillar proteins denaturation, is induced by pressure shift freezing, similar to the effect of pressure on protein structures: high-pressure treatment (200 MPa) of sea bass muscle (Urrutia et al., 2007). As shown in Figure 9, cooking process was caused a total denaturation of proteins as comparison with pressure shift freezing. Bower (1987) showed the proteins were completely denatured under cooking process at 80°C.

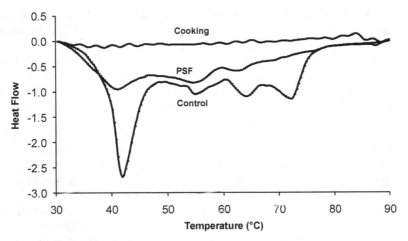

Fig. 9. Typical DSC thermograms of Atlantic salmon fillets (Alizadeh, 2007).

5. Conclusion

The quality of frozen foods is closely related to the size and distribution of ice crystals. Existence of large ice crystals within the frozen food tissue could result in mechanical damage, drip loss, and thus reduction in product quality. This chapter offers once again the advantage of pressure shift freezing process, which is widely used in the industry. Pressure shift freezing (200 MPa) process produced a large amount of small and regular intracellular ice crystals that can improved the microstructure of ice crystals (size, formation and location). The pressure shift freezing was responsible of a partial protein denaturation, which is reflected by an increase in texture. The total change of colour was observed after the freezing / thawing and cooking processes. The integration of results showed that the pressure shift freezing provides an interesting alternative compared to conventional freezing.

6. References

Alizadeh, E. (2007). Contribution à l'étude des procédés combinés de congélation – décongélation sur la qualité du saumon (*salmo salar*); impact des procédés haute pression et conventionnels. Thèse de Doctorat, Ecole Polytechnique de l'Université de Nantes, France, 77–150.

Alizadeh, E.; Chapleau, N.; De lamballerie, M. & Le bail, A. (2007a). Effects of different freezing processes on the microstructure of Atlantic salmon (*Salmo salar*) fillets. Innovative Food Science and Emerging Technologies, 8, 493–499.

Alizadeh, E.; Chapleau, N.; De lamballerie, M. & Le bail, A. (2007b). Effects of different freezing and thawing processes on the quality of Atlantic salmon (*Salmo salar*) fillets. Journal of Food Science: Food Engineering and Physical Properties, 72 (5), E279-E284.

Alizadeh, E.; Chapleau, N.; De lamballerie, M. & Le bail, A. (2009). Effect of Freezing and Cooking Processes on the Texture of Atlantic Salmon (*Salmo Salar*) Fillets.

Proceedings of the 5th CIGR Section VI International Symposium on Food Processing, Monitoring Technology in Bioprocesses and Food Quality Management (pp: 262-269), Potsdam, Germany, 31 August - 02 September 2009.

Awad, A.; Powrie, W. D. & Fennema, O. (1969). Deterioration of fresh-water whitefish muscle during frozen storage at -10°C. Journal of Food Science, 34, 1-9.

Awonorin, S. O. & Ayoade, J. A. (1992). Texture and eating quality of raw- and thawed-roasted turkey and chicken breasts as influenced by age of birds and period of frozen storage. Journal of Food Service System, 6, 241.

Bello, R. A.; Luft, J. H. & Pigott, G. M. (1982). Ultrastructural study of skeletal fish muscle after freezing at different rates. Journal of Food Science, 47, 1389-1394.

Bowers, J. A.; Craig, J. A.; Kropf, D. H. & Tucker, T. J. (1987). Flavor, color, and other characteristics of beef longissimus muscle heated to seven internal temperatures between 55 and 85°C. Journal Food Science, 52, 533-536.

Bridgman, P. W. (1912). Water in the liquid and five solid forms under pressure. Proceedings of the American Academy of Arts and Sciences, 47, 439-558.

Burke, M. J.; George, M. F. & Bryant R. G. (1975). Water in plant tissues and frost hardiness. In *Water relations of foods*, R. B. Duckworth, London, Academic Press, 111-135.

Cheftel, J. C.; Levy, J. & Dumay, E. (2000). Pressure-assisted freezing and thawing: principles and potential applications. Food Reviews International, 16, 453-483.

Cheftel, J. C.; Thiebaud, M. & Dumay, E. (2002). Pressure-assisted freezing and thawing: a review of recent studies. High Pressure Research, 22, 601–611.

Chevalier, D.; LeBail, A.; Chourot, J. M. & Chantreau, P. (1999). High pressure thawing of fish (Whiting): Influence of the process parameters on drip losses. Lebensmittel-Wissenschaft und-Technologie, 32, 25-31.

Chevalier, D.; Sequeira-Munoz, A. ; Le Bail, A. ; Simpson, B. K. & Ghoul, M. (2000). Effect of freezing conditions and storage on ice crystal and drip volume in turbot (*Scophthalmus maximus*): evaluation of pressure shift freezing vs. air-blast freezing. Innovative Food Science and Emerging Technologies, 1(3), 193-201.

Chevalier, D.; Sequeira-Munoz, A.; Le Bail, A.; Simpson, B. K. & Ghoul, M. (2001). Effect of freezing conditions and storage on ice crystal and drip volume in turbot (*Scophthalmus maximus*), evaluation of pressure shift freezing vs. air-blast freezing. Innovative Food Science and Emerging Technologies, 1, 193–201.

Chourot, J. M.; Lemaire, R.; Cornier, G. & LeBail, A. (1996). Modelling of high-pressure thawing. High Pressure Bioscience and Biotechnology, 439-444.

Delgado, A. E. & Sun, D. W. (2001). Heat and mass transfer models for predicting freezing processes – a review. Journal of Food Engineering, 47 (3),157–174.

Dyer, W. J. (1951). Protein denaturation in frozen and stored fish. Food Research, 16, 522–527.

Fennema, O. R.; Powrie, W. D. & Marth, E. H. (1973). Nature of the Freezing Process. In *Low Temperature Preservation of Foods and Living Matter*, M. Dekker, New York, 151-222.

Institut International du Froid (1986). Recommendations for the processing and handling of frozen foods. IIF, Paris, France, 32-39.

Jiang, S. T. (2000). Effect of proteinases on the meat texture and seafood quality. Food Science and Agricultural Chemistry, 2, 55-74.

Kalichevsky, M. T.; Knorr, D. & Lillford P. J. (1995). Potential food applications of high-pressure effects on ice-water transitions. Trends in Food Science & Technology, 6(8), 253-259.

Le Bail, A.; Chevalier, D.; Mussa, D. M. & Ghoul, M. (2002). High pressure freezing and thawing of foods: a review. International Journal of Refrigeration, 25(5), 504-513.

Ledward, D. A. (1998). High pressure processing of meat and fish. Fresh novel foods by high pressure. Biotechnology and Food Research, 165-175.

Li, B. & Sun, D. W. (2002). Novel methods for rapid freezing and thawing of foods— a review. Journal of Food Engineering, 54 (3),175-182.

Mackie, I. M. (1993). The effect of freezing on fish proteins. Food Review International, 9, 575-610.

Martino, M. N.; Otero, L.; Sanz, P.D. & Zaritzky, N.E. (1998). Size and location of ice crystals in pork frozen by high-pressure-assisted freezing as compared to classical methods. Meat Science, 50(3), 303-313.

Miller, A. J.; Ackerman, S. A. & Palumbo, S. A. (1980). Effects of frozen storage on functionality of meat for processing. Journal of Food Science, 45, 1466-1471.

Murakami, T.; Kimura, I.; Yamagishi, T. & Fujimoto, M. (1992). Thawing of frozen fish by hydrostatic pressure. High Pressure Biotechnology, 224, 329-331.

Ngapo, T. M.; Babare, I. H.; Reynolds, J. & Mawson, R. F. (1999). Freezing and thawing rate effects on drip loss from samples of pork. Meat Science, 53, 149-158.

Offer, G. (1984). Progress in the biochemistry physiology and structure of meat. In R. Lawrie (Ed.), Proceedings of the 30th European meeting of meat science (4th ed.) (pp. 173-241). Oxford: Elsevier.

Okamoto, A. & Suzuki, A. (2002). Effects of high hydrostatic pressure-thawing on pork meat. Trends in High Pressure Bioscience and Biotechnology, 19, 571-576.

Otero, L.; Martino, M.; Zaritzky, N.; Solas, M. & Sanz, P. D. (2000). Preservation of microstructure in peach and mango during highpressure- shift freezing. Journal of Food Science, 65(3), 466-470.

Palka, K. & Daun, H. (1999). Changes in texture, cooking losses, and miofibrillar structure of bovine M. semitendinosus during heating. Meat Science, 51, 237-243.

Pham, Q. T. (1987). Calculation of bound water in frozen food. Journal of Food Science, 52(1), 210-212.

Phan, Q. T. & Mawson, R. F. (1997). Moisture migration and ice recrystalization in frozen foods. In Quality in Frozen Food, M. C. Erickson, Y. C. Hung, New York, Chapman and Hall, 67-91.

Rouillé, J.; LeBail, A.; Ramaswamy, H. S. & Leclerc, L. (2002). High pressure thawing of fish and shellfish. Journal of Food Engineering, 53, 83-88.

Saenz, C.; Sepulveda, A. E. & Calvo, C. (1993). Color changes in concentrated juices of prickly pear (Opuntia ficus indica) during storage at different temperatures. Lebensmittel-Wissenschaft und-Technologie, 26, 417-421.

Schubring, R. (1999). DSC studies on deep frozen fishery products. Thermochimica Acta, 337, 89-95.

Sequeira-Munoz, A.; Chevalier, D.; Simpson, B. K.; Le Bail, A. & Ramaswamy, H. S. (2005). Effect of pressure-shift freezing versus air-blast freezing of carp (Cyprinus carpio) fillets: a storage study. Journal of Food Biochemistry, 29, 504-516.

Shaevel, M. L. (1993). Manufacturing of frozen prepared meals. In *Frozen Food Technology*, C. P. Mallett, London, Blackie Academic & Professional, 270–302.

Sikorski, Z.; Olley, J. & Kostuch, S. (1976). Protein changes in frozen fish. Critical Reviews in Food Science and Nutrition, 8, 97–129.

Urrutia, G.; Arabas, J.; Autio, K.; Brul, S.; Hendrickx, M.; Ka꜒kolewski, A.; Knorr, D.; Le Bail, A.; Lille, M.; Molina-Garcı́a, A. D.; Ousegui, A.; Sanz, P. D.; Shen, T. & Van Buggenhout, S. (2007). SAFE ICE: Low-temperature pressure processing of foods: Safety and quality aspects, process parameters and consumer acceptance. Journal of Food Engineering, 83, 293–315.

Zhu, S.; Le Bail, A. & Ramaswamy, H. S. (2003). Ice crystal formation in pressure shift freezing of Atlantic salmon (*Salmo salar*) as compared to classical freezing methods. Journal of Food Processing and Preservation, 27(6), 427–444.

Zhu, S.; Ramaswamy, H. S. & Simpson, B. K. (2004). Effect of high-pressure versus conventional thawing on colour, drip loss and texture of Atlantic salmon frozen by different methods. Lebensmittel-Wissenschaft und-Technologie, 37, 291-299.

5

Novel Fractionation Method for Squalene and Phytosterols Contained in the Deodorization Distillate of Rice Bran Oil

Yukihiro Yamamoto and Setsuko Hara
Seikei University
Japan

1. Introduction

To obtain the valuable constituents from natural products, "fractionation" is very important step in industrial processing. In addition, how it is efficient, how it is green for the environment, animals, nature and human, has been required for its standard. Especially in food industry, there is limitation of usage of organic solvent, which is well adopted for extraction of hydrophobic constituents from natural products. Although, hexane or ethanol has used for extraction in food industry at least in Japan, it is often less efficient than chloroform or methanol for extraction of low polar constituents such as lipids. In addition, it is difficult to remove residual solvent. Safe and simple extraction or fractionation methods have been required. In this point, supercritical fluid extraction method may be useful. In this chapter, after introducing about the key words of our study, we will present about - novel fractionation method for squalene and phytosterols contained in the deodorization distillate of rice bran oil.

2. Rice bran and rice bran oil

Rice bran, ingredient of rice bran oil (RBO), is co-product of milled rice containing pericarp, seed coat, perisperm and germ. In Japan, 1,000,000 t/year of rice bran have produced. One third is used for production of RBO and the residue is for feeding stuff, fertilizer, Japanese pickle and etc. Rice bran contains ~20% of RBO, and in RBO, unsaponifiables such as squalene and phytosterol are specifically highly contained compare to other seed oils (Table 1). Fatty acid in RBO is mainly composed by oleic acid (C18:1) and linoleic acid (C18:0). Palmitic acid (C16:0) is contained higher compare to other edible oils. In addition lower content of linolenic acid (C18:3) makes RBO stable for oxidation.

It is known that RBO exerts cholesterol lowering effect, for example, Chou et al. reported that RBO diet improves lipid abnormalities and suppress hyperinsulinemic responses in rats with streptozotocin/nicotinamide-induced type 2 diabetes (Chou et al., 2009). Physiological function of RBO is well documented in several reviews (Cicero and Gaddi, 2001; Jariwalla, 2001; Sugano et al., 1999).

In the process of produce RBO, crude RBO is steam-distilled for its flavor. The volatile component is called "deodorization distillate of RBO". The content of deodorization distillate of RBO is 0.15-0.45% varied with its condition of distillation. The deodorization distillate of RBO is viscous and typical smelled liquid. Besides diacylglyceride and free fatty acid, deodorization distillate of RBO has nearly 40% of unsaponifiable substances such as squalene, phytosterols and tocopherol. Particularly it contains ca 10% of squalene in deodorization distillate of RBO.

	RBO	Soy been	Canola	Corn
Unsaponifiables/Oil (%)	2.31	0.46	0.87	0.96
Fatty acid composition (%)				
C16:0	16.4	10.5	4.2	10.4
C18:0	1.6	3.9	2.0	1.9
C18:1	42.0	23.3	60.8	27.5
C18:2	35.8	53.0	20.6	57.2
C18:3	1.3	7.6	9.2	1.2
Others	2.9	1.7	3.2	1.8

Table 1. Chemical properties and fatty acid compositions of major edible oils.

3. Squalene

Squalene is widely found in marine animal oils as a trace component, and has been extensively studied their preventive effect in many diseases such as cardiovascular diseases and cancer (Esrich et al., 2011; Smith, 2000). Recently it is also attracted attention as food factor (Bhilwade et al., 2010). Since it has been well known that the liver oil of some varieties of shark (*Squalidae* family), especially those inhabiting the deep sea, is rich in squalene, this substance has been fractionated from shark liver oil. On the other hand, squalene obtained from shark liver oil has not been fully utilized recently on humane grounds, due to the unstable supply of shark liver oil as an industrial material, the characteristic smell of fish oil, and the large variation of the constituents. Then, attention has shifted to squalene of plant origin, and the application of this type of squalene to cosmetics, medicine and functional foods has been attempted. For example, squalene originating from olive oil is being produced in Europe. However, the production is poor to make up for the sagging production of squalene from shark liver oil. Therefore, the possibility of extracting squalene from RBO is being investigated. Furthermore, because squalene is easily oxidized owing to its structural character having a lot of carbon double bond (Fig. 1.), novel fractionation method conducting in mild condition is required.

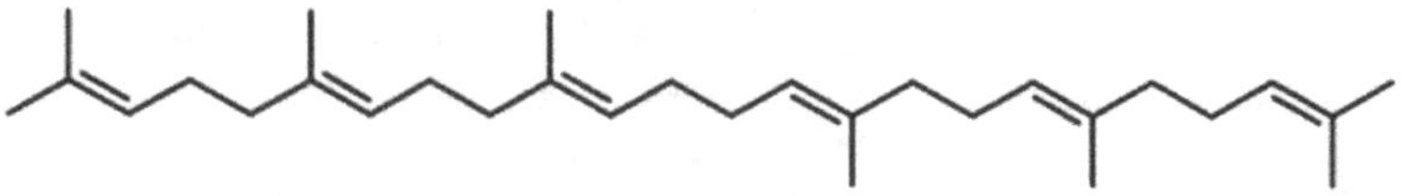

Fig. 1. Chemical structure of squalene.

4. Phytosterol

The main phytosterols in RBO are β-sitosterol, campesterol, stigmasterol and isofucosterol (Fig. 2.), and the content in phytosterols were ca 50%, 20%, 15%, 5%, respectively. It is well known that phytosterols inhibit cholesterol absorption on small intestine which results cholesterol-reducing activity of phytosterols (Gupta et al., 2011; Niijar et al., 2010; Malinowski & Gehret, 2010). In addition, the sterol content of RBO is specifically high compare to other major oils from plant- ca 1% in RBO, and 0.2-0.5% in soy bean, canola, and corn oil. These phytosterols are effective substances for utilization for functional foods called foods for specified health use.

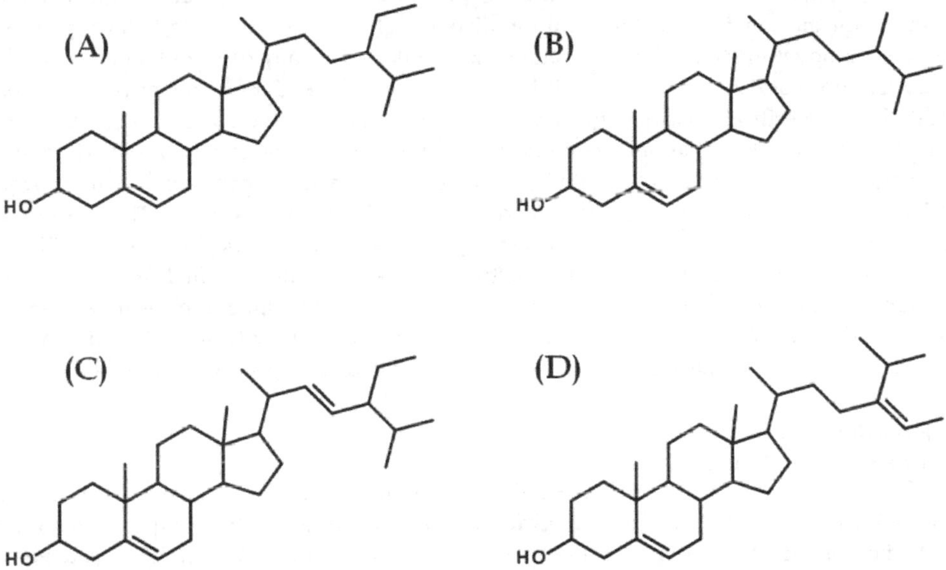

Fig. 2. Chemical structures of main phytosterols in RBO; (A) β-sitosterol, (B) campesterol, (C) stigmasterol, (D) isofucosterol.

5. Supercritical fluid extraction and supercritical fluid chromatography

Supercritical fluid is any substance at a temperature and pressure above its critical point, where distinct liquid and gas phases do not exist. It can effuse through solids like a gas, and dissolve materials like a liquid. There are mainly two techniques, supercritical fluid extraction (SFE) and supercritical fluid chromatography (SFC). For high fractionation selectivity, combination of SFE and SFC is recently adopted. Following advantages of SFE or SFC are known.

- Carried at low temperature-thermal denaturation of substances is poorly occurred.
- Inactive gas (CO_2)-denaturation and oxidation are almost ignored.
- Supercritical fluid is whole diminished after extraction.
- Bland, innocuous and harmless- applicable for food industry.

Carbon dioxide and water are the most commonly used supercritical fluids, being used for decaffeination of coffee been (Khosravi-Darani, 2010), extraction of eicosapentaenoic acid and docosapentaenoic acid from fish oil (Higashidate et al., 1990) and fractionation of terpenes from lemon peel oil (Yamauchi & Saito, 1990), respectively. For the detail of the application of supercritical fluid for variety of field including food industry, several reviews are available (Herrero et al., 2010; Khosravi-Darani, 2010; Zhao & Jiang, 2010).

6. Novel fractionation method for squalene and phytosterols contained in the deodorization distillate of rice bran oil

Although there are already some reports on methods of concentrating squalene originating from plants, the thermal and oxidative deterioration of squalene poses significant problems for its use in foods and medicines. In addition, solvent fractionation bears some difficulties related to proper solvent selectivity during processing, the elimination of the residual solvent from the squalene fraction, and the disposal of the waste fluid. In this chapter, we presented a novel method of concentrating the squalene and phytosterols contained in the deodorization distillate of rice bran oil by combining SFE, SFC and solvent fractionation. Under the supercritical condition, gas acquires unique characteristics: upon a slight pressure change, it becomes more dense, less viscous and more soluble (Xiao-Wen, 2005). Therefore, supercritical fluid has the excellent ability to separate and extract specific components from a mixture. Since the critical temperature and pressure of carbon dioxide are 31.1°C and 72.8 kg/cm², respectively, there are many advantages to using supercritical carbon dioxide for extraction at room temperature: the extract undergoes no thermal and oxidative deterioration, no remnant gas is generated in the extract at atmospheric pressure and etc as described above. SFC is determined here that the method extracting from the solution mixed with silica gel as a stationary phase, not capillary column SFC or packed column SFC.

6.1 Experimental
6.1.1 Samples and reagents

The deodorization distillate of RBO used to concentrate squalene and phytosterols was supplied from Boso Oil & Fat Co., Ltd., (Funabashi, Japan). The composition of the deodorization distillate was as follows: squalene, 8.5%; phytosterols, 3.5% (sitosterol: 2.05%, stigmasterol: 0.69%, campesterol: 0.76%); tocopherol (Toc), 1.6%; triacylglycerol (TG), 58.0%; diacylglycerol (DG) + fatty acids (FA), 26.5%. Squalene standard of special grade and 2-propanol of HPLC grade were purchased from Wako Pure Chemical Industries, Ltd. (Osaka, Japan), and all other reagents used were of special grade or first grade, as commercially available, and used without further purification, unless otherwise specified.

6.1.2 Condensation of squalene from the deodorization distillate

6.1.2.1 Condensation of squalene by SFE

To fractionate squalene from the deodorization distillate, SFE with supercritical carbon dioxide was employed. SFE makes it possible to extract the target compound by changing the density of supercritical carbon dioxide under certain conditions of temperature and/or pressure. In the present experiment, SFE was carried out with an SFC apparatus, model 880-81 (JASCO, Tokyo, Japan), and a 10- or 50-mL stainless vessel. The extractor was kept in a thermostatic oven, model CH-201 (Scinics Corporation, Tokyo, Japan) and temperatures ranging from 30 to 80°C and pressures ranging from 80 to 220 kg/cm² were examined at an extraction time of 2 h and a supercritical carbon dioxide flow rate of 7 mL/min. The squalene concentration in the extract was determined by TLC-FID analysis with hexane as the primary developing solvent and benzene/ethyl acetate (9:1, v/v) as the secondary developing solvent.

6.1.2.2 Condensation of squalene by SFC

The optimal conditions for squalene concentration by SFE were determined to be as follows: extraction temperature, 30°C; supercritical carbon dioxide pressure, 100 kg/cm²; flow rate, 7 mL/min. However, squalene was only concentrated to 25%, and the other components in the extract had higher polarity, such as TG, DG and FA. Therefore, 1 to 3-fold higher quantities of activated silica gel as compared with the deodorization distillate were placed in the extraction vessel to adsorb the components with higher polarity, and then squalene was extracted and/or concentrated under the optimal conditions listed above by means of SFC as shown in Fig. 3.

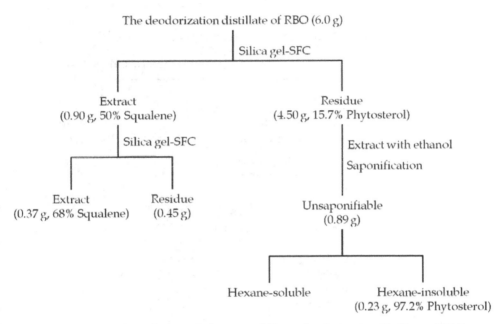

Fig. 3. Fractionation of squalene and phytosterols from deodorization distillate of RBO.

6.1.3 Condensation of phytosterols from the deodorization distillate

6.1.3.1 Condensation of phytosterols by SFC

Since phytosterols were concentrated in the residue after the extraction of squalene by SFC, further concentration of the phytosterols contained in 6 g of residue was attempted. In addition, the residue recovered as a result of SFE was also subjected to SFC to concentrate the phytosterols adsorbed onto the silica gel. In this case, the concentration ratio of phytosterols was compared with the amounts of residual phytosterols before and after the further extraction of squalene for 2 h, because phytosterols remained in the residue.

6.1.3.2 Condensation of phytosterols by solvent fractionation

The residue remaining in the vessel packed with silica gel were extracted with ethanol and then saponified by refluxing for 4 h with 3 mL of 25% potassium hydroxide aqueous solution and 40 mL each of ethanol and hexane. After the saponification, the reactant was separated into a hexane layer and a hydrated ethanol layer.

6.1.3.3 Condensation of phytosterols from the unsaponifiable components of the deodorization distillate

Another means of concentrating phytosterols was also examined, as shown in Figs. 4: the phytosterols were concentrated after the saponification of the deodorization distillate.

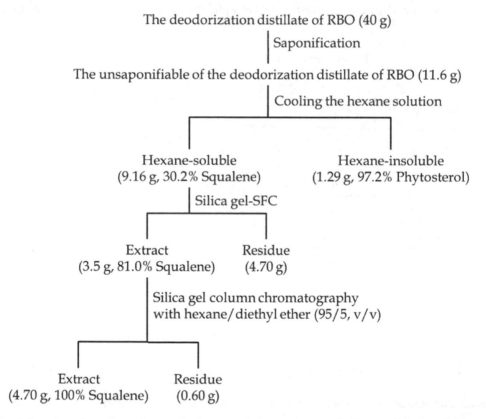

Fig. 4. Fractionation of squalene and phytosterols from unsaponifiable components of the deodorization distillate of RBO.

6.1.3.4 Preparation of highly purified squalene

To obtain more highly purified squalene, 6 g of the extract, which contained squalene having 81% purity and recovered by SFC of the unsaponifiable components, were column-chromatographed with hexane/diethyl ether (95:5, v/v) as an eluent by the procedure shown in Fig. 4.

6.1.4 Analysis

To confirm the composition of the deodorization distillate of rice bran oil, TLC-FID analysis was carried out on a Iatroscan, model MK-5 (Mitsubishi Kagaku Iatron Inc., Tokyo, Japan) equipped with a Chromatopac C-R6A (Shimadzu Corp., Kyoto, Japan). The deodorization distillate, squalene, canola oil, oleic acid, phytosterols and Toc as standard samples were loaded on silica gel sintering quartz rods (Chromarod S-III, Mitsubishi Kagaku Iatron Inc., Tokyo, Japan). Developing solvents consisting of hexane as a primary solvent to detect squalene and benzene/ethyl acetate at a ratio of 9:1 (v/v) as a secondary solvent to detect the other components were used. The fatty acid composition of the lipids was analyzed by GLC. The constituent fatty acids were methyl-esterified (Jham et al., 1982), and a fused silica capillary column, CBP1-S25-050 (0.32 mm x 25 m; Sinwa-kagaku Co., Ltd.), was connected to a GLC (model GC-18A; Shimadzu Corp., Kyoto, Japan) for constant-temperature GLC analysis. The column temperature was set at 200°C, helium was used as the carrier gas, and FID was used as the detector. A normal-phase column, Finepak SIL-5 (4.6 × 250 mm; JASCO Corp.), was connected to an HPLC (model BIP-1; JASCO), and a mobile-phase solvent that consisted of hexane and 2-propanol at a ratio of 124:1 was allowed to flow through the column at a rate of 0.7 mL/min. A fluorescence detector (model FP-2020 Plus, JASCO) with excitation at 295 nm and emission at 325 nm was used. For the quantitative analysis of Toc, 2,2,5,7,8-pentamethyl-6-chromanol (PMC) was used as the internal standard, and the calibration curves prepared beforehand for the Toc isomers [for a-Toc, y=0.4508x-0.0927 (R2=0.9682); for b-Toc, y=0.7795x-0.3017 (R2=0.9488); for g-Toc, y=1.2532x-0.7692 (R2=0.9887); and for δ-Toc, y=1.2454x-0.9934 (R2=0.9752)] were used.

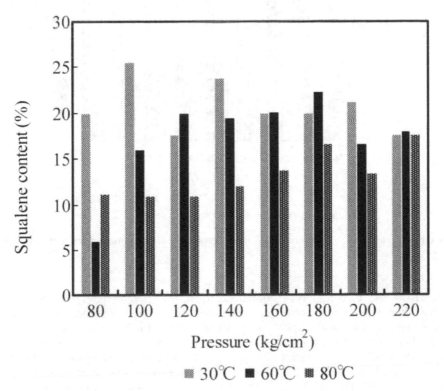

Fig. 5. Changes in squalene content under various conditions of SFE.

6.2 Results and discussion

6.2.1 Condensation of squalene from the deodorization distillate

6.2.1.1 Condensation of squalene by SFE

As shown in Fig. 5, squalene was concentrated to an average concentration of 25% with quantitative recovery under the following conditions: extraction temperature, 30°C; pressure, 100 kg/cm²; extraction time, 2 h. It was confirmed that squalene could be concentrated to 3 times the original content (8.5%) in the deodorization distillate with peroxide value (PV) of 3.0 meq/kg.

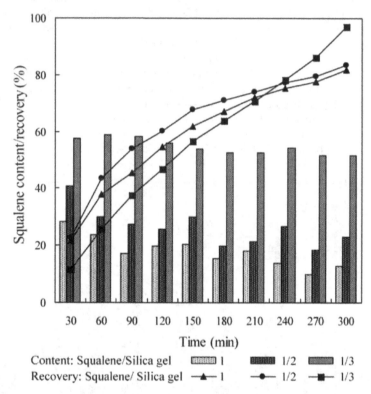

Fig. 6. Changes in the squalene content and recovery at various amounts of activated silica gel in SFC.

6.2.1.2 Condensation of squalene by SFC

As shown in Fig. 6, the higher squalene content was obtained with more than 95% recovery when 3 times more silica gel was added to the deodorization distillate and the squalene was extracted with supercritical carbon dioxide for 5 h. The extract obtained by silica-gel SFC contained, on average, 50% squalene and 38% components with higher polarity. Then, SFC of the extract obtained by the first SFC was repeated after mixing fresh silica gel. As a result, it was proved that squalene could be concentrated to 68% in the extract with PV of 3.1 meq/kg by repeating SFC with 3-fold more silica gel under the conditions stated earlier. The PV of intact squalene was 3.0 meq/kg as mentioned above.

From these results, it is suggested that the squalene was not oxidized under these conditions. Therefore, it was found that the present silica gel-SFC, with the addition of silica gel as a stationary phase into the supercritical vessel to create a chromatographic system, had a higher selectivity than mere SFE. The present silica gel-SFC is expected to become a very useful technique for concentrating squalene from the deodorization distillate of RBO as shown in Fig. 3.

6.2.2 Condensation of phytosterols from the deodorization distillate

6.2.2.1 Condensation of phytosterols by SFC

The composition of the residue with PV of 3.0 meq/kg recovered by SFC with silica gel packed as a stationary phase under the conditions of 30°C, 100 kg/cm^2, 7 mL/min and 5 h was 10.4% phytosterols, 3.9% Toc, 48.6% TG, and 37.1% DG + FA. In addition, the residue recovered from the procedure described in 6.1.2.1 contained 17.3% phytosterols under the following conditions: 30°C, 220 kg/cm^2, 7 mL/min and 7 h. From these results, it was considered that SFC did not suit the separation of phytosterols from a mixture of phytosterols, TG, DG and FA, which have nearly the same polarities, although SFC was suitable for the extraction of compounds with lower polarity, such as a squalene. Then, we examined solvent fractionation to concentrate the phytosterols from the deodorization distillate.

6.2.2.2 Condensation of phytosterols by solvent fractionation

The residual fraction shown in Fig. 3 contained 15.7% phytosterols in 4.5 g of recovered residue. The 4.5 g of residue remaining in the vessel packed with silica gel were extracted with ethanol and then saponified by refluxing for 4 h with 3 mL of 25% potassium hydroxide aqueous solution and 40 mL each of ethanol and hexane. After the saponification, the reactant was separated into a hexane layer and a hydrated ethanol layer. The unsaponifiable components thus obtained in the hexane layer were then cooled to obtain 0.23 g of crystalline phytosterols with 97.3% purity. As described above, squalene was fractionated by SFC with silica gel packed into the vessel, and phytosterols were highly concentrated from the residue by solvent fractionation. Therefore, it is considered that the combination of silica gel-SFC and solvent fractionation was a very effective means of obtaining both components with higher purity. This method, however, is rather time-consuming and costly, because SFC has to be repeated in order to concentrate the squalene, and the residue has to be extracted from the silica gel in the SFC column to concentrate the phytosterols.

6.2.2.3 Condensation of phytosterols from the unsaponifiable components of the deodorization distillate

After the saponification of the deodorization distillate (40 g) by refluxing for 4 h with 3 mL of 25% potassium hydroxide aqueous solution, 11.6 g of unsaponifiable components were recovered. Then, hexane was added to the components, and the crystalline phytosterols were recovered from the hexane-insoluble fraction under cooling. By a series of processes, 9.16 g of hexane-soluble fraction and 1.29 g of hexane-insoluble fraction were obtained and analyzed by TLC-FID. As a result, it was found that the phytosterols were concentrated to 97.2% in the hexane-insoluble fraction as shown in Table 2.

Fraction	Hexane soluble (9.16 g)	Hexane insoluble (1.29 g)
Less polar components	11.9	0
Squalene	30.2	0
Phytosterol	14.1	97.2
TG	17.5	2.8
DG	26.3	0
FA		
PV (meq/kg)	3.5	3.8

Table 2. Composition of the hexane-soluble and hexane- insoluble fractions by solvent fractionation (%).

In hexane soluble faction, saponifiables such as TG, DG and FA were contained. In this study, the condition of saponification was not finely examined. By controlling the reflux time and temperature or the concentration of potassium hydrate, TG, DG and FA could be well saponified.

It was confirmed that a combination of saponification and solvent fractionation of the deodorization distillate is an effective means of concentrating phytosterols. Since squalene was concentrated to 30.2% in the hexane soluble fraction, this fraction were subjected to SFC with silica gel under the following conditions to obtain higher purity squalene: flow rate of supercritical carbon dioxide, 3 or 7 mL/min; extraction pressure, 80-140 kg/cm². As results, it was found that higher squalene recovery tended to be obtained at faster flow rates and higher pressures. Furthermore, the squalene content in the extract reached 81.0%. From these results, it is considered that the deodorization distillate which is usually discarded as waste can be utilized for sources of functional components. In addition, the comparison of Fig. 3 with Fig. 4 indicates that the solvent fractionation of unsaponifiable components of the deodorization distillate is a practicable and convenient method of concentrating phytosterols and squalene. The combination of solvent fractionation and SFC developed in the present work is deemed to be an effective and safe means of fractionating squalene and phytosterols, which can then be used as additives in cosmetics and functional foods.

6.2.2.4 Preparation of highly purified squalene

The 3.50 g of extract containing 81.0% squalene obtained from the SFC were further purified by column chromatography with hexane/diethyl ether (95:5, v/v). As a result, 2.55 g of squalene with 100% purity and PV of 4.0 meq/kg could be obtained with 500 mL of eluate.

7. Conclusion

In this chapter, a novel method of fractionating squalene and phytosterols contained in the deodorization distillate of RBO without any oxidative rancidity was established by the combination of solvent fractionation and SFC after saponification of the deodorization distillate. Although there are some industrial production methods which are patented (Hirota & Ohta, 1997; Tsujiwaki et al., 1995; Ando et al., 1994) of squalene or squalane from the deodorization distillate of RBO, those methods have to perform many processes such as saponification, solvent fractionation, distillation, hydrogenation, and final molecular distillation to avoid the oxidative rancidity of squalene, or another is a cultivation method with yeast extracts for 6 days at 30°C. A Japan patent (Kohno, 2002) for the production for phytosterols are released from Kao Corporation, in which phytosterols are concentrated to

90-94% purity from crude phytosterols (purity: ca. 80%) with hydrocarbon solvents. Commercial squalenes obtained from shark liver oil, olive oil, and rice bran oil are now on sale as 1,000-1,500 yen/kg, 2,500 yen/kg, and 15,000 yen/kg, respectively. The market prices of phytosterols are 3,500-15,000 yen/kg based on their purities. Therefore, the present method has some merits such as a fewer operation process, time-saving, no oxidative rancidity and continuous production of the two functional components. In addition, there is a strong possibility of lower prices production than existent methods, since carbon dioxide used as a supercritical gas is costly but recyclable. It was found that the present method very safely and effectively fractionates the functional components contained in deodorization distillate, which is usually regarded as waste.

8. References

Ando, Y., Watanabe, Y. & Nakazato, M. (1994). Japan patent. 306387.

Bhilwade, HN., Tatewaki, N., Nishida, H. & Konishi, T. (2010). Squalene as Novel Food Factor. *Current Pharmaceutical Biotechnology*, Vol. 11 (No. 8): 29-36.

Chou, TW., Ma, CY., Cheng, HH. & Gaddi, A. (2009). A Rice Bran Oil Diet Improves Lipid Abnormalities and Suppress Hyperinsulinemic Responses in Rats with Streptozotocin/Nicotinamide-Induced Type 2 Diabetes. *Journal of Clinical Biochemistry and Nutrition*, Vol. 45 (No. 1): 29-36.

Cicero, AF. & Gaddi, A. (2001). Rice Bran Oil and Gamma-Oryzanol in the Treatment of Hyperlipoproteinaemias and Other Conditions. *Phytotheraphy research : (PTR)*, Vol. 14 (No. 4): 277-289.

Escrich, E., Solanas, M., Moral, R. & Escrich, R. (2011). Modulatory Effects and Molecular Mechanisms of Olive Oil and Other Dietary Lipids in Breast Cancer. *Current Pharmaceutical Design*, Vol. 17 (No. 8): 813-830.

Gupta, AK., Savopoulos, CG., Ahuja, J. & Hatzitolios, AI. (2011). Role of phytosterols in lipid-lowering: current perspectives. *QJM : Monthly Journal of the Association of Physicians*, Vol. 104 (No. 4): 301-308.

Herrero, M., Mendiola, JA., Cifuentes, A. & Ibáñez, E. (2010). Supercritical Fluid Extraction: Recent Advances and Applications. *Journal of Chromatography A*, Vol. 1217 (No. 16): 2495-2511.

Higashidate, S., Yamauchi, Y. & Saito, M. (1990). Enrichment of Eicosapentaenoic Acid and Docosahexaenoic Acid Esters from Esterified Fish Oil by Programmed Extraction-Elution with Supercritical Carbon Dioxide. *Journal of Chromatography A*, Vol. 515 (No. 31): 295-303.

Hirota, Y. & Ohta, Y. (1997). Japan patent. 176057.

Jarowalla, RJ. (2001). Rice-Bran Products: Phytonutrients with Potential Applications in Preventive and Clinical Medicine. *Drugs Under Experimental and Clinical Research*, Vol. 27 (No. 1): 17-26.

Jham, GN., Teles, FFF. & Campos, LG (1982). Use of Aqueous HCl/MeOH as Esterification Reagent for Analysis of Fatty Acid Derived from Soybean Lipids. *Journal of the American Oil Chemists Society*, Vol. 59 (No. 3): 132-133.

Khosravi-Darani, K. (2010). Research Activities on Supercritical Fluid Science in Food Biotechnology. *Critical Reviews in Food Science and Nutrition*, Vol. 50 (No. 6): 479-488.

Khono, J. (2002). Japan patent. 316996.

Malinowski, JM. & Gehret, MM. (2010). Phytosterols for Dyslipidemia. *American Journal of Health-System Pharmacy : AJHP : Official Journal of the American Society of Health-System Pharmacists*, Vol. 67 (No. 14): 1165-1173.

Niijar, PS., Burke, FM., Bloesch, A. & Rader, DJ. (2010). Role of Dietary Supplements in Lowering Low-Density Lipoprotein Cholesterol: a review. *Journal of Clinical Lipidology*, Vol. 4 (No. 4): 248-258.

Smith. (2000). Squalene: Potential Chemopreventive Agent. *Expert Opinion on Investigational Drugs*, Vol. 9 (No. 8): 1841-1848.

Sugano, M., Koba, K. & Tsuji, E. (1999). Health Benefits of Rice Bran Oil. *Anticancer Research*, Vol. 10 (No. 5A): 3651-3657.

Tsujiwaki, Y., Yamamoto, H. & Minami, K. (1995). Japan patent. 327687.

Xiao-Wen, W. (2005). Leading Technology in the 21st Century "Supercritical Fluid Extraction".*Shokuhin to Kaihatsu*, Vol. 40: 68-69.

Yamauchi, Y. & Saito, M. (1990). Fractionation of Lemon-Peel Oil by Semi-Preparative Supercritical Fluid Chromatography. *Journal of Chromatography*, Vol. 505 (No. 1): 237-246.

Zhao, HY. & Jiang, JG. (2010). Application of Chromatography Technology in the Separation of Active Components from Nature Derived Drugs. *Mini Reviews in Medicinal Chemistry*, Vol. 10 (No. 13): 1223-1234.

6

Micro and Nano Corrosion in Steel Cans Used in the Seafood Industry

Gustavo Lopez Badilla[1],
Benjamin Valdez Salas[2] and Michael Schorr Wiener[2]
[1]UNIVER, Plantel Oriente, Mexicali, Baja California
[2]Instituto de Ingenieria, Departamento de Materiales, Minerales y Corrosion,
Universidad Autonoma de Baja California
Mexico

1. Introduction

The use of metal containers for food preservation comes from the early nineteenth century, has been important in the food industry. This type of packaging was developed to improve food preservation, which were stored in glass jars, manufactured for the French army at the time of Napoleon Bonaparte (XVIII century), but were very fragile and difficult to handle in battlefields, so it was decided the produce metal containers (Brody et al, 2008). Peter Durand invented the metallic cans in1810 to improve the packaging of food. In 1903 the English company Cob Preserving, made studies to develop coatings and prevent internal and external corrosion of the cans and maintain the nutritional properties of food (Brody, 2001). Currently, the cans are made from steel sheets treated with electrolytic processes for depositing tin. In addition, a variety of plastic coatings used to protect steel from corrosion and produce the adequate brightness for printing legends on the outside of the metallic cans (Doyle, 2006). This type of metal containers does not affect the taste and smell of the product; the insulator between the food and the steel, is non-toxic and avoid the deterioration of the food. The differences between metal and glass containers, as well as the negative effects that cause damage to the environment and human health are presented in Table 1.

The wide use of steel packaging in the food industry, from their initial experimental process, has been very supportive to keep food in good conditions, with advantages over other materials such as glass, ceramics, iron and tin. The mechanical and physicochemical properties of steel help in its use for quick and easy manufacturing process (Brown, 2003). At present, exist a wide variety of foods conserved in steel cans, but in harsh environments, they corrode. Aluminum is used due to its better resistance to corrosion, but is more expensive. With metal packaging, the food reaches to the most remote places of the planet, and its stays for longer times without losing its nutritional properties, established and regulated for health standards by the Mexican Official Standards (NOM). The difference between using metal cans to glass (Table 1) indicate greater advantages for steel cans (Finkenzeller, 2003). In coastal areas, where some food companies operate, using steel cans, three types of deterioration are detected: atmospheric corrosion, filiform corrosion and microbiological corrosion. Even with the implementation of techniques and methods of protection and use of metal and plastic coatings, corrosion is still generated, being lower with the use of plastics (Lange et al, 2003). Variaitons of humidity and temperature deteriorate steel cans (Table2).

1.1 Steel

Steel is the most used metal in industrial plants, for its mechanical and thermal properties, and manufacturing facility. It is an alloy of iron and carbon. Steel manufacturing is a key part of the Mexican economy. Altos Hornos is the largest company in Mexico, with a production of more than 3,000000 tons per year, located in Monclova, Coahuila, near the U.S. border (AHMSA, 2010). Steel is used in the food industry, especially in the packaging of sardines and tuna (Lord, 2008).

PROPERTIES AND UNPROPERTIES		NEGATIVE EFFCTS	
METAL	GLASS	METÁL	GLASS
Resist the irregular handling and transport	Fragile and easily broken	Generation of filiform and microbiological corrosion	Cause spots of black color
Hermetically sealed	Not sealed; air enters	Bad sealed, creates rancidity by microbiological corrosion	High percentage of microorganisms by poor seal
Good shelf life without refrigeration at room temperature	Necessity of refrigeration of marine food	At warm and cold temperatures, foods lose their nutritional properties	At warm and cold temperatures, foods lose their nutritional properties
Accessibility manufacture	Manufacturing process complex by its fragility	By bad handling and the internal deterioration of the coating, generates filiform corrosion	Cover deformation generates gas food deterioration
No frequent supervision	Frequent supervision	Susceptible to atmospheric corrosion in indoor and outdoor environments	Broken pieces of glass are mixed with food, generating health damage
Easy recycling	Difficult to recycle	Sterilization time is 20 minutes	In sterilizing process, glass cans remains in hot water for 10 to 15 minutes and can generate bacteria

Table 1. Differences between metallic cans and glass containers in the food industry and their effect on health and environment

1.2 Metallic cans
The steel cans consist of two parts: body and ring or three parts: body, joint and ring (Figures 1a and 1b).

When a steel can is not properly sealed, it is damaged by drastic variations of humidity and temperature creating microorganisms, which cause an injury on the health of consumers (Cooksey, 2005). Every day millions of cans are produced, the companies express their interest in research studies to improve their designs. There are two main types of steel cans: tin plated and plastic coated. Plastic coatings have good resistance to compression, and the resistance to corrosion is better than the tin plate. Since the oxide layer that forms on the container surface is not completely inert, the container should be covered internally with a health compatible coating (Nachay, 2007).

Corrosion	Climatic factors	Coatings	
		Metallic	Plastic
Atmospheric External	High levels of humidity and temperature	In aggressive environments, is generated external and internal damage of steel cans	Originates stains and bad appearance without damage
Filiform Internal	Low levels of humidity	In harsh environments, are generated cracks under the coatings and, is formed the filiform corrosion	No formation of cracks in coatings as in the steel cans
Microbiological Internal	High levels of humidity	Dense black spots are formed by OH- and rancidity	Isolated black spots

Table 2. Corrosion types in coated metallic cans used in the food industry

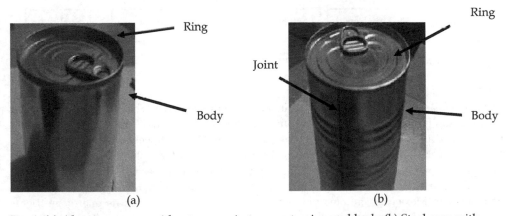

(a)   (b)

Fig. 1. (a) Aluminum cans without seams, in two parts: ring and body (b) Steel cans with seams: body, joint and rings

1.3 Production stages
The manufacture stages in a food industry are shown in Figure 2 (Avella, 2005):

Washing: Cans are cleaned thoroughly to remove the bacteria that could alter the food nutritional value.

Blanching: The product is subjected to hot water immersion to remove the enzymes that produce food darkening and the microorganisms that cause rancidity.

Preparation: Before placing food in the can the non-consumable parts of the sardine and tuna are removed, then the ingredients to prepare the food in accordance with the consumption requirements are added.

Packaging: The food is placed in the can, adding preservatives such as vinegar, syrup, salt and others to obtain the desired flavor.

Air removal: The can pass through a steam tunnel at 70 ° C, to avoid bad taste and odor.

Sealing: by soldering or with seams.

Sterilization: It is of great importance for the full elimination of microorganisms that might be left over from the previous stages, when the can is treated at temperatures of 120 ° C.

Cooling. Once sterilized the cans are cooled under running cold water or cold water immersion, from the outside without affecting the food quality.

Labeling. On the can label are placed legends with product ingredients, expiration dates and lot numbers of production.

Packaging, is made to organize the food steel cans in boxes.

Food technology specialists considers, that an adequate manufacturing process of canned foods, helps to keep certain products up to several months and years, as the case of milk powder to nine months, some vegetables and meat foods two and up to five years. A diagram summarizing all these stages is displayed in Figure 2.

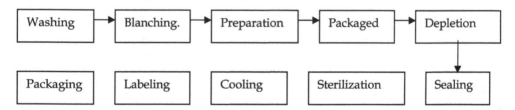

Fig. 2. Manufacturing steps in a food industry.

1.4 Sea food industry in Mexico

The main coastal cities in Mexico, with installed companies that fabricate metallic cans for sardines and tuna conservation are Acapulco, Guerrero, Ciudad del Cabo in the State of Baja California Sur; Ensenada, Baja California, Campeche, Campeche, Mazatlan, Sinaloa, Veracruz, Veracruz (Bancomext, 2010). The sardine is a blue fish with good source of omega-3, helping to lower cholesterol and triglycerides, and increase blood flow, decreasing the risk of atherosclerosis and thrombosis. Due to these nutrition properties, its widely consumed in Mexico; it contains vitamins B12, niacin and B1, using its energy nutrients (carbohydrates, fats and proteins) as a good diet. This food is important in the biological processes for formation of red blood cells, synthesis of genetic material and production of sex hormones. Tuna is an excellent food with high biological value protein, vitamins and minerals. It has minerals such as phosphorus, potassium, iron, magnesium and sodium and vitamins A, D, B, B3 and B12, which are beneficial for the care of the eyes and also provides folic acid to pregnant women. Fat rich in omega-3, is ideal for people who suffer from cardiovascular disease (FAO, 2010).

1.5 Atmospheric corrosion

Atmospheric corrosion is an electrochemical phenomenon that occurs in the wet film formed on metal surfaces by climatic factors (Lopez et al, 2011, AHRAE, 1999). One factor that determines the intensity of damage in metals exposed to atmosphere is the corrosive chemical composition in the environments. The sulphur oxides (SO_X), nitrogen oxides (NO_X), carbon oxide (CO) and sodium chloride (NaCl) that generates chloride ions (Cl^-), are the most common corrosive agents. The NaCl enters to the atmosphere from the sea; SO_X, NO_X and CO, is emitted by traffic vehicle. The joint action of the causes of pollution and weather determine the intensity and nature of corrosion processes, acting simultaneously, and increasing their effects. It is also important to mention other factors such as exposure conditions, the metal composition and properties of oxide formed, which combined, have an influence on the corrosion phenomena (Lopez, 2008). The most important atmospheric feature that is directly related to the corrosion process is moisture, which is the source of electrolyte required in the electrochemical process. In spite of existing corrosion prevention and protection systems as well as application of coatings in steel cans the corrosion control, is not easy in specific climatic regions, especially in marine regions. Ensenada which is a marine region of Mexico on the Pacific Ocean has a marine climate with cold winter mornings around 5 °C and in summer 35° C. Relative humidity (RH) is around 20% to 80%. The main climate factors analyzed were humidity, temperature and wind to determine the time of wetness (TOW) and the periods of formation of thin films of SO_X and Cl^- which were analyzed to determine the corrosivity levels (CL) in outdoors and indoors of seafood industry plants (Lopez et al, 2010).

1.6 Corrosion of steel cans

Corrosion of tinplate for food packaging is an electrochemical process that deteriorates the metallic surfaces (Ibars et al, 1992). The layer of tin provides a discontinuous structure, due to their porosity and mechanical damage or defects resulting from handling the can. The lack of continuity of the tin layer allows the food, product to be in contact with the various constituents of the steel, with the consequent formation of galvanic cells, inside of the cans. The presence of solder alloy used in the conventional container side seam is a further element in the formation of galvanic cells. Corrosion of tin plate for acidic food produces the dissolution of tin and hydrogen gas formation resulting from the cathodic reaction that accumulate in the cans. At present, the problems arising from the simultaneous presence of an aggressive environment, mechanical stress and localized corrosion (pitting) are too frequent (CGB, 2007).

1.7 Coatings

The food in steel can is protected by a metallic or plastic coating regulated by the FDA (Food Drugs Administration, USA) that does not generate any health problems in consumers (Weiss et al, 2006). The coating is adhered on the metal plate and its function is due to three main features:

- Thermal and chemical resistance assures the protection of the steel surface when a food produces a chemical attack by rancidity, changing the food taste.
- Adherence. The coating is easily attached to the inside can surface.
- Flexibility. Resistance to mechanical operations that modify the structure of the can, in the manufacturing process, such as molding shapes and bad handling.

Currently, new materials and coatings are analyzed to fit them to food variety, beverages and other canned products (Table 3). The coatings used in the food industry are organosol type, with high solids content, creating dry films with thickness 10 to 14 g/m², for manufacturing or recycling, allowing large deformation (Soroka, 2002). To improve the strength of steel, two layers of epoxy-phenolic are applied, in the organosol film. If the food suffers decomposition, it generates deformation in the can (Yam et al, 2005). Coatings are applied on the cans on the inside and outside. Since the early twentieth century, coatings manufacturers have supported the food and beverage industries, using oleoresins resins, phenolic and later in 1935, was applied vinyl coating in the beer cans. Later comes the epoxy-phenolic coatings, organosoles, acrylic and polyester (Ray, 2006).

CLASIFICACTION	
COATINGS	**DEFINITION**
Protection in indoors of cans	They are in contact with the packaged product and are called "health coatings"
Exterior coatings pigmented	Are used to underlie the decorative printing of packaging, called "white enamel" or "white lacquer. "
Exterior coatings transparent	Are used to underlie print, called "coatings of hitch. " Protects the printing inks of defectives manipulations, known as "clear coatings".
FUNCTIONS	
Protects the metal of the food	
Protects the product from contamination when steel parts from the can, are detached	
Facilitates the production	
Provides a basis for decoration	
Acts as a barrier against external corrosion and abrasion	
CHARACTERISTICS	
Must be compatible with the packaged product and to resist their aggressiveness	
Have a high adhesion to the tin or other metal	
Are free of toxic substances	
Not affect the organoleptic characteristics of the packaged product	
Not contain any items prohibited by the health legislation	
Is resistant of the sterilization and / or treatment, that is subject to packaging the product	
Adequately support to the welding of the body in two and three pieces of containers	
TYPES	
METALLIC	**PLASTICS**
Tin compound	Oleoresins, phenolic, epoxy, vinyl, acrylic, polyester
COMPOSITION	
COATING MATERIAL	**COMPOSITION**
Acrylic Polyester Copolymer	The polyester is inserted in the acrylic
Polyester resin	The resin is unsaturated
Crosslinker	The polyester contains acrylic acid, or maleic anhydride with styrene.

Table 3. Coatings used in the food industry

2. Materials and methods

2.1 Climate factors

The climate is composed of several parameters; RH and temperature are the most important factors in the damage of steel cans. Scientists that analyze the atmospheric corrosion, consider that the grade of deterioration of steel cans is due to the drastic changes in the humidity and temperature in certain times of the year, as expressed in ISO 9223 (ISO 9223, 1992). Managers and technicians of companies and members of health institutions in Mexico are concerned in some periods of the year, by the quality of seafood contained in steel cans (Moncmanova; 2007).

2.2 Corrosion testing

Pieces of steel rolls were prepared for corrosion testing simulating steel cans, which were exposed at indoor conditions of seafood plants for periods of one, three, six and twelve months in Ensenada, following the ASTM standards G 1, G 4, G 31 (ASTM, 2000). The results were correlated with RH, TOW and temperature parameters. The concentration levels of SO_X and Cl^- were evaluated with the sulfatation technique plate (SPT) and wet candle method (WCM), (ASTM G 91-97, 2010; ASTM G140-02, 2008). The industrial plants of seafood in this city are located at distances at 1 km to 10 from the sea shore. Steel plates used to fabricate steel cans with dimensions of 3 cm. X 2 cm. and 0.5 cm of width, were cleaned by immersion in an isopropyl alcohol ultrasound bath for 15 minutes (ISO 11844-1, ISO 11844-2, Lopez et al, 2008). Immediately after cleaning the steel probes were placed in sealed plastic bags, ready to be installed in the test indoor and outdoor sites. After each exposure period the steel specimens were removed, cleaned and weighed to obtain the weight loss and to calculate the corrosion rate (CR).

2.3 Examination techniques

The corrosion products morphology was examined by the scanning electron microscope (SEM) and the Auger Electron Spectroscopy (AES) techniques.

- SEM. Used to determine the morphology of the corrosion products formed by chemical agents that react with the steel internal an external surface. The SEM technique produces very high-resolution images of a sample surface. A wide range of magnifications is possible, from about 10 times to more than 500,000 times. The SEM model SSX-550 was used; revealing details less than 3.5 nm, in size from 20 to 300,000 magnifications and 0.5 V to 30kV by step.
- AES. It determines the chemical composition of elements and compounds in the steel cans and rolls, and analyzes the air pollutants deposited on the steel. With this technique we knew in detail, quickly and with a good precision, the structural form and location of corrosion at surface level which determined the type of corrosion (Clark,et al, 2006). AES analysis was performed in Bruker Quantax and ESCA / SAM 560 models, and the bombarding were obtained when samples with a beam of electrons with energy of 5keV. We made a clean surface of steel specimens analyzed with an ion beam with energy Ar $^+$ 5keV and current density of 0.3 uA / cm^3 to remove CO_2 from the atmosphere (Asami et al, 1997). The sputtering process indicates the type of film formed on the metallic surface of steel and the corrosion on separated points such as pitting corrosion.

2.4 Numerical analysis

A mathematical correlation was made applying MatLab software to determine the CL in indoors of seafood industry in Ensenada in summer and winter (Duncan et al, 2005). With this simulation we find out the deterioration grade of steel probes, correlating the climate factors (humidity and temperature) and air pollutants (CO, NO_X and SO_X), with the corrosion rate (CR).

3. Results

The generation of corrosion in steel cans is promoted by the formation of the thin film of corrosion products in their surface and the exposition of chlorides and sulfides. The seafood industry is concerned with the economic losses caused by bad appearance of the containers and the loss of nutritional properties of sardine and tuna.

3.1 Deterioration of steel cans

Levels of humidity and temperature bigger than 75% and 35 °C accelerated the CR. In summer the CR was higher after one year. For temperatures in the range from 25 °C to 35 °C, and RH level of 35% to 75%, the CR was very high. Furthermore, in winter, at temperatures around 10 °C to 20 °C and RH levels from 25% to 85%, water condensates on the metal surface and the CR increases very fast. Variations of RH in the range from 25% to 75% and temperatures from 5 °C to 30 °C, and the concentration levels of air pollutants such as sulfides and chlorides, which exceeds the permitted levels of the air quality standard (AQS), increase the corrosion process. In the autumn and winter, corrosion is generated by a film formed uniformly on the steels (Lopez et al, 2010). Exposition to SO_2 indicates more damage, compared with the effect of the chlorides on the steel surface. The maximum CR representing the deterioration with steel exposed to SO_2 was in winter for the high concentration levels of RH and the minimum was in spring. The major effect of Cl- on the deterioration of metallic surface occurred in winter and the minimum was in spring, same with the exposition of SO_2 (Table 4).

Climate factors	Sulphur oxide (SO_2) RH[a] T[b] C[c] CR[d]				Chloride (Cl-) RH[a] T[b] C[c] CR[d]			
Spring								
Max	76.3	21.4	0.24	68.8	73.5	25.6	234	59.3
Min	24.3	13.5	0.20	35.7	24.3	13.3	122	24.8
Autumn								
Max	82.7	29.8	0.31	145.7	78.9	30.6	267	136.7
Min	28.4	20.7	0.18	109.8	16.7	15.8	187	99.8
Winter								
Max	88.9	23.4	0.51	205.6	84.3	31.2	299	178.9
Min	23.2	15.6	0.36	144.6	22.1	13.7	197	122.3

[a] RH. Relative Humidity (%), [b] T. Temperature (°C), [c] C. Concentration Level of Air Pollutant (ppm), [d] CR- Corrosion rate (mg/m2.year).; Source. TPS and WCM.

Table 4. Effect of RH, temperature and air pollutants on the CR of steel (2010)

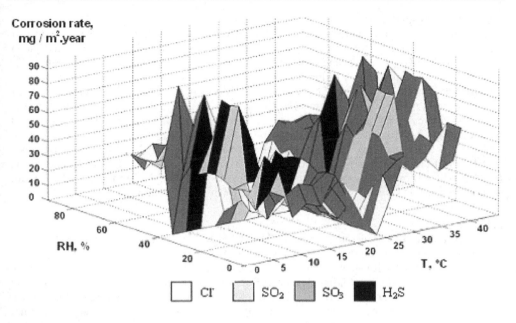

Fig. 3. CL of steel during summer exposition in Ensenada.

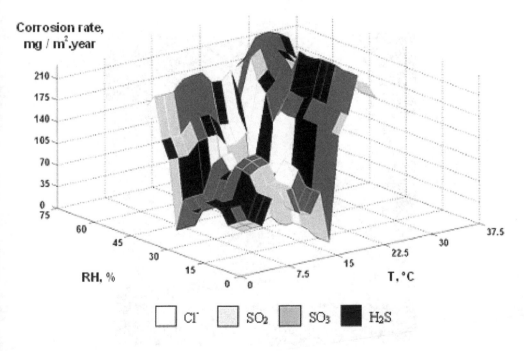

Fig. 4. CL of steel during winter exposition in Ensenada.

3.2 Corrosivity analysis

A computer model of atmospheric corrosion has been used to simulate the steel exposed to air pollutants: Cl^-, SO_2, NO_2, O_3 and H_2S from a thermoelectric station located between Tijuana and Ensenada. RH was correlated with the major CR was 35% to 55% with temperatures of 20°C to 30°C. In summer CR was different than in winter, and in both environments (Figures 3 and 4). Air pollutants such as Cl^-, NO_2 and sulfide penetrate through defects of the air conditioning systems. Figure 3 shows the CL analysis of indoors in summer, indicating the level 1, as the major aggressive environment and levels 4 the low aggressiveness grade which generate high deterioration grade of this type of materials. Some sections of the Figure 4, represents the different grades of aggressiveness, with high areas of level 1 and 2 but levels 3 and 4 exists in less percentage. RH and temperature ranges were from 25% to 80% and 20 ° C to 30° C with CR from 30 mg / m².year to 100 mg / m².year with RH and temperatures from 40% a 75% and 20 ° C to 35 ° C, with CR from 10 to 160 mg / m².year.

3.3 SEM analysis

The steel samples of 1, 3, 6 and 12 months show localized corrosion with small spots during the summer period and more corroded areas with uniform corrosion in the winter. Air pollutants that react with steel surface form corrosion products, in some zones of steel cans and rolls with chloride ions (light color) and other with sulfides (dark color), as shown in the AES analysis. Some corrosion products in the internal of steel cans appeared on the surface contaminating the sardine (Figures 5 and 6). Various microorganisms and microbial metabolites are human pathogens in sardine and tuna conserved in steel cans were detected (Figures 7 and 8). According to the most common source of these organisms, they can be grouped as follows:

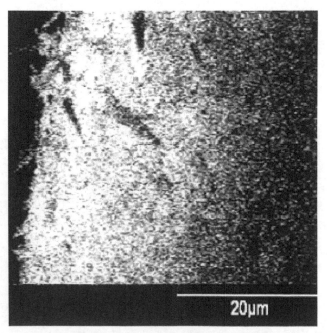

Fig. 5. Sardine contaminate with tin plate steel corrosion products.

Fig. 6. Filiform corrosion formed in internal of tin plate steel cans

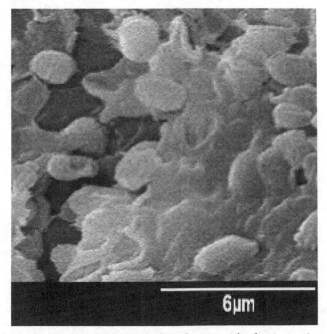

Fig. 7. Microbiological corrosion in internal of steel cans with plastic coatings.

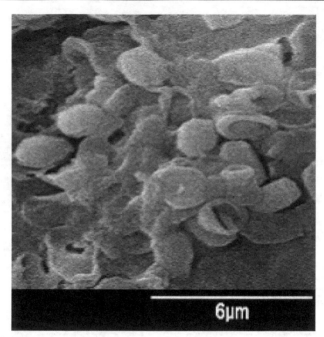

Fig. 8. Microbiological corrosion in internal of steel cans with tin coatings.

1. Endogenous. Originally present in the food before collection, including food animal, which produces the zoonoses diseases, transmitted from animals to humans in various ways, including through the digestive tract through food.
2. Exogenous. Do not exist in the food at the time of collection, at least in their internal structures, but came from the environment during production, transportation, storage, industrialization. Fungi are uni-or multicellular eukaryotic type, their most characteristic form is a mycelium or thallus and hyphae that are like branches.

3.4 AES examination

AES analyses were carried out to determine the corrosion products formed in indoor and outdoor of the steel cans. Figure 9a show scanning electron micrograph (SEM) images of areas selected for AES analysis covered by the principal corrosion products which are rich in chlorides and sulfides in tin plate steel cans evaluated. The Auger map process was performed to analyze punctual zones, indicating the presence of Cl- and S^{2-} as the main corrosive ions present in the steel corrosion products. The Auger spectra of steel cans was generated using a 5keV electron beam (Clark et al, 2006), which shows an analysis of the chemical composition of thin films formed in the steel surface (Figure 9b). The AES spectra of steel cans in the seafood plants show the surface analysis of two points evaluated in different zones of the steel probes. The peaks of steel appear between 700 and 705 eV, finding the chlorides and sulfides. In figure 10, the spectra reveals the same process as in figure 9 wit plastic coatings, with variable concentration in the chemical composition. In the two regions analyzed, where the principal pollutant was Cl- ion. In the region of steel surface were observed different concentrations of sulfide, carbon and oxygen, with low levels concentrations of H$_2$S, which damage the steel surface.

The standard thickness of 300 nm of tin plate and plastic coatings of internal and external of steel cans was determined by the AES technique with the sputtering process.

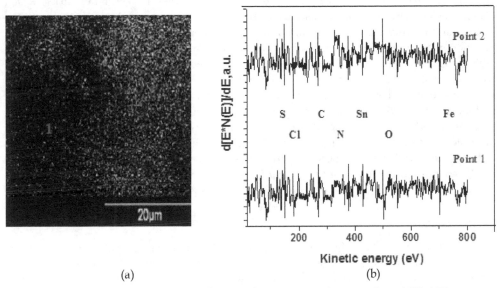

(a)   (b)

Fig. 9. Corrosion products of tin plated steel: (a) SEM microphotograph and (b) AES analysis, three months exposure.

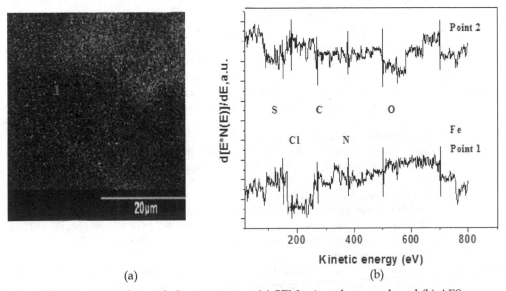

(a)   (b)

Fig. 10. Corrosion products of plastic coatings: (a) SEM microphotograph and (b) AES analysis, three months exposure.

4. Conclusions

Corrosion is the general cause of the destruction of most of engineering materials; this destructive force has always existed. The development of thermoelectric industries, which generates electricity and the increased vehicular traffic, has changed the composition of the atmosphere of industrial centers and large urban centers, making it more corrosive. Steel production and improved mechanical properties have made it a very useful material, along with these improvements, but still, it is with great economic losses, because 25% of annual world steel production is destroyed by corrosion. The corrosion of metals is one of the greatest economic losses of modern civilization. Steel used in the cannery industry for seafood suffer from corrosion. The majority of seafood industries in Mexico are on the coast, such as Ensenada, where chloride and sulfide ions are the most aggressive agents that promote the corrosion process in the steel cans The air pollutants mentioned come from traffic vehicles and from the thermoelectric industry, located around 50kms from Ensenada. Plastic coatings are better than tin coating because, on the plastic coatings do not develop microorganisms and do not damage on the internal surface.

5. References

AHRAE; Handbook; Heating, Ventilating and Ari-Conditioning; applications; *American Society of Heating*, Refrigerating and Air-Conditioning Engineers Inc.; 1999.

Altos Hornos de Mexico, Acero AHMSA para la industria petrolera y de construccion; www.ahmsa.com, consulted, may 2011.

Annual Book of ASTM Standards, 2000, Wear and Erosion: Metal Corrosion, Vol. 03.02.

Asami K., Kikuchi M. and Hashimoto K.; An auger electron spectroscopic study of the corrosion behavior of an amorphous $Zr_{40}Cu_{60}$ alloy; *Corrosion Science*; Volume 39, Issue 1, January 1997, Pages 95-106; 1997.

ASTM G140-02; Standard Test Method for Determining Atmospheric Chloride Deposition Rate by Wet Candle Method; 2008.

ASTM G91-97; Standard Practice for Monitoring Atmospheric SO2 Using the Sulfation Plate Technique (SPT); 2010.

Avella M, De Vlieger JJ, Errico ME, Fischer S, Vacca P, Volpe MG.; Biodegradable starch/clay nanocomposite films for food packaging applications. Food Chem; 93(3):467-74; 2005.

BANCOMEXT, Datos de producción pesquera en México;

Brody A, Strupinsky ER, Kline LR. Odor removers. In: Brody A, Strupinsky ER, Kline LR, editors. Active packaging for food applications. Lancaster, Pa.: Technomic Publishing Company, Inc. p 107-17; 2001.

Brody Aaron L., Bugusu Betty, Han Jung h., Sand Koelsh, Mchugh Tara H.; Innovative Food Packing Solutions; Journal of Food Science; 2008.

Brown H, Williams; Packaged product quality and shelf life. In: Coles R, McDowell D, Kirwan MJ, editors. Food packaging technology. Oxford, U.K.: Blackwell Publishing Ltd. p 65-94; 2003.

Canning Green Beans (CGB); Ecoprofile of Truitt Brothers Process; Institute for Environmental Research and Education; 2007.

Clark A. E., Pantan C. G, Hench L. L; Auger Spectroscopic Analysis of Bioglass Corrosion Films; *Journal of the American Ceramic Society*; Volume 59 Issue 1-2, Pages 37-39; 2006.

Cooksey K.; Effectiveness of antimicrobial food packaging materials. Food Addit Contam 22(10):980-7; 2005.

Doyle ME. ; Nanotechnology: a brief literature review. Food Research Institute Briefings [Internet]; http://www.wisc.edu/ fri/briefs/FRIBrief Nanotech Lit Rev.pdf; 2006.

FAO, Corporate Repository Report; consulted in

Finkenzeller K.; RFID handbook: fundamentals and applications. 2nd ed. West Sussex, U.K.: JohnWiley & Sons Ltd. 452 p.; 2003.

http://www.fao.org/documents/en/Fisheries%20and%20aquaculture%20manage ment%20and%20conservation/topicsearch/3, 2011.

http://www.financierarural.gob.mx/informacionsectorrural/Documents/Sector% 20pesquero/SectorPesqueroM%C3%A9xicoFR07.pdf, consulted, june 2011.

Ibars JR, Moreno DA, Ranninger C.; Microbial corrosion of stainless steel; Microbiologia. Nov;8(2):63-75; 1992.

ISO 11844-1:2006. Corrosion of metals and alloys - Classification of low corrosivity of indoor atmospheres- Determination and estimation of indoor corrosivity. ISO, Geneva, 2006.

ISO 11844-2:2005. Corrosion of metals and alloys - Classification of low corrosivity of indoor atmospheres - Determination and estimation attack in indoor atmospheres. ISO, Geneva, 2005.

ISO 9223:1992, Corrosion of metals and alloys, Corrosivity of Atmospheres, Classification.

Lange J, Wyser Y.; Recent innovations in barrier technologies for plastic packaging — a review. Packag Technol Sci 16:149-58.; 2003.

Lopez B. Gustavo, Valdez S. Benjamin, Schorr W. Miguel, Zlatev R., Tiznado V. Hugo, Soto H. Gerardo, De la Cruz W.; AES in corrosion of electronic devices in arid in marine environments; *AntiCorrosion Methods and Materials*; 2011.

Lopez B.G.; Ph.D. Thesis; Caracterización de la corrosión en materiales metálicos de la industria electrónica en Mexicali, B.C., 2008 (Spanish).

Lopez B.G.; Valdez S. B.; Schorr M. W.; "Spectroscopy analysis of corrosion in the electronic industry influeced by Santa Ana winds in marine environments of Mexico"; INTECH Ed. INDOOR AND OUTDOOR POLLUTON, 4; Edited by Jose A. Orosa, Book, 2011.

Lord JB.; The food industry in the United States. In: Brody AL, Lord J, editors. Developing new food products for a changing market place. 2nd ed. Boca Raton, Fla.: CRS Press. p 1-23; 2008.

Moncmanova A. Ed. ; Environmental Deterioration of Materials, WITPress, pp 108-112; 2007.

Nachay K. ; Analyzing nanotechnology. Food Tech 61(1):34-6; 2007.

Ray S, Easteal A, Quek SY, Chen XD; The potential use of polymer-clay nanocomposites in food packaging. Int J Food Eng 2(4):1-11; 2006

Soroka, W, "Fundamentals of Packaging Technology", Institute of Packaging Professionals (IoPP), ISBN 1-930268-25-4; 2002.

Walsh, Azarm, Balachandran, Magrab, Herold & Duncan Engineers Guide to MATLAB, Prentice Hall, 2010, ISBN-10: 0131991108.

Weiss J, Takhistov P, McClements J.; Functional materials in food nanotechnology; J. Food
       Science; 71(9):R107–16; 2006.
Yam KL, Takhistov PT,Miltz J.; Intelligent packaging: concepts and applications; J Food Sci
       70(1):R1–10; 2005.

7

Nanotechnology and Food Industry

Francisco Javier Gutiérrez, Mª Luisa Mussons,
Paloma Gatón and Ruth Rojo
Centro Tecnológico CARTIF. Parque Tecnológico de Boecillo,
Valladolid
España

1. Introduction

Human population will reach 9,100 million by 2,050, which supposes an increase of 34% respect present situation. This growth will occur in emerging countries mainly. As a consequence of that, there will be an increase in global demand for foods, feed and energy. Initial estimation on the increment of this world demands are in the order of 70%. Accordingly, the pressure on resources (in special water and crops) will be higher. World surface devoted to crop production will be increased, in order to meet demands on food energy and other industrial uses and so, the environmental impact is some areas could be high.

In order to obtain commodities and other feedstock in a sustainable way it is necessary to improve the current working methods and control the environmental impact, acting on:

- Water management and use,
- Agriculture,
- Animal exploitation and, in general,
- Food processing.

In a broad context, some factors affecting living standards are the water availability and food. In fact, life expectancy and health are determined by both. Considering the expected demand on food (and water) as well as the global situation, several technological challenges should be overcome to make a rational use of resources possible. In this sense, nanotechnology could suppose a great tool in solving that situation.

Where is the interest of nanotechnology? Which are the possibilities of nanotechnology in relation to foods and food production? Is it possible to improve crops using nanotechnology? Which are the advantages concerning water and use management? How is it envisaged to achieve those potential benefits at a global scale?

A general view on the nanotechnology concerning foods and water is presented in this chapter and some answers to the former questions are proposed.

According to Chaudhry et al., 2011, in food nanotechnology different categories can be specified:

- When food ingredients are processed to form nanostructures.
- When additives are used in nanocapsules (or reduced somehow to nanometric size).
- When nanomaterials are used in surfaces, contact surface development, packaging materials intelligent packaging and nanosensors.

- When nanomaterials are used in the development of new pesticides, veterinarian drugs or other agrochemical aimed at production improvement.
- When nanomaterials are used in the removal of unwanted substances from foods and water.

The state of the nanotechnology in relation with food production and water use is presented in this chapter through some examples. Why the nanotechnology shows a great application potential will be explained presenting some of the most recent findings. Therefore, a general view on how nanotechnology could be a solution for the improvement of methods used in the food production and water uses is provided.

1.1 Basic notions on nanotechnology

Nanotechnology can be briefly defined as 'the engineering of very small systems and structures'. Actually, nanotechnology consists in a set of technologies than can be developed and used in several activities and agro-food sector it is not an exception.

The main feature of nanotechnology is defined by the size of the systems of work: the 'nanoscale'. At this scale the matter presents properties which are different (and new) than the observed at macroscopic level. These properties emerge as a consequence of the size of the structures that produces the material and their interactions. Somehow, in nanotechnology, atomic and molecular forces turn on determinant factors over other effects of relevance at higher scales. Other properties and possibilities of nanotechnology, which have great interest in food technology, are high reactivity, enhanced bioavailability and bioactivity, adherence effects and surface effects of nanoparticles.

According to the former, which is the magnitude order defining nanoscale?. Although there is a consensus about considering this as a range between 10 and 100 nm, there is growing evidence on particles and systems of several hundred of nanometres with activity that can be related with a typical behaviour of nanomaterial (Ashwood et al., 2007).

The definition 'working under 100 nm' could be considered a common approach when defining nanotechnology, but this approach could be too restrictive in the agro-food sector, as well as when the effects on health and environment are considered.

At present several definitions about the term 'nano' are presented by organisations at international level. For instance, Commonwealth Scientific and Industrial Research Organisation (CSIRO) or the Food and Drug Administration (FDA) define nanomaterial as that material under 1,000 nm.

According to the ISO/TS 8004-1:2010 norm, nanotechnology is: 'The application of scientific knowledge to control and utilize matter in the nanoscale, where properties and phenomena related to size or structure can emerge'. And 'nanoscale' is defined as: 'Size range from approximately 10 nm to 100 nm'.

Regardless of the former, in this chapter nanoscale will be considered when the working range is less than 1,000 nm. This decision comes from the fact that a clear distinction can be made when considering the production of foods and water use:

- Application could be devoted to be a part of the food that will be not absorbed. (i.e. packaging)
- Application is designed to be an ingredient or additive that will be consumed.

So, since the final aim is the search for a kind of functionality (mainly physiological type) through the control of the size of the matter and, considering the fact that particles of several hundred of nanometres showing activity can be found, it seem reasonable to extent the definition up to 1,000 nm.

Despite that the potential of nanotechnology has been already recognised and applied in other industrial sectors (electronic, medical, pharmaceutical energetic sector and material sciences), the application on the agro-food sector has been limited up to now. The most promising applications of nanotechnology in foods includes: Enhancement of activity and bioavailability of nutrients and activity principles of foods, improvement of organoleptic features (colour, flavour), better consistence of food matrix, new packaging development, food traceability, safety and food monitoring during transport and storage.

In the case of the water use, the high nanoparticle activity allows the use of new purification techniques and removal of unwanted substances. In agriculture, the increase of bioavailability and the particles behaviour could boost the reduction of the side-effects in environment.

2. Potential applications of nanotechnology in the agro-food industries

2.1 Can nanotechnology enhance the access to the crops?

The pressure on crops for agricultural production will increase in the future (both the area available for cultivation and the consumption of water required for this purpose). This is related to the fact that soil is used not only for food production but also for other products of industrial interest such as biofuels. To give some figures, the annual cereal production will have to increase from 2.1 billion to 3 billion while the annual meat production will have to double.

The scientific community trust in nanotechnology as a tool which could help to solve the challenge faced by the farmer: get highly productive crops while minimizing the use of synthetic chemicals. Despite the promising use of this technology in agriculture, most applications are still under research: nanotechnology and the genetic improvement of crops and production of more selective, effective and easier to dose plant protection products (FAO / WHO, 2009; Nair et al., 2010). Nanotechnology will likely have something to say about the search of alternative energy sources and cleaning and decontamination of water or soil resources. Researches also work in the application of nanotechnology in surveillance and control systems to determine when is the best moment to harvest or to monitor crop safety (Joseph & Morrison, 2006).

But the crops will not only be benefited from the potential advantages of this technology, but also could be used as organic producers of nanoparticles.

2.1.1 Nanotechnology and crops genetic improvement

This technology, combined with others such as biotechnology, can make genetic manipulation of plants easier. It allows that nanoparticles, nanofibers or nanocapsules are used as vectors of new genetic material instead of conventional viral vectors. These new vehicles could carry a larger number of genes as well as substances able to trigger gene expression (Miller & Senjen, 2008; Nair et al., 2010) or to control the release of genetic material throughout time.

Chitosan is one of the most studied non-viral gene vectors since its positive charge density allows the condensation of DNA by electrostatic interaction, protecting the entrapped genes from nuclease action. However, chitosan nanocarriers have still not replaced conventional vectors. These vehicles are often associated with low transfection efficiencies. Authors like Zhao et al., 2011 are recently working on the enhancement of this property through modification of chitosan nanocomplexes with an octapeptide. They have condensed DNA into spherical nanoparticles of around 100-200 nm size with higher transfection efficiencies and lower cytotoxicity than those of DNA complexed with unmodified chitosan. Their transfection capacity varies depending on cell type. Authors like Mao et al., 2001 when they used as a genetic vehicle chitosan nanoparticles, found higher levels of gene expression in human kidney cells and bronchial cells than in cancerous cells. Wang et al., 2011 worked with *Jatropha curcas* callus cells and demonstrated the integration of DNA carried by this type of nanoparticles in their genome.

Chitosan nanoparticles are quite versatile, as well as their transfection efficiency can be modified, they can be PEGylated in order to control the release of genetic material as time goes by (Mao et al., 2001). This effect of time controlled genetic material release can be achieved by encapsulating pDNA into poly (DL-lactide-co-glycolide) particles. Specifically speaking, Cohen et al., 2000 achieved sustained release of pDNA over a month. Nonetheless the major advantage of nanobiotechnology is the simultaneous delivery of both DNA and effector molecules to specific sites. This effect has been achieved in intact tobacco and maize tissues when using gold-capped mesoporous silica nanoparticles (MSNs) as plasmid DNA transfers (Nair et al., 2010).

2.1.2 Production of nano-agrochemicals

Nanotechnology can help significantly to improve crop management techniques and it seems to be particularly useful in the use and handling of agrochemicals. Usually, only a very small amount of a given active compound is needed for treatment. Nowadays larger amount of chemicals, which are applied several times, are used in the field and nanotechnology could help to increase efficacy and to prevent losses. These losses can be explained for several reasons: UV degradation, hydrolysis, microbiota interaction or leaching. This way of handling chemicals brings negative effects to the environment (soil degradation, water pollution and side effects in other species). It is postulated that active substances in their nano form will allow to formulate more effective products, with time controlled action, active only under certain environmental conditions and against specific organisms, or able to reach and act on specific sites inducing changes in plant metabolism.

Research in this field is carried out by groups such as Wang et al., 2007 who formulated beta-cypermethrin nanoemulsions. They were more effective than commercial micro-emulsions of the same compound. The surface-functionalized silica nanoparticle (SNP) developed by Debnath et al., 2011 (about 15-30 nm) caused 90% mortality in *Sitophilus oryzae* when comparing its effectiveness with conventional silica (> 1 μm). Qian et al., 2011 achieved 2 weeks of validamycin sustained release when nano-sized calcium carbonate (nano-CC) was used. Guan et al., 2010 encapsulated imidacloprid with a coating of chitosan and sodium alginate through layer-by-layer self-assembly, increasing its speed rate in soil applications.

In the synthesis of products active only under specific environmental conditions work among others, Song et al., 2009 group, who showed that triazophos can be effectively protected from hydrolysis in acidic and neutral media by including it in a nano-emulsion, while its release was very easily achieved in alkaline media. Other examples of selective chemicals are the nanoparticles functionalized with the aim of being absorbed through insects' cuticle. The protective wax layer of insects is damaged and leads them to death by desiccation. This approach is safe for plants and entails less environmental damage. Nair et al., 2010 reported that certain nanoparticles are able to reach sap, and Corredor et al., 2009 found that they can move towards several sites in pumpkins. In other words, nanoparticles have the ability to act in sites different from their application point.

Substances in their nano form can affect cellular metabolism in a specific manner; for instance Ursache-Oprisan et al., 2011 reported an inhibition in chlorophyll biosynthesis caused by magnetic nanoparticles. But nanomaterials not only influence plants, but also animals like *Eisenia fetida* earthworms which avoid silver nanoparticles enriched soils as was evidenced by Shoults-Wilson et al., 2011.

2.1.3 Alternative energy sources and nanotechnology

One of the biggest challenges in this century is the search for new and feasible energy sources different from fuel, nuclear or hydroelectric power. Alternative energy should maintain socio-economic development without jeopardizing the environment. According to U.S. Energy Information Administration, more than 90% of the total energy produced during the first eight months of 2010 in the USA was obtained from coal, gas, nuclear, oil and wood. This energetic mix finds explanation in the fact that renewable energies (solar, wind, geothermal and tidal) are still non cost-effective.

Although, it is not the aim of this chapter to talk about the connection between energy and agriculture, it is worth to mention that agricultural sector requires important energy inputs (direct but also indirect). We think any step forward in the production; storage or use of new energy sources could be easier and faster with the help of nanotechnology. Then society will be moving towards a more sustainable, affordable and less dependent on fuel agriculture.

2.1.4 Water and soil resources remediation

Furthermore being vehicle for active substances (pesticides, plant growth regulators or fertilizers), nanoparticles can also be synthesized with a catalytic oxidation-reduction objective. The latter application would reduce the amount of these active substances present in the environment and also the time during which it is exposed to their action (Knauer & Bucheli, 2009). This section focuses on the application of reduction-oxidation catalytic nanoparticles as soil decontaminants, while the role of nanotechnology in water remediation is developed in section 2.3. Nanotechnology and water supplies.

The research focus is twofold in this issue: first, try to accelerate the degradation of residual pesticides in the soil, and secondly, improving these pollutants' detection and quantification methods.

Between those who try to accelerate the decomposition of these contaminants in soil, are Shen et al., 2007 who synthesized magnetic Fe_3O_4-C18 composite nanoparticles (5-10 nm) more effective than conventional C18 materials as cleaner substances of organophosphorous pesticides. Zeng et al., 2010 proved that TiO_2 nanoparticles can enhance organophosphorous and carbamates' degradation rate (30% faster) in crop fields.

Currently, several authors are working in pesticide analysis methods' optimization. Magnetic composite nanoparticle-modified screen printed carbon electrodes (Gan et al.,

2010), electrodes coated with multiwalled carbon nanotubes (Sundari & Manisankar, 2011) or nano TiO2 (Kumaravel & Chandrasekaran, 2011) are some examples of nanomaterials which have being successfully used. They all act as sensors of pesticides. As opposed to more conventional ones, these methods are more sensitive and selective.

2.1.5 Crop monitoring systems

Nanotechnology is contributing very much in the development of sensors with applications in fields such as: agriculture, farming or food packaging. In agriculture, nano-sensors could make real-time detection of: humidity, nutrient status, temperature, pH or pesticides, pollutants and pathogens presence in air, water, soil or plants. All these collected data could help to save agro-chemicals and to reduce waste production (Baruah & Dutta, 2009; Joseph & Morrison, 2006).

2.1.6 Particle farming

One of the cornerstones of nanotechnology is the synthesis of nanoparticles and their self-assembly (Gardea et al., 2002) since the methods used until now are very expensive and some of them involve the use of hazardous chemical reagents.

Alternative nanoparticles production processes are continually sought in order to make them more easily scalable and affordable. One of these routes of synthesis under study is known as "particle farming" and involves the usage of living plants or their extracts as factories of nanoparticles. This process opens up new opportunities in the recycling of wastes and could be useful in areas such as cosmetics, food or medicine.

The latest research in this field focus on the synthesis of gold and silver nanoparticles with various plants: *Medicago sativa* (Bali & Harris, 2010; Gardea et al., 2002), *Vigna radiata*, *Arachis hypogaea*, *Cyamopsis tetragonolobus*, *Zea mays* , *Pennisetum glaucum*, *Sorghum vulgare* (Rajani et al., 2010), *Brassica juncea* (Bali & Harris, 2010; Beattie & Haverkamp, 2011) or extracts from *B. juncea* and *M.sativa* (Bali & Harris, 2010), *Memecylon edule* (Elavazhagan & Arunachalam, 2011) or *Allium sativum* L. (Ahamed et al., 2011).

Depending on the nanoparticle's nature, specie of plant or tissue in which they are stored, metal nanoparticles of different shapes and sizes can be obtained. However, all these processes share the advantages of being simple, cost-effective and environmentally friendly.

Apart from the potential benefits of nanotechnology in agricultural sector (described throughout this section), it also involves some risks. Farmers' chronic exposure to nanomaterials, unknown life cycles, interactions with the biotic or abiotic environment and their possible amplified bioaccumulation effects, should be seriously considered before these applications move from laboratories to the field.

2.2 Nanotechnology and animal production

Livestock contribute 40 percent of the global value of agricultural output and support the livelihoods and food security of almost a billion people (FAO, 2009). Rapidly rising incomes and urbanization, combined with underlying population growth, are driving demand for meat and other animal products in many developing countries, being the annual growth rate 0,9% in developed countries and 2,7% annual worldwide rate. In the last decade, per capita consumption in developing countries is nearly twice and this tendency keeps on growing surpassing 100 million tonnes of produced meat in Asia in 2007 for instance (FAO, 2009).

Therefore, some of the challenges the animal production sector will have to deal with are:

- Look for an environmentally friendly sustainable production system that contributes to maintain environment preventing further degradation (livestock is responsible for 18 percent of global greenhouse gas emissions, (FAO, 2009) and livestock grazing occupies 26 percent of the earth's ice-free land surface), this joint to the increasing supply of meat products to a growing population (food demand is foreseen to be twice by 2050 due to the socio-economic and population growths, FAO, 2009). However, this growing population has not always higher economic resources (FAO estimates that between 2003–05, 75 million more people were added to the total number of undernourished and high food prices share part of the blame and this number increased in 2007 up to 923 million (FAO,2008).
- Diseases control in animal-food production sector, since these diseases are extended quickly now than in the past due to the market globalisation, so this point is a key to preserve human health and food safety (according to a study carried out in USA, UK and Ireland during the last decade approximately 20% of the retired food for food safety came from the meat sector, 12% from food processed products and 11% from vegetable and fruits sectors (Agromeat, 2011). Animal diseases reduce production and productivity, disrupt local and national economies, threaten human health and exacerbate poverty being essential its minimisation.
- An accurate traceability of products derived from the meat sector in an international growing market, which would warrant the product identity and avoid possible food fraud in this sector.

In this way, nanotechnology will be able to solve these problems in the animal production sector. In order to explain this issue, we will speak of the different areas within the animal production where nanotechnology can give support and provide some important solutions. These fields can be classified within 5 categories (Kuzma, 2010) being currently all of them under research and development.

- Pathogen detection and removal
- Veterinary medicine
- Feed improvement and waste remediation
- Animal breeding and genetics
- Identity preservation and supply-chain tracking

2.2.1 Pathogen detection and removal

This first category includes the use of nanodetectors not only to detect pathogens but to bind and remove them. An example will be the case of S. *typhi* detection in chickens skins, which uses magnetic particles functionalised with antibodies that after binding the pathogen, removes them by means of the introduction of magnetic forces. Other case is the study against foot-and mouth disease (FMD) virus in chickens, with the development of surfaces based on nanostructured gold films with topography matched to that of the size of FMDV. These surfaces are functionalized with a single chain antibody specific for FMDV in such a way that liquid crystals will uniformly anchor on these surfaces. FMDV will bind to these surfaces in such a way that it will give rise to easily visualized changes in the appearance of liquid crystals anchored on these surfaces and so virus presence is detected (Platypus Technologies LLC, 2002).

The benefit of this application would allow improving human health reducing the risk of diseases derived from food consumption and allow great benefits for animal health, likewise an economic advantage when producing. However, this technology is not still very developed and it is necessary studies to guarantee the lack of toxicity of these nanoparticles. Also, nanotechnology is used in the diagnosis of animal diseases control based on microfluidic, microarray, electronic and photo-electronic, integrated on-chip and nanotechnology together with analytical systems, which enable the development of point-of-care analysers (Bollo, 2007).

2.2.2 Veterinary medicine

In the veterinary medicine field, nanoparticles to deliver animal drugs like growth hormones (somatotropine) in pigs through PGLA nanocapsules and the improvement in animal vaccination are two of the application examples. Among vaccination products, some polystyrene nanoparticles bound to antigens are experimentally being proved in sheep (Scheerlinck et al., 2005) and vaccines against *Salmonella enteriditis* based on an immunogenic subcellular extract obtained from whole bacteria encapsulated in nanoparticles made with the polymer Gantrez (Ochoa et al., 2007).

Animal production and food safety will be increased and this will lead to cheaper and greater availability of meat products. However, some of these measures could not be successfully applied in some countries, as it would be the case of hormones supply to animals, which is not accepted as good practice by consumers or even banned in some countries.

2.2.3 Feed Improvement and waste remediation

In the category of feed improvement and waste remediation are being studied the use of nanoparticles in feed for pathogen detection. These particles would bind pathogens in the gut of poultry preventing its colonization and growth and avoiding its presence in waste.

Other example would be the use of nanoparticles to detect chemical and microbiological contamination in feed production. Advantages of both applications are clear: Farms would be more environmentally friendly and alternatives for using of antibiotics to face pathogens will be proposed. The former option could be very interesting due to the growth of antibiotic resistance found in animals and humans. However, life cycle of these particles must be verified in order to warrant that these particles are not harmful for the animal and that they don't end up on animal-derived food products or food chain.

Toxicology must be studied and environmental exposure, likewise the effect that animal wastes containing these particles would have. In feed production case, workers safety must be studied and the effect of leftovers remaining in farms and fields.

Within this part, we can also mention the use of polymeric nanocapsules to carry bioactive compounds (such as proteins, peptides, vitamins, etc.) in such a way that they will be able to pass through the gastrointestinal mucus layer and/or epithelial tissue providing or increasing the absorption of these bioactive compounds (Luppi et al., 2008; Plapied et al., 2011), improving the availability of these nutrients and therefore, the quality of animal production.

2.2.4 Animal breeding and genetics

Genetic animal therapy is one of the possible applications in a near future. Currently, studies of genes delivery in animal cells for selection and temporal expression are being carried out in order to determine specific features of livestock (Mc Knight et al., 2003).

Animal health and breeding would be improved as well as animal production. For instance, some kind of genetic diseases could be avoided and genetic configuration could be determined in a way that animals will be more fertile. However, before more studies about toxicity, safety and legislation must be made. Moreover, public perception and current legal framework related to genetic modification would have to change a lot, since society opposition is still rather difficult to save.

2.2.5 Identity preservation and supply-chain tracking

Concerning traceability and identity preservation chips of DNA are being studied in order to detect genes in feed and food, likewise nanobarcodes to get feed (Jayarao et al., 2011) and animal traceability from farm to fork. This will aid to maintain animals healthy but also to improve public health, since feed traceability would help to prevent and avoid possible food related illnesses, being able to find quickly the source of such diseases and even to prevent them.

In conclusion, the potential benefit of these technologies would allow a diminution cost and an improvement in the animal production sector as much in quality as in quantity, the waste reduction, increasing efficiency in the use of resources and rising safety in animal derived food products.

Possible risks related to human and animal health and also the environmental impact have to be overcome before applying of these technologies. Other matters such as exploitation rights, intellectual property and legal issues derived from the use of these inventions and patents must be considered too.

2.3 Nanotechnology and water supplies

In this decade, the demand of water supplies is rapidly growing and the competition for water resources is being every day a bigger concern. Currently, 70% of the obtained fresh water in the entire world is used in agriculture, while 20% is used in the industry and 10% are devoted to municipal uses (FAO, Agotamiento de los Recursos de Agua dulce).

The access to inexpensive and clean water sources is an overriding global challenge mainly due to:

- The global climate change that will contribute to fresh water scarcity.
- The constant growth in world population. Water sources demand has exceeded by a factor of two or more the world population growth, (FAO, Agotamiento de los Recursos de Agua dulce valoración ambiental: Indicadores de Presión Estado Respuesta), this means that the expected population growth will be pressing water resources. In order to have a rough idea, the daily drinking-water requirements per person are 2-4 litres. However, it takes 2000 - 5000 litres of water to produce a person's daily food, if this is added to the increasing population growth... (FAO. FAOWATER. Water at a Glance: The relationship between water, agriculture, food security and poverty.)
- The raising industrial and municipal water waste. Every day, 2 million tons of human waste is disposed of in water courses. Worldwide total water waste volume was estimated in more than 1.500 km3 in 1995, knowing that every litre of waste water contaminates 8 litres of freshwater, 12.000 km3 of water resources from the planet are estimated not being available for its use. If this figure maintains its growth at the same

rate than the population is increasing (by 2050, population will reach 9.000 million people), the planet will lose every year around 18.000000 km3 of water resources (UNESCO, 2007).

- The large quantities of water used to produce increasing amounts of energy from traditional sources.
- The intensive water use in agricultural (i.e., it takes 1-3 tonnes of water to grown 1kg of cereal; FAO. FAOWATER. Water at a Glance: The relationship between water, agriculture, food security and poverty) and livestock activities. Added to this, it is the fact that the foreseen urban population growth will increase pressure on local water resources, while food production demand is rising in remote areas to support the industrial animal production systems connected to cities.

These are some of the reasons why alternatives should be thought for obtaining and recycling water and in this point is where nanotechnology could provide innovative solutions.

Currently, there are several nanotechnology research lines concerning water treatment. Some of these examples are mentioned next:

- Magnetic nanoadsorbents, which can transform high polluted water in water for human consumption, sanitary uses and irrigation. Magnetic nanoadsorbents can be used to bind pollutants such as arsenic or petroleum and then could be easily eliminated by a magnet (Mayo et al., 2007; Wolfe, 2006).
- Water desalination by means of carbon nanoelectrodes. Water for human consumption could be obtained in areas, which are rich in high salinity water, using less economic and energy resources. An example would be the use of capacitive deionization (CDI), which is based on high-surface-area electrodes and that when are electrically charged can adsorb ionic components from water (Oren, 2008).
- Biofilms, which create surfaces that avoid bacterial adhesion and which would be self-cleaning, preventing fouling and hindering biocorrosion and pathogens in water distribution, connexion and storage systems. (Brame et al., 2011).
- Nanomaterials with catalytic and phototacalytic activities able to eliminate bacteria, pesticides, antibiotics and hormonal disruptors, likewise chlorine of organic solvents like trichloroethylene (hypercatalysis) from the environment. Remediation of groundwater contaminated with organochlorine compounds and other compounds can be done, having impact not only on water purification systems for drinkable water but on human health. Among some of the applications would be nanocomposites of dioxide of titanium, which in presence of UV-visible and solar light have bactericide and anti-viral properties. Other application would be the hypercatalysis with palladium covered by gold nanoparticles that would eliminate organic solvents like trichloroethylene (Clean Water Project: WATER DETOXIFICATION USING INNOVATIVE vi-NANOCATALYSTS, FP7 Collaborative Project, ENV NMP227017, 2009-2012; Nutt et al., 2005; Senthilnathan & Ligy, 2010).
- Membranes systems based on metal nanoparticles, which prevent fouling to purify water in places where water quality is not good for human consumption and in water waste, reducing the number of treatment steps; nevertheless several drawbacks must be solved before applying it. Among these problems to be solved would be the use of metal-based nanoparticles, which implies a potential selection for antibiotic resistance. These multi-functional membranes would be self-cleaning, antifouling

performance and could inactivate virus and eliminate organic matter. An example would be reverse osmosis membranes and its spacers with nanosilver coating, which would improve bacteria removal and antifouling performance (Yang at al., 2009; Zodrow et al., 2009).

- Nanosponge for rainwater harvesting, a combination of polymers and glass nanoparticles that can be printed onto surfaces like fabrics to soak up water. This application can be very beneficial for tropical climate countries where humidity is high and there could be a more efficient water mist catching (Grimshaw, 2009).

- Nanomesh waterstick, a straw-like filtration device made of carbon nanotubes placed on a flexible and porous material. This stick cleans water as someone drinks and eliminates virus and bacteria. It has been already used in Africa (Seldom laboratories) and is currently available in the market. Speaking about active carbon, this one has been used together with nanofibres, in point-of-use water purifying systems. They are some kind of biodegradable tea-bags-like to filtrate water. Once, they have been used the nanofibres will disintegrate in liquids after a few days (developed by E. Cloete; Universidad de Pretoria).

- Nanoparticles with DNA able to track hydrological flows, which will allow to know water sources and if the water is polluted or not (Walter et al., 2005).

The key point is that these technologies would allow to treat processing and residual water from food industry and recycle and reuse them, contributing to the economy, environmental improvement and water security of local communities and industries.

This is essential in areas where water scarcity is a problem, drinkable water is difficult to obtain and water is considered as a precious resource.

Although the operation cost of some of these installations and facilities would be higher than the operation costs of the traditional facilities, the initial investment would be cheaper (construction costs), moreover life cycle of some of these materials is unknown and these are only some of issues that must be studied before trying to implement anything, so there is still a long way to cover before some of these applications can be real facts.

2.4 Nanotechnology in food processing

Up to now in this chapter, the nanotechnology applications have been referred to water use and commodity production but, nanotechnology can be a way to get better manufacturing food process (or even a completely new and different one). Actually, food sector could be facing a paradigm shift. As commented by Pray & Yaktine, 2009, nowadays foods are structured using a recipe (o formulation) whereas two simultaneous processes occur:

- Formation the structure (i.e: by means of phase creation, reactions, biopolymer transformation …)
- Stabilization of the system (i.e. vitrification, crystallization, network formation…)

On the other side, nanotechnology affords the possibility of structure foods from the base elements (in an approach derived from its constituents self-assembly). Rather than the use of a recipe, nanotechnology raises the opportunity of using molecules as starting material, so that new interaction are achieved which in turn generates the required properties. The paradigm shift resides in the fact the food formulation will based on the use of "food matrix precursors or structural elements". To achieve this possibility it will be necessary the development of new knowledge and techniques.

Some examples on the potentiality of the former approach is the development of new textures and flavours, the possibility of the design of low dense, low calorie foods but nutritionally enhanced. Those foods can be aimed at a nutrition adapted to different life-style and consumer conditions (i.e. obesity cases).

Two points are commented here to give a general view of the multiple possibilities of nanotechnology use in food processing:

- Food processing issues: mixing, component stability, safety, etc.
- Intrinsic food features: texture, flavour, taste masking, availability and delivery, etc.

Among the processing issues the most of the applications are related with the use of nanoparticles and nanocapsules. These particles enhance the products functionality, and they are responsible of that enhancement because they perform a protection function on:

- The contained active principle. Nanoparticles avoid the degradation produced by the surrounding environment of the food or by the manufacturing process. Gökmen et al., 2011, have reported that omega 3 was successfully nanoencapsulated and used in bread making. In addition of the positive effect in taste masking, thermooxidation during baking was reduced. As a consequence of that, the production of other further degradation by-products (i.e. acrylamide) was strongly reduced.
- The food itself. This happens when the active principle is used with a technological purpose. For instance nanoparticles containing essential oil have been used to improve antimicrobial activity in juices (Donsì et al., 2011). In this case, the addition of small amounts of nanoparticles containing terpenes to an orange and pear juice delayed or avoided the microbial growth (depending on the concentration). Organoleptic properties were maintained.

One advanced aspect of the use of nanotechnology is the possibility of acting straight onto the food structure. Nanostructured foods aimed at producing better texturized, flavoured and other properties can be obtained. In particular it can be highlighted the possibility of producing cream-like foods such as: ice-creams, spreads and low fat dressings. Leser et al., 2006, comment how polar lipids (monoglycerides and phospholipids) can be used with this purpose. Some of those polar lipids are capable of create nanostructures that finally result in crystalline phases. These structures can be used as a base for the development of spreads taking advantage on the texture properties emerging from interaction between nanostructures clusters. The manufacturing process will depend on water relative quantity, stirring, and the system temperature evolution.

The contribution of particles to the stability of foams and emulsions is another aspect of interest in the use of nanotechnology in foods. Most of the foods are (or have been during processing) dispersions like emulsions or foams. Several examples could be derived from bakery, confectionery, meat products, dressings and spreads. Another example is the prepared foods. In general, foam and emulsion stability can be improved in the presence of nanoparticles and nanostructures.

The macroscopic properties of dispersion can be improved by controlling the formation of nanostructures in the interphase (Rodríguez et al., 2007). The application of new techniques to food development such as Brewster angle microscopy (BAM), atomic force microscopy (AFM) y imaging ellipsometry (IE), will allow the study and application of those possibilities to processing.

Dickinson, 2010 presents a review about the use of inorganic nanoparticles (silica), fat crystals and protein nanoparticles. Main conclusions to highlight are:

- "Natural biopolymer structural assemblies are obviously attractive as (nano)particle building blocks"
- Polysaccharides (i.e. starch or cellulose) represent can be considered as a cheap and a quickly source of nanomaterial for food uses. Anyhow, it is required some type of modification to increase hydrophobicity.
- Proteins represent the best potential derived from the amphiphilic behaviour as well as due to their carrying capacity (i.e. casein nanotubes)
- The interaction between protein and polysaccharides is of great interest also. For instance, the interaction obtained from the interaction of caseinate and Arabic gum or, the heat degraded beta-lactoglobulin aggregates and stabilised by pectin. One important implication of this type of interaction could be the complexing between casein and a charged polysaccharide which can be used in the maintenance of solubility and functionality under acidic pH conditions

Another relevant aspect in food processing is the time. In this case, when considering the interaction between food components in a structure, the time required to reach a position should be considered. "In order to interact, different components of a food structure must come into position at the right time." (Pray & Yaktine, 2009) For example, the foam formation is a kinetic process that is initiated at the nanoscale in miliseconds, after that foam is stabilised at macroscopic level in the order of minutes. The capability of nanoparticles and nanosystems to arrange in time is then another issue to consider. In this sense, Sánchez & Patino, 2005 have reported how the diffusion speed of caseinate to the liquid-air interphase affects to the foam formation and, how depending on the concentration (above 2%), interphase saturation occurs and appear a micelle formation for sort times.

Besides processing features, a possible breakthrough in food manufacturing is the enhancement of the bioavailability. Nanoemulsions and nanoparticles suppose a delivery vehicle capable of override two great problems:

- Conservation in the food previous to the consumption and the stomach degradation
- The absorption of the compounds by the organism

In the last case, it should be considered that the main absorption route of foods is the intestine and nanotechnology could improve that somehow. Direct absorption of nanoparticles is controlled by size and surface chemistry of the particle. Generally nanoparticles can be done by an active transport mechanism or by a passive transport (Acosta, 2009).

In the first case (active transport) absorption happens by means of specific channels that present the epithelial intestinal cells. In this case absorption is controlled by the hormonal system and by homeostatic effect derived from blood levels of compounds. According to the last control mechanism, once a level is surpassed for an ingredient, it will be not absorbed and the excess is accumulated or excreted. Selective absorption can be conditioned by specific receptors in the cell surface (enterocytes and M cells). Those nanoparticles with a specific surface composition can be absorbed by this way. Generally, the M cells present a better permeability. For example, there is a great interest in the development of lecithin nanoparticles since it could be an improvement in the absorption of isoflavones by this route (Acosta, 2009).

Passive transport happens by means of diffusion though epithelial tissue. The particle has to be fixed to intestinal mucosa, and from there, to contact the cell. The absorption speed is determined by a concentration gradient. In an interesting work, Cattania et al., 2010, studied the use *in vivo* of lipidic core nanocapsules. Those nanocapsules are fixed to gastrointestinal lumen and act as active principle reservoirs. Nanocapsules delivered the active principle once they were fixed by diffusion. As relevant conclusion of this work it is pointed that there is important differences between evaluation models used *(in vivo* and *ex vivo),* and the live model system complexity cannot be predicted by a simpler model, at least at this time.

In fact, nanomaterial dosimetry it is not clear at all. Despite this, there is a general consensus about the relation of particle size and bioavailability in the sense that, as far the size decreases, the bioavailability is improved. A size reduction under 500 nm produces a higher absorption of the active principle and more particle uptake regardless of the system composition. However, above 500 nm, the bioavailability depends on the system (Acosta, 2009). So, it is possible to obtain more effect from a food additive with the use of nanotechnology (with lower content in an active principle) than with the use of the traditional approach (i.e. microencapsulation).

Food properties can be enhanced this way but there is a necessity for new test and assay methods to evaluate the state of the particle in the food (or structure) and the effect on the organism and health.

Ultimately, nanotechnology application on foods supposes a great challenge but requires solving several aspects in the near future. Once those barrier have been overcome, manufacturing of foods adapted to especial necessities will be obtained, as well as the improvement in efficiency, action in the organism and even getting a better experience in the consume.

2.5 Food packaging

Most of the nanotechnology applications in food industry are in the packaging sector. It is expected that by 2015 this sector will be 19% of nanotechnology food applications.

This is mainly due to the big development of nanotechnology in this field, to a higher acceptance by consumers of the use of this technology in packaging than in food as ingredients and to the normative requirements, which are less restrictive than for the enforced current food legislation. All this contributes to the fact that there are many more applications in this area than in others of food sector, together with the necessity of a more sustainable, lighter and stronger at the same time, efficient and intelligent packages. These packages would be able to provide safer products with more quality and at the same time maintaining the products in the best possible conditions and with a longer shelf-life.

During the last decade, the use of polymers as material for food packaging has incredibly increased due to its advantages over the use of traditional materials.

In polymers world market, which has been increased of about 5 million tons in 1950 to almost 100 million tons at the present time, 42% corresponds to packaging and containers (Silvestre et al., 2011).

In the following section, some nanotechnology derived applications of materials in contact with food are mentioned. These applications are currently available or at research level.

2.5.1 Nanomaterials to improve packaging properties

As it was previously mentioned, polymers provide packaging with strength, stiffness, oxygen barrier, humidity, protection against certain food compounds attacks and flexibility (Sánchez-García et al., 2010b).

The possibility of polymers improving these features in food packaging by means of nanoparticles addition has allowed the development of a huge variety of polymers with nanomaterials in its composition (Azeredo, 2009; Bradley et al., 2011; Lagaron & López-Rubio, 2011).

The use of nanocomposites for food packaging not only protects food but also increases shelf-life of food products and solves environmental problems reducing the necessity of using plastics. Most of packaging materials are not degradable and current biodegradable films have poor barrier and mechanical properties, so these properties need to be considerably improved before these films can replaced traditional plastics and aid to manage worldwide waste problem (Sorrentino et al., 2007).

With the introduction of inorganic particles as clay in the biopolymer matrix (Bordes et al., 2009; 48, Utracki et al., 2011), numerous advantages are reached.

Natural structure of clay in layers at nanoscale level makes that when clay is incorporated to polymers, gas permeation will be restricted and product will be anti UV radiation proof. In addition, mechanical properties and thermal stability of package are improved.

Polymers made up of nanoclay are being made from thermoplastics reinforced with clay nanoparticles. At present, there are a huge number of nanoclay polymers available on the market. Some well-known commercial applications are beer and soft drinks bottles and thermoformed containers.

Other examples that can be mentioned are nitrure of naotitanium, which is use to increase mechanical strength and as aid in processing and dioxide of titanium to protect against UV radiation, in transparent plastics. Among these applications the oxide of nanozinc and nanomagnesium are expected to be affordable and safer solutions for food packaging in a near future (Lepot et al., 2011; Li et al., 2010).

Also carbon nanotubes (Sánchez-García et al., 2010a) or nanoparticles of SiO_2 have been used for improving mechanical and barrier properties of several polymeric matrices (Vladimiriov et al., 2006).

The use as reinforcement elements of biodegradable cellulosic nanowhiskers and nanostructures obtained by electrospinning (Goffin et al., 2011; López-Rubio et al., 2007; Siqueira et al., 2009; Torres-Giner et al., 2008) must be highlighted too. Finally, to mention the use of biological nanofillers to strength bioplastics has the added value of generating formulations of complete biological base. These nanofillers have a high surface-mass ratio, an excellent mechanical strength, flexibility, lightness and in some cases, even they are edible, since they can be made from food hydrocolloids.

2.5.2 Active packaging

Active packaging is thought to incorporate components that liberate or absorb substances in the package or in the air in contact to food. Up to now, active packaging has being mainly developed for antimicrobiological applications, nevertheless other promising applications include oxygen captation, ethylene elimination, CO_2 absorption / emission, steam resistances and bad odours protection, liberation of antioxidants, preservatives addition, additives or flavours.

Nanoparticles more used in active packaging development are nanomaterials of metals and oxide of metals in antimicrobial packaging. Nanosilver use in packaging helps to maintain healthy conditions in the surface of food avoiding or reducing microbial growth. However, its action is not as a preservative even though, it is a biocide (Morones et al, 2005; Travan et al., 2009). Based on these properties, a big number of food contact materials, which inhibit microorganisms' growth have been created (i.e. plastic containers and bags to store food).

2.5.3 Intelligent packaging

Nanotechnology can be also applied in coatings or labels of packaging providing information about the traceability and tracking of outside as well as inside product conditions through the whole food chain. Some examples of these applications are: leak detections for foodstuffs packed under vacuum or inert atmosphere (when inert atmosphere has been ruptured some compounds change of colour warming consumers that air has come inside in where should be an inert atmosphere) (Mills & Hazafy, 2009); temperature changes (freeze–thaw–refreezing, monitoring of cold chain by means of silicon with nanopores structure), humidity variations through the product shelf-life or foodstuffs being gone off (unusual microbial presence).

Currently, sensors based on nanoparticles incrusted in a polymeric matrix isolated to detect and identify pathogens transmitted by food are being studied. These sensors work producing a specific pattern of answer against each microorganism (Yang et al., 2007). Technology called "Electronic tongue" must be underlined, too. It is made up of sensor arrays to signal condition of the foodstuffs. The device consists of an array of nanosensors extremely sensitive to gases released by spoiling microorganisms, producing a colour change which indicates whether the food is deteriorated.

DNA-based biochips are also under development, which will be able to detect the presence of harmful bacteria in meat, fish, or fungi affecting fruit (Heidenreich et al., 2010).

3. Nanotechnology challenges

As described throughout part 2, the implementation of nanotechnology in the food industry offers a wide range of opportunities to improve farm management, livestock waste, processing and food packaging. According to Helmut Kaiser Consultancy, this market was valued at USD 2.6 bn in 2003, doubled in 2005 and is expected to soar to USD 20.4 bn in 2015 (Groves & Titoria, 2009).

Despite these figures, nanotechnology has a lot of work to do in the food industry compared to its implementation in other fields such as health and fitness, home and garden or automotive. These were the three categories with the largest number of nanoproducts in March 2011 (Project on Emerging Nanotechnologies, 2011).

According to the same source, in 2010 there were sold a total of 1317 different products based on nanotechnology. This figure is small compared with the R & D investment and shows that nanotechnology commercialization is still in its infancy not only in the food sector.

The main common themes addressed by all company surveys related to commercialising nanotechnology, are: high processing costs, problems in the scalability of R & D for prototype and industrial production and concerns about public perception of environment, health and safety issues (Palmberg et al., 2009).

At the same time, as research on new and different applications of nanotechnology is carried out, others should be done with the aim of developing reliable and reproducible instrumental techniques for detecting, quantification and characterization of new materials in environmental, food and human samples. It will be necessary to study: different absorption pathways, exposure levels, metabolism, acute and chronic toxicity and its short or long term bioaccumulation. The knowledge gained in all these areas is essential to sketch a realistic and effective nanotechnology regulatory framework.

3.1 Scientific and technological challenges

There is currently a research boom in nanotechnology; both companies and universities are increasing their efforts to study the human health and environmental effects of exposure to nanomaterials. During last years, it has been shown that these materials can affect biological behaviours at the cellular, sub-cellular and protein levels due to its high potential to cross cell membranes. Some of these effects are not at all desirable, turning to be even toxic. Despite the efforts, conventional toxicity studies need to be updated to nanoscale. These new methods must define scenarios and routes of human exposure (so far there are only few studies involving oral routes), consider the behaviour of nanomaterials in watery environment and conditions that may influence its aggregation state and association with toxicants. Also, they must select a model organism to test toxicity.

In order to carry out such toxicity studies, it is necessary to implement new analytical methods able to detect the presence of very small quantities of nanomaterials in both environmental and food samples. This issue arouses so much interest that scientific journals such as Trends in Analytical Chemistry have published two special issues about characterization, analysis and risks of nanomaterials in environmental and food samples. Its papers emphasize that these are complex samples and therefore their analysis often involves four stages: (1) Sample preparation, (2) Imaging by means of different microscope techniques, (3) Separation and (4) Characterization by measuring size, size distribution, type, composition or charge density by, between others, light scattering techniques. Anyway it is very important to take into account that nanoparticles can change its structure and composition as a function of the medium and treatment. That is why resulting sample after its preparation may differ from the original one determining the reliability and conclusions of the whole analysis (Peters et al., 2011).

3.2 Socio-economic challenges

After years of public and private economic investments in R & D, nanotechnology in return is thought to develop new and more environmental friendly and efficient production methods in order to supply a growing population with commodities, and new and safer products with enhanced properties, and to generate qualified jobs as well as scientific advances.

In other words, one of the biggest challenges which nanotechnology faces up to is the ability to create an industrial and business scope. It is thought that nanotechnology will have an important impact on employment sooner than later, despite the fact that not all consultancies agree in their expectations. It depends, for instance, on nanomaterials' definition. The American NSF (National Science Foundation) estimated in 2001 that 2 million workers will be needed in nanotechnology based companies by 2015. According to LuxResearch in its 2004 report, 10 million jobs related to nanotechnology will have been created worldwide by 2014 (Palmberg et al., 2009). Although all long-term forecasts share that they were made in a buoyant and optimistic economic scenario (before 2008 crisis), and they should be considered with caution.

Related to its geographical distribution and according to NSF, by 2015 45% of new jobs will be generated in USA and 30% in Japan. Other agencies think that job layout will change and countries like China, India or Russia will become more important (Seear et al., 2009).

Another aspect to consider when studying nanotechnology influence on employment is the workers´ health. Professionals are not immune to the new materials' effects on health and could show symptoms related to chronic expositions (Seear et al., 2009). This fact would have repercussion on economic status of health systems.

In order to achieve private initiative investment in this sector, it is strictly necessary that a stable and effective regulatory framework exists, but also channels to inform and educate the public about what this technology is, its advantages, disadvantages but also the risks it may involve.

3.2.1 Development of an effective and specific legislation

The current implementation of nanotechnology in food industry does not count with a specific legislation. Although it does not mean that new food products, ingredients, surfaces or materials intended to come into contact with food are not obliged to pass safety controls before entering the market.

According to European Commission, the scope of the current legislation is wide enough to deal with new technologies (Commission of the European Communities, 2008). It should be applied what is established in one or other normative depending on the nanotechnology implementation and on the resultant product (ingredients, additives, packages...).

Against public agencies' opinion, other society sectors like Friends of the Earth defend that it is necessary to develop new nano-specific legislation which consider engineered nanomaterials as new substances with characteristic risks and properties and different from those associated with the same substance in its bulk form (Miller & Senjen, 2008).

These rules should be observed by any food producer, whether using nanotechnology or not. Although it is true that this technology implementation in the agro-food sector goes further than the observance of the regulations, codes or acts which appears in the table. It also concerns such different topics as: workplace health and security, water quality, wastes management, pesticides or animal health.

Implementation of this current legal framework is quite complex. It is necessary for a nanotechnology regulation to work properly in food industry that public agencies define precisely: (1) Nanomaterials, (2) An international regulatory body, (3) Detection, characterization and quantification methods and (4) Exposure and risk assessment of the new products.

The main problem is the lack of agreement between the most important agencies and international bodies in the legal definition of engineered nanomaterials. Council of the European Union defines it as follows: "any intentionally produced material that has one or more dimensions of the order of 100 nm or less or is composed of discrete functional parts, either internally or at the surface, structures, agglomerates or aggregates, which may have a size above the order of 100 nm to the nanoscale include: (i) those related to the large specific surface area of the materials considered; and or (ii) specific physico-chemical properties that are different from those of the nanoform of the same material" in the proposal for the novel foods amending Regulation (EC) No. 258/97 (Council of the European Union, 2009).

Use	European Union	United States of America	Australia & New Zealand.
General Food Safety	Regulation (EC) No.178/2002.	Federal Food, Drug and Cosmetic Act (the FDC Act).	Australian and New Zealand Food Standards Code (the Food Standards Code)
Novel Foods and Novel Food Ingredients	Regulation (EC) No.258/97.		Part 1.3. of the Food Standards Code
Food additives	Regulation (EC) No.1331/2008 Regulation (EC) No.1332/2008. Regulation (EC) No.1333/2008. Regulation (EC) No.1334/2008.		
Packaging and Food Contact Materials (FCMs)	Regulation (EC) No.1935/2004. Regulation (EC) No.450/2009.	Federal Food, Drug and Cosmetic Act (the FDC Act) in its Title 21, Chapter 9.	Standard 1.4.3 of the Food Standards Code

Table 1. Summary of the food related legislation in the UE, USA, Australia and New Zealand.

Institutions like International Union of Food Science & Technology (IUFoST) or House of Lords advice that engineered nanomaterials' legal definition shouldn't be based on size alone and recommend that it should refer in an explicit manner to its functionality. Other way, if the size threshold is fixed in 100 nm, producers could declare that their goods only contain particles with dimensions of 101 nm, avoiding the established safety controls (House of Lords, Science and Technology Committee, 2010a; Morris, 2010).

Overlapping between the different international regulatory entities or agencies is another difficulty related to nanotechnology implementation control. It is because this technology can be used in many and different fields and also because its resulting products should compete in a global market, this is why it is so important to define what body is going to organize the trade. Once this challenge has been overcome, trade barriers will be reduced and there will be a free movement of goods. Codex Alimentarius organized an expert meeting in June 2009 where they thought about the use of nanotechnology in the food and agriculture sectors and its potential food safety implications (FAO/WHO, 2009). On the other hand, entities such as Organisation for Economic Co-operation and Development (OECD) can also act as arbitrator in this issue on an international scale. Regarding this body,

Friends of the Earth Australia pointed out that many countries are not represented at the OECD, in particular developing nations (House of Lords, Science and Technology Committee, 2010a). There are also people who think that an international regulatory agency is unnecessary and agreements between countries are enough.

Apart from establishing nanomaterials' definition and deciding which organization is going to coordinate the international trade of nanotechnology food products, it is also necessary to standardize protocols and reliable detection, characterization and quantification methods of nanomaterials in food samples. Otherwise, any written regulation would be limited *per se* (Institute of Food Science & Technology, 2009).

Currently, food safety legislation in western countries is designed with the aim of offering the highest health guaranties. Each agency has established its own pre-market approval assessments for new products, additives, flavourings, enzymes and materials intended to come into contact with food. Any approved application will be included in a positive list which authorizes its use under certain concentrations and foods. If the application fails the assessment, (neither the company nor the authorities in charge are able to prove the substance's safety), its marketing will be denied.

Normally, every substance approved for human consumption is associated with a tolerable intake (expressed in concentration units). The nanomaterials inclusion in those lists will make necessary to change this units, because their effects are quite different from those in the macro-scale (Gergely et al., 2010). This is exactly the reason why it is strictly necessary the exposure and risk assessment of every new nanotechnology implementation in food industry.

3.2.2 Public perception generation from a critical approach

Public perception is a key point in the development of any new technology, since without public acceptation any opportunity for development, even if scientific-technological perspectives are gorgeous, would be vanished. The good news is that public perception can be created, changes and evolutions (an example of this would be genetically modified food products).

Nevertheless, for a true and lasting public perception, this one has to be created by and inside consumers of the mentioned technology, has not to be something only imposed from outside and for it some requirements must be accomplished (Magnuson, 2010; Yada, 2011).

Among the requirements would be the simple and transparent access to information, education related to nanotechnology to acquire the necessary knowledge for allowing society benefits and risks identification, as well as management of risk control by independent and reliable organisations, knowing cost impact and who will pay for its implementation, assessment of environmental impact and finally and more important, freedom of choice. This means allowing users of this technology to choose and decide consciously if they want to consume products in which nanotechnology has been used or not and for it again, here we are, the right to be informed.

Citizens' participation in committees and forums, where people give their opinion and can be informed in these matters and "nano" mandatory labelling would be some of the pending issues (Miller & Senjen, 2008), happily the path starts to be tracked in this direction.

Currently public perception of nanotechnology faced two problems: on one hand, the technical unknowledge of the subject and on the other hand the exaggerated expectatives arisen, which show it as the solution for all the problems together with the rejection to this excessive idealized vision.

The repulse to these exaggerated views, fear of nanotechnology being uncontrolled and becoming a threat, the fact that nanotechnology is a difficult concept to understand for consumers due to its complex nature and the small size of what is being treated (it's not something that can be seen just looking at it) contribute to the fact that there is still too many things to be done in this field.

According to a public perception document of nanotechnology published by the Food Standard Agency (Food Standards Agency, 2011), its success is conditioned by several factors. Next, some of the factors mentioned in this and other reports are shown:

- Use: On one side, consumers in general are more favourable to use nanotechnology in other sectors than Food sector (House of Lords, Science and Technology Committee, 2010b).This is due to food is not only perceived as from the functional point of view but related to health, environment, science, etc. not to mentioning personal and familiar habits in each home. On the other side, within Food sector it seems that public is more likely favourable to accept nanotechnologies in fat and salt reduction meals (issues that directly affect to its health) without taste and texture damage, than to accept it for new tastes and textures development (Joseph & Morrison, 2006).

- Physical proximity with food: As some authors have said consumers are more favourable to accept the use of nanotechnology in packaging than the use of the same as an additional food ingredient. Moreover, they better understand the advantages related to packaging use of nanotechnology (extended shelf-life of the product, intelligent packaging that shows when the product has gone off, etc.) (Harrington & Dawson, 2011). This was shown in a study carried out by Siegrist et al., 2007, 2008 that evaluated public perception of different kinds of food products. Results in 153 people interviewed were that packaging derived from nanotechnology was perceived as more beneficial that food modified with nanotechnology. The use of this technology is perceived as more acceptable if it is outside food. Furthermore, even if age didn't seem to be a determining factor, it was observed that people older than 66 years old was more favourable to consider the use of the nanotechnology in packaging, not being significant differences with respect to the other age groups.

- Concerns about unknowns in some issues: its effectiveness, human health risks, and regulation (40% people interviewed in a study carried out by the Woodrow Wilson Centre for Scholars, WWCS show their concerns with these issues), testing and research for safety (12% show their worries), environmental impact (10%), short and long term side-effects in food and food chain (7%), control and regulatory concerns, etc. As reported by the WWCS (Macoubrie, 2005) 40% of the study participants relieves that regulatory agencies shouldn't be trust, from 177 participants 55% thought that voluntary standards applied by industry weren't enough to assess nanotechnology risks. However, after receiving a bit more of information, when they were asked to ban this technology until more studies of potential risks were carried out 76% of people considered that this measure would be exaggerated. When they were asked about how government and industry could increase their trust in nanotechnology, 34% answered that increasing safety tests before the product will be on the market and 25% said that providing information to consumers supporting them to make an informed choice. Other suggestion was tracking risks of products on the market. These proposals of improvement of public information and consumers' education would allow them to make better choices and gain trust on industry and government, since lack of information is one the main mechanisms that breed suspicions and lack of trust. This coincides with the opinions shown in other reports, for example the one of FSA "Nanotechnology and food. TNS-BMRB Report" in which the consumers' acceptance is conditioned by transparent information transmission and the reliability in the involved authorities (Food Standards Agency, 2011).

- Information sources (Dudo et al., 2011; House of Lords, Science and Technology Committee, 2010a): Sources and means from which public obtain information conditions in part social perception, being more likely favourable to accept or defeat a new technology. Means from which consumers have obtained more information are mainly television and radio (22%) and from other people (20%) (Macoubrie, 2005). This probably will mean that those that had acquired their knowledge through television and radio have a general knowledge about nanotechnology and not a view as much scientific-technological as if they had acquired this knowledge through journals or specialised papers.

- Socio-demographic and cultural factors (Rollin et al., 2011): According to these authors women are less optimistic than men (33% vs. 49%), and slightly less supportive (53% vs. 59%); religious people were less likely to a favourable perception; the age is not a significant influencing factor in perception, although older people (more than 66 years old) are less likely favourable to the use of nanotechnology.

- Finally, different results were found when nanotechnology perception was studied in different countries. For instance, in 2005 in Europe, 44% of Europeans had heard about nanotechnology. In Europe, acceptance seems to be increasing. In 2002, only 29% agreed on the future positive impact of nanotechnology, and 53% answered "don't know", while in 2005, almost half (48%) considered that nanotechnology will have positive effects on their way of life in the next 20 years. In 2006, over half of Europeans interviewed (55%) support the development of nanotechnology as they perceived this technology as useful to society and morally acceptable. However, in USA in 2002 consumers were more optimistic about nanotechnology (50% optimistic) than Europeans (29% optimistic). Nevertheless, by 2005, European, US and Canadian citizens were equally optimistic about nanotechnology. Europeans were more concerned about the impact of nanotechnology on the environment and were less confident in regulation than North Americans.

- Public general interests in nanotechnology applications. Among these interest must be cited the following ones, 31% in medical applications, 27% better consumers products (i.e. less toxic paint coatings, rubbish bags that biodegrade, etc..), 12% general progress (i.e. qualitative and quantitative advance in human knowledge, improvement in communications, etc.), 8% environmental protection, 6% in food safety and 4% in energy, economy, and electronics, finally 3% shows interest in army and militar security (Macoubrie, 2005).

- Attitude towards risks-benefits balance. This point had a big influence in the consumers' acceptance or not of this technology. When expected risks were rather lower than benefits public were more likely favourable to accept this technology, this could explain packaging issue too. Personal situation influences perception of the risks-benefits balance, for instance people with diseases like obesity, hypertension or diabetes usually prone to see higher benefits in the applications to mitigate these diseases than risks in their possible applications (Food Standards Agency, 2011).

None previous studies have been found about how nanotechnology´s perception conditions behaviours regarding eating and food buying. These are very important facts, if a more useful and commercial approach of the real acceptance of this technology wants to be known, but to get it, first the access to this information must be simple and transparent.

4. Conclusions

Throughout this chapter a review on how the world is facing a situation where in the near future access to foods and water will be one of the main problems for a great part of the world has been described. The pressure on environment, efficiency on production systems and population growth will require new and imaginative solutions to answer those problems. A great part of these solutions could require a technological leap or a breakthrough to achieve the final result. In this sense, nanotechnology could result a great opportunity for that.

As previously mentioned, nanotechnology shows solutions for foods manufacturing and production as well as for the water management. These approaches raise technical possibilities that could help to solve the situation in real context. Despite this, it will be needed to invest more time, financial resources and technological means to achieve widespread nanotechnology application in the sectors described here. For this reason some years will be still required to see a general market application.

Challenges of the nanotechnology are technological and social. Technological challenges will require new analytical techniques so that we can understand how the things are actually working at the nanoscale. On the other hand, and according to the evidence, new evaluation methods for the determination of potential environmental and health effects of using nanotechnology are required. This way the responsible and safe use of the nanotechnology will be possible.

Concerning the social aspects, the achievement of a global consensus about the use of nanotechnology will be necessary so that some limits were respected, (at least from a health and safety point of view).

So, technological and scientific considerations are not enough for the development of nanotechnology. A great part of the possibilities and potential applications of nanotechnology in the near future will depend upon public acceptance. Society (as a whole) must evaluate critically and objectively nanotechnology.

5. Acknowledgement

This work has been granted by the Science and Innovation Office of Spanish Government (2008-2011) within National Programme of Applied Research, Applied Research-Technology Centre Subprogramme (DINAMO Project: *Development of Nanoencapsulates for Nutritional Use* , AIP-600100-2008-23).

6. References

Acosta, E. (2009). Bioavailability of nanoparticles in nutrient and nutraceutical delivery. *Current Opinion in Colloid & Interface Science,* Vol. 14, pp. 3–15, ISSN 1359-0294

Agromeat (2011). Calidad Alimentaria: Cuáles son las causas principales para retirar alimentos del mercado., In: *Agromeat website,* 4 July 2011, Available from: <http://www.agromeat.com/index.php?idNews=115714>

Ahamed, M.; Khan, M. A. M.; Siddiqui, M. K. J.; AlSalhi, M. S. & Alkorayan, S. A. (2011). Green synthesis, characterization and evaluation of biocompatibility of silver nanoparticles. *Physica E-Low-Dimensional Systems & Nanostructures,* Vol. 43, No. 6, (April 2011), pp. 1266-1271, ISSN 1386-9477

Ashwood, P.; Thompson, R.; Powell, J. (2007). Fine Particles That Adsorb Lipopolysaccharide Via Bridging Calcium Cations May Mimic Bacterial Pathogenicity Towards Cells. *Experimental Biology and Medicin,* Vol. 232, pp. 107-117, ISSN 1535-3702

Australia New Zealand Food Standards Code. (2003). In: *Food Standards Australia New Zealand (FSANZ) website,* 8 June 2011, Available from: <http://www.foodstandards.gov.au/foodstandards/foodstandardscode.cfm>

Azeredo, H. M. C. (2009). Nanocomposites for food packaging applications. *Food Research International,* Vol. 42, pp. 1240-1253, ISSN 0963-9969

Bali, R. & Harris, A. T. (2010). Biogenic Synthesis of Au Nanoparticles Using Vascular Plants. *Industrial & Engineering Chemistry Research,* Vol. 49, No. 24, (December 2010), pp. 12762-12772, ISSN 0888-5885

Baruah, S. & Dutta, J. (2009). Nanotechnology applications in pollution sensing and degradation in agriculture: a review. *Environmental Chemistry Letters,* Vol. 7, No. 3, (September 2009), pp. 191-204, ISSN 1610-3653

Beattie, I. R. & Haverkamp, R. G. (2011). Silver and gold nanoparticles in plants: sites for the reduction to metal. *Metallomics,* Vol. 3, No. 6, (2011), pp. 628-632, ISSN 1756-5901

Bollo, E. (2007). Nanotechnologies applied to veterinary diagnostics. Conference 60th Annual Meeting of the Italian-Society-of-Veterinary-Sciences, Terrasini, Italy, Aug-2007 ,*Veterinary Research Communications,* Vol. 31, Supplement 1, pp. 145-147

Bordes, P.; Pollet, E. & Avérous, L. (2009). Nano-biocomposites: biodegradable polyester/nanoclay systems. *Progress in Polymer Science,* Vol. 34, pp. 125-155, ISSN 0079-06700

Bradley, E. L.; Castle, L. & Chaudhry, Q. (2011). Applications of nanomaterials in food packaging with a consideration of opportunities for developing countries. *Trends in Food Science & Technology,* doi:10.1016/j.tifs.2011.01.002

Brame, J.; Li, Q. & Álvarez, P. J. J. (2011). Nanotechnology-enabled water treatment and reuse: emerging opportunities and challenges for developing countries. *Trends in Food Science & Technology,* doi:10.1016/j.tifs.2011.01.004

Cattania, V. B.; Almeida, Fiel L. A.; Jager, A.; Jager, E.; Colomé, L. M.; Uchoa, F.; Stefani, V.; Dalla, Costa T.; Guterres, S. S. & Pohlmann A. R. (2010). Lipid-core nanocapsules restrained the indomethacin ethyl ester hydrolysis in the gastrointestinal lumen and wall acting as mucoadhesive reservoirs. *European Journal of Pharmaceutical Sciences,* Vol. 39, No. 1-3, pp. 116–124, ISSN 0928-0987

Chaudhry, Q. & Castle, L. (2011). Food applications of nanotechnologies: an overview of opportunities and challenges for developing countries. *Trends in Food Science & Technology*, doi:10.1016/j.tifs.2011.01.001

Clean Water Project: Water Detoxification Using Innovative vi-Nanocatalysts, FP7 Collaborative Project, ENV NMP227017. (2009-2012), In: *Project website*, 04 August 2011, Available from: <http://www.photocleanwater.eu/>

Cohen, H.; Levy, R. J.; Gao, J.; Fishbein, I.; Kousaev, V.; Sosnowski, S.; Slomkowski, S. & Golomb, G. (2000). Sustained delivery and expression of DNA encapsulated in polymeric nanoparticles. *Gene Therapy*, Vol. 7, No. 22, (November 2000), pp. 1896-1905, ISSN 0969-7128

Commission of the European Communities. (2008). Communication from the Commission to the European Parliament, the Council and the European Economic and Social Commmittee. Regulatory Aspects of Nanomaterials. [SEC (2008) 2036], In: *European Commission website* , 20 June 2011, Available from: <http://ec.europa.eu/nanotechnology/pdf/comm_2008_0366_en.pdf>

Corredor, E.; Testillano, P. S.; Coronado, M. J.; González-Melendi, P.; Fernández-Pacheco, R.; Marquina, C.; Ibarra, M .R.; de la Fuente, J. M.; Rubiales, D.; Pérez-de-Luque, A. & Risueño, M.C. (2009). Nanoparticle penetration and transport in living pumpkin plants: in situ subcellular identification. *BMC Plant Biology*, Vol. 9, (April 2009), ISSN 1471-2229

Council of the European Union (June 2009). Proposal for a Regulation of the European Parliament and of the Council on novel foods an amending Regulation (EC) No XXX/XXXX [common procedure] (LA) (First reading), In: *Council of the European Union website*, 8 June 2011, Available from: <http://register.consilium.europa.eu/pdf/en/09/st10/st10754.en09.pdf>

Debnath, N.; Das, S.; Seth, D.; Chandra, R.; Bhattacharya, S. C. & Goswami, A. (2011). Entomotoxic effect of silica nanoparticles against Sitophilus oryzae (L.). *Journal of Pest Science*, Vol. 81, No. 1, (March 2011), pp. 99-105, ISSN 1612-4758

Dickinson, E. (2010). Food emulsions and foams: stabilization by particles. *Current Opinion in Colloid & Interface Science*, Vol. 15, pp. 40-49, ISSN 1359-0294

Donsì, F.; Annuziata, M.; Sessa, M. & Ferrari, G. (2011). Nanoencapsulation of essential oils to enhance their antimicrobial activity in foods / LWT. *Food Science and Technology*, Vol. 44, pp. 1908-1914

Dudo, A.; Choi, D-H. & Scheufele, D. A. (2011). Food nanotechnology in the news: coverage patterns and thematic emphases during the last decade. *Appetite*, Vol.56, No.1, pp. 78-89, ISSN 0195-6663

Elavazhagan, T. & Arunachalam, K.D., (2011). Memecylon edule leaf extract mediated green synthesis of silver and gold nanoparticles. *International Journal of Nanomedicine*, Vol. 6, (2011), pp. 1265-1278, ISSN 1178-2013

FAO (2008). El estado de la inseguridad alimentaria en el mundo 2008: Los precios elevados de los alimentos y la seguridad alimentaria: amenazas y oportunidades, In: *Depósito de documentos de la FAO website*, 01 August 2011, Available from: <ftp://ftp.fao.org/docrep/fao/011/i0291s/i0291s00.pdf>

FAO (2009). El Estado mundial de la agricultura y la alimentación 2009: La ganadería, a examen, In: *Depósito de documentos de la FAO website*, 01 August 2011, Available from: <http://www.fao.org/docrep/012/i0680s/i0680s.pdf>

FAO. Agotamiento de los Recursos de Agua Dulce Valoración Ambiental: Indicadores de Presión Estado Respuesta, In: *FAO website*, 04 August 2011, Available from: <http://www.fao.org/ag/againfo/programmes/es/lead/toolbox/Indust/DepWatEA.htm>

FAO. Agotamiento de los Recursos de Agua Dulce, In: *FAO website*, 04 August 2011, Availablefrom:<http://www.fao.org/ag/againfo/programmes/es/lead/toolbox/Indust/DFreWat.htm>

FAO. FAOWATER. Water at a Glance: The relationship between water, agriculture, food security and poverty. In: *FAO website*, 04 August 2011, Available from: <http://www.fao.org/nr/water/docs/wateataglance.pdf>

FAO/WHO (2009). Expert Meeting on the Application of Nanotechnologies in the Food and Agriculture Sectors: Potential Food Safety Implications, In: *FAO website*, April 2011, Available from: <http://www.fao.org/ag/agn/agns/files/FAO_WHO_Nano_Expert_Meeting_Report_Final.pdf>

Federal Food, Drug, and Cosmetic Act (FD&C Act) Chapter IV: Food. (March 2005). In: *FDA U.S. Food and Drug Administration website*, 8 June 2011, Available from: <http://www.fda.gov/RegulatoryInformation/Legislation/FederalFoodDrugand CosmeticActFDCAct/default.htm>

Food Standards Agency (2011). FSA Citizens Forums: Nanotechnology and food. In: *TNS-BMRB Report. JN 219186*, April 2011, Available from: <http://www.food.gov.uk/multimedia/pdfs/publication/fsacfnanotechnologyfood.pdf>

Gan, N.; Yang, X.; Xie, D. H.; Wu, Y. Z. & Wen, W. G. (2010). A Disposable Organophosphorus Pesticides Enzyme Biosensor Based on Magnetic Composite Nano-Particles Modified Screen Printed Carbon. *Sensors*, Vol. 10, No. 1, (January 2010), pp. 625-638, ISSN 1424-8220

Gardea-Torresdey, J. L.; Parsons, J. G.; Gomez, E.; Peralta-Videa, J.; Troiani, H. E.; Santiago, P. & Yacaman, M. J. (2002). Formation and growth of Au nanoparticles inside live alfalfa plants. *Nano Letters*, Vol. 2, No. 4, (April 2002), pp. 397-401, ISSN 1530-6984

Gergely, A.; Chaudry, Q. & Bowman, D. M., (2010). Regulatory perspectives on nanotechnologies in foods and food contact materials, In: *International Handbook on Regulating Nanotechnologies*, Hodge, G.A., Bowman, D.M., & Maynard, A.D., pp. 321-341, Edward Elgar Publishing Limited, Retrieved from <http://www.steptoe.com/assets/attachments/Gergely%20et%20al%202010%20R egulatory%20perspectives%20on%20nanotech%20in%20foods.pdf>

Goffin, A. L.; Raquez, J. M.; Duquesne, E.; Siqueira, G.; Hbibi, Y.; Dufresne, A. & Dubois, P. (2011). From interfacial ring-opening polymerization to melt processing of cellulose nanowhisker-filled polylactide-based nanocomposites. *Biomacromolecules*, Vol. 12, No. 7, pp. 2456-2465, ISSN 1525-7797

Gökmen, V.; Mogol, B.; Lumaga, R.; Fogliano, V.; Kaplun, Z. & Shimoni, E. (2011). Development of functional bread containing nanoencapsulated omega-3 fatty acids. *Journal of Food Engineering*, Vol. 105, pp. 585–591, ISSN 0260-8774

Grimshaw, D. (2009). Nanotechnology for clean water: facts and figures, In: *SciDevNet. Science and Development Network website*, 02 August 2011, Available from: <http://www.scidev.net/en/new-technologies/nanotechnology/features/ nanotechnology-for-clean-water-facts-and-figures.html>

Groves, K. & Titoria, P. (2009). Nanotechnology for the food industry. *Nano the Magazine for Small Science*, Vol. 13, (August 2009), pp. 12-14, ISSN 1757-2517

Guan, H. A.; Chi, D.F.; Yu, J. & Li, H. (2010). Dynamics of residues from a novel nano-imidacloprid formulation in soyabean fields. *Crop protection*, Vol. 29, No. 9, (September 2010), pp. 942-946, ISSN 02161-2194

Harrington, R. & Dawson, F. (2011). Nano-based packaging more acceptable to consumers, In: *FoodProductionDaily.com website*, Available from: <http://www. foodproductiondaily.com/Quality-Safety/Nano-based-packaging-more-acceptable-to-consumers>

Heidenreich, B.; Pohlmann, C.; Sprinzi, M. & Gareis, M. (2010). Detection of Escherichia coli in meat with an electochemical biochip. *Journal of Food Protection*, Vol. 73, No. 11, pp. 2025-2033, ISSN 0362-028X

House of Lords, Science and Technology Committee. (2010a). Nanotechnologies and Food. 1st Report of Session 2009-10 Volume I: Report, In: *UK Parliament website*, 9 June 2011, Available from: <http://www.publications.parliament.uk/pa/ld200910 /ldselect/ldsctech/22/22i.pdf>

House of Lords, Science and Technology Committee. (2010b). Nanotechnologies and Food. 1st Report of Session 2009–10 Volume II: Minutes of Evidence, In: *UK Parliament website*, 9 June 2011, Available from: <http://www.publications.parliament. uk/pa/ld200910/ldselect/ldsctech/22/2202.htm>

Institute of Food Science & Technology (2009). The House of Lords Science and Technology Committee Inquiry into the Use of Nanotechnologies in the Food Sector 2009. Written evidence submitted by the Institute of Food Science & Technology (IFST), In: *Institute of Food Science & Technology website*, 10 June 2011, Available from: <http://www.ifst.org/documents/submissions/holselectcttenanotechnologyinqui r.pdf>

Jayarao, B.; Wolfgang, D. R.; Keating, C. D.; Van Saun, R. J. & Hovingh, E. (2011). Nanobarcodes-based bioassays for tracing consignments of meat and bone meal wiyhin the project use nanobar technologies to trace commodities for contamination and/or purity, In: *Veterinary and Biomedical Sciences*, 03 August 2011 Available from: <http://vbs.psu.edu/extension/research/projects/nanobarcodes-based-bioassays-for-tracing-consignments-of-meat-and-bone-meal, College of Agricultural Sciences web 2011>

Joseph, T. & Morrison, M. (2006). Nanotechnology in Agriculture and Food. Nanoforum Report, In: *Nanoforum.org European Nanotechnology Gateway website*, April 2011, Available from: <http://www.nanoforum.org/dateien/temp/nanotechnology %20in%20agriculture%20and%20food.pdf>

Knauer, K. & Bucheli, T. (2009). Nano-materials-the need for research in agriculture. *Agrarforschung*, Vol. 16, No. 10, (October 2009), pp. 390-395, ISSN 1022-663X

Kumaravel, A. & Chandrasekaran, M. (2011). A biocompatible nano TiO(2)/nafion composite modified glassy carbon electrode for the detection of fenitrothion. *Journal of Electroanalytical Chemistry*, Vol. 650, No. 2, (January 2011), pp. 163-170, ISSN 1572-6657

Kuzma, J. (2010). Nanotechnology in animal production—Upstream assessment of applications. *Livestock Science*, Vol. 130, pp. 14–24, ISSN 1871-1413

Lagaron, J. M. & López-Rubio, A. (2011). Nanotechnology for bioplastics: opportunities, challenges and strategies. *Trends in Food Science & Technology*, doi:10.1016/j.tifs.2011.01.007

Lepot, N.; Van Bael, M. K.; Van den Rul, H.; D'Haen, J.; Peeters, R.; Franco, D. & Mullens, J. (2011). Influence of incorporation of ZnO nanoparticles and biaxial orientation on mechanical and oxygen barrier properties of polypropylene films for food packaging appliations. *Journal of Applied Polymer Science*, Vol. 120, No. 3; pp. 1616-1623, ISSN 0021-8995

Leser, M. E.; Sagalowicz, L.; Michel, M. & Watzke, H. J. (2006). Self-assembly of polar food lipids. *Advances in Colloid and Interface Science*, pp. 123– 126, 125– 136, ISSN 0001-8686

Li, X. H.; Xing, Y. G.; Li, W. L.; Jiang, J. H. & Ding, Y. L. (2010). Antibacterial and physical properties of poly(vinylchoride)-based film coated with ZnO nanoparticles. *Food Science and Technology International*, Vol. 16, No. 3, pp. 225-232, ISSN 1082-0132

Lopez-Rubio, A.; Lagaron, J. M.; Ankerfors, M.; Lindstrom, T.; Nordqvist, D. & Mattozzi, A. (2007). Enhanced film forming and film properties of amylopectin using micro-fibrillated cellulose. *Carbohydrate Polymers*, Vol. 68, No. 4, pp. 718-727

Luppi, B.; Bigucci, F.; Cerchiara, T.; Mandrioli, R.; Di Pietra, A. M. & Zecchi, V. (2008). New environmental sensitive system for colon-specific delivery of peptidic drugs. *International Journal of Pharmaceutics*, Vol. 358, pp. 44–49, ISSN 0378-5173

Macoubrie, J. (2005). Informed Public Perceptions of Nanotechnology and Trust in Government. The Project on Emerging Nanotechnologies at the Woodrow Wilson International Center for Scholars, In: *Woodrow Wilson International Center for Scholars website*, 05 August, 2011, Available from: <http://www.pewtrusts.org/uploadedFiles/wwwpewtrustsorg/Reports/Nanotechnologies/Nanotech_0905.pdf>

Magnuson, B. (2010). State of the science on safety/toxicological assessment of nanomaterials for food applications Presentation. Cantox (Health Science International). IFT International Food Nanoscience Conference, Chicago, Illinois (USA), Ontario, Canada. July 17, 2010

Mao, H-Q.; Roy, K.; Troung-Le, V.L.; Janes, K. A.; Lin, K. Y.; Wang, Y.; August, J. T. & Leong, K. W. (2001). Chitosan-DNA nanoparticles as gene carriers: synthesis, characterization and transfection efficiency. *Journal of Controlled Release*, Vol. 70, No. 3, (February 2001), pp.399-421, ISSN 0168-3659

Mayo, J. T.; Yavuz, C.; Yean, S.; Cong, L.; Shipley, H.; Yu, W.; Falkner, J.; Kan, A.; Tomson, M. & Colvin, V.L., (2007). The effect of nanocrystalline magnetite size on arsenic

Removal. *Science and Technology of Advanced Materials*, Vol. 8, pp. 71–75, ISSN 1468-6996

Mc Knight, T. E.; Melechko, A. V.; Griffin, G.D.; Guillorn, M.A.; Merkulov, V.I.; Serna, F.; Hensley, D.K.; Doktycz, M.J.; Lowndes, D.H. & Simpson, M.L. (2003). Intracellular integration of synthetic nanostructures with viable cells for controlled biochemical manipulation. *Nanotechnology*, Vol. 14, No. 5, pp. 551-556, ISSN 0957-4484

Miller, G. & Senjen, R. (2008). Out of the laboratory and on to our plates. Nanotechnology in food & agriculture, In: Friends of the Earth, Australia, Europe & U.S.A., In: *Friends of the Earth Europe website*, 31 May 2011, Available from: <http://www.foeeurope.org/activities/nanotechnology/Documents/Nano_food_report.pdf>

Mills, A. & Hazafy, D. (2009) Nanocrystalline SnO_2-based, UVB-activated, colourimetric oxygen indicator. *Sensors and Actuators B-Chemical*, Vol. 136, No. 2, pp. 344–349, ISSN 0925-4005

Morones, J. R., Elechiguerra, J. L., Camacho, A., Holt, K., Kouri, J. B., Ramirez, J. T. & Yacaman, M.J. (2005). The bactericidal effect of silver nanoparticles. *Nanotechnology*, Vol. 16, No. 10, pp. 2346-2353, ISSN 0957-4484

Morris, V. (2010). Nanotechnology and Food, In: *International Union of Food Science & Technology IUFoST Scientific Information Bulletin August 2010 Update*, 9 June 2011, Available from:
http://www.iufost.org/sites/default/files/docs/IUF.SIB.NanotechnologyandFoodupdate.pdf>

Nair, R.; Varguese, S. H.; Nair, B. G.; Maekawa, T.; Yoshida, Y. & Kumar, D. S. (2010). Nanoparticulate material delivery to plants. *Plant Science*, Vol. 179, pp. 154-163

Nutt, M. O.; Hughes, J. H. & Wong, M. S. (2005). Designing Pd-on-Au bimetallic nanoparticle catalysts for trichloroethene hydrodechlorination. *Environmental Science & Technology*, Vol. 39, pp. 1346-1353, ISSN 0013-936X

Ochoa, J.; Irache, J. M.; Tamayo, I.; Walz, A.; DelVecchio, V.G. & Gamazo, C. (2007). Protective immunity of biodegradable nanoparticle-based vaccine against an experimental challenge with Salmonella Enteritidis in mice , *Vaccine*, Vol. 25, No. 22, pp. 4410-4419, ISSN 0264-410X

Oren, Y. (2008). Capacitive deionization (CDI) for desalination and water treatment — past, present and future (a review). *Desalination*, Vol. 228, pp. 10–29, ISSN 0011-9164

Palmberg, C.; Dernis, H. & Miguet, C. (2009). Nanotechnology: an overview based on indicators and statistics. statistical analysis of science, technology and industry, In: *Organisation for Economic Co-operation and Development (OECD) website*, 18 July 2011, Available from: <http://www.oecd.org/dataoecd/59/9/43179651.pdf>

Peters, R.; Dam, G.; Bouwmeester, H.; Helsper, H.; Allmaier, G.; Kammer, F.; Ramsch, R.; Solans, C.; Tomaniová, M.; Hajslova, J. & Weigel, S. (2011). Identification and characterization of organic nanoparticles in food. *Trends in Analytical Chemistry*, Vol. 30, No. 1, (January 2011), pp. 100-112, ISSN 0167-2940

Plapied, L.; Duhem, N.; des Rieux, A. & Préat, V. (2011). Fate of polymeric nanocarriers for oral drug delivery. *Current Opinion in Colloid & Interface Science*, Vol. 16, pp. 228–237, ISSN 1359-0294

Platypus Technologies LLC (2002). Detection of FMDV by Anchoring Transitions of Liquid Crystals. Program: 2002/SBIR, Award ID:57257, In: *SBIR/STTR Small Business Innovation Research Small Business Technology Transfer*, 05 August 2011, Available from: <http://sba-sbir-qa.reisys.com/sbirsearch/detail/277343>

Pray, L. & Yaktine, A. (2009). Nanotechnology in Food Products: Workshop Summary. *Food Forum; Institute of Medicine*, ISBN 978-0-309-13772-0

Project on Emerging Nanotechnologies (2011). Inventory of nanotechnology-based consumer products, In: *Project on Emerging Nanotechnologies website*, July 2011, Available from

http://www.nanotechproject.org/inventories/consumer/analysis_draft/>

Qian, K.; Shi, T. Y.; Tang, T.; Zhang, S. L.; Liu, X. L. & Cao, Y. S. (2011). Preparation and characterization of nano-sized calcium carbonate as controlled release pesticide carrier for validamycin against Rhizoctonia solani. *Microchimica Acta*, Vol. 173, No. 1-2, (April 2011), pp. 51-57, ISSN 0026-3672

Rajani, P.; SriSindhura, K.; Prasad, T. N. V. K. V.; Hussain, O. M.; Sudhakar, P.; Latha, P.; Balakrishna, M.; Kambala, V.; Reddy, K. R.; Giri, P. K.; Goswami, D. K.; Perumal, A. & Chattopadhyay, A. (2010). Fabrication of Biogenic Silver Nanoparticles Using Agricultural Crop Plant Leaf Extracts. *Proceedings of International Conference on Advanced Nanomaterials and Nanotechnology*, ISBN 978-0-7354-0825-8, Guwahati, December 2009

Rodríguez, J. M.; Lucero, A.; Rodríguez, M. R.; Mackie, A. R.; Gunning, A. P.; Morris, V. J. (2007). Some implications of nanoscience in food dispersion formulations containing phospholipids as emulsifiers. *Food Chemistry*, Vol. 102, pp. 532–541, ISSN 0308-8146

Rollin, F.; Kennedy, J. & Wills, J. (2011). Review: Consumers and new food technologies. *Trends in Food Science & Technology*, Vol. 22, Issue. 2-3, pp. 99-111, ISSN 0167-7799

Sánchez, C. C. & Patino, J. M. R. (2005). Interfacial, foaming and emulsifying characteristics of sodium caseinate as influenced by protein concentration in solution. *Food Hydrocolloids*, Vol.19 407–416, ISSN 0268-005X

Sánchez-García, M. D.; Lagaron, J. M. & Hoa, S. V. (2010a). Effect of addition of carbon nanofibers and carbon nanotubes on properties of thermoplastic biopolymers. *Composites Science and Technology*, Vol. 70, No. 7, pp. 1095-1105, ISSN 0266-3538

Sánchez-García, M. D.; López-Rubio, A. & Lagaron, J. M. (2010b). Natural micro and nanobiocomposites with enhanced barrier properties and novel functionalities for food biopackaging applications. *Trends in Food Science & Technology*, Vol. 21, pp. 528-536, ISSN 0167-7799

Scheerlinck, J. P. Y.; Gloster, S.; Gamvrellis, A.; Mottram, P.L. & Plebanski, M. (2005). Systemic immune responses in sheep, induced by a novel nano-bead adjuvant. *Vaccine*, Vol.24, No. 8, pp. 1124-1131, ISSN 0264-410X

Seear, K.; Petersen, A. & Bowman, D. (2009) The Social and Economic Impacts of Nanotechnologies: A Literature Review, In: *Department of Innovation, Industry, Science and Research of Australian Government website*, 25 July 2011, Available from:<http://www.innovation.gov.au/Industry/Nanotechnology/NationalEnabli

ngTechnologiesStrategy/Documents/SocialandEconomicImpacts_LiteratureRevie
w.pdf>

Seldon laboratoires. Personal Water treatment: Waterstick, In: *Source: Seldon website*, 04
August 2011, Available from: <http://www.seldontechnologies.com/products.
htm>

Senthilnathan, J. & Ligy, P. (2010). Removal of mixed pesticides from drinking water system
using surfactant-assisted nano-TiO_2. *Water Air and Soil Pollution*, Vol. 210, pp. 143-
154, ISSN 0049-6979

Shen, H. Y.; Zhu, Y.; Wen, X. E. & Zhuang, Y.M. (2007). Preparation of Fe3O4-C18 nano-
magnetic composite materials and their cleanup properties for organophosphorus
pesticides. *Analytical and Bioanalytical Chemistry*, Vol. 387, No. 6, (March 2007), pp.
2227-2237, ISSN 1618-2642

Shoults-Wilson, W. A.; Zhurbich, O. I.; McNear, D. H.; Tsyusko, O. V.; Bertsch, P. M. &
Unrine, J. M., (2011). Evidence for avoidance of Ag nanoparticles by earthworms
(Eisenia fetida). *Ecotoxicology*, Vol. 20, No. 2, (March 2011), pp. 385-296, ISSN 0963-
9292

Siegrist, M.; Cousin, M-E.; Kastenholz, H. & Wiek, A. (2007). Public acceptance of
nanotechnology foods and food packaging: The influence of affect and trust.
Appetite, Vol. 49, pp. 459-466, ISSN 0195-6663

Siegrist, M.; Stampfli, N.; Kastenholz, H. & Keller, C. (2008). Perceived risks and perceived
benefits of different nanotechnology foods and nanotechnology food packaging.
Appetite, Vol. 51, pp. 283-290, ISSN 0195-6663

Silvestre, C.; Duraccio, D. & Cimmino, S. (2011). Food packaging based on polymer
nanomaterials. Progress in Polymer Science,
doi:10.1016/j.progpolymsci.2011.02.003

Siqueira, G.; Bras, J. & Dufresne, A. (2009). Cellulose whiskeys versus microfibrils: influence
of the nature of the nanoparticle and its surface functionalization on the thermal
and mechanical properties of nanocomposites. *Biomacromolecules*, Vol. 10, No. 2, pp.
425-432, ISSN 1525-7797

Song, S. L.; Liu, X. H.; Jiang, J. H.; Qian, Y. H.; Zhang, N. & Wu, Q. H. (2009). Stability of
triazophos in self-nanoemulsifying pesticide delivery system. *Colloids and Surfaces
A-Physicochemical and Engineering*, Vol. 350, No. 1-3, (October 2009), pp. 57-62, ISSN
0927-7757

Sorrentino, A.; Gorrasi, G. & Vittoria, V. (2007). Potential perspectives of bionanocomposites
for food packaging applications. *Trends in Food Science & Technology*, Vol. 18, pp.
84-95, ISSN 0167-7799

Sundari, P. A. & Manisankar, P. (2011). Development of Nano Poly (3-methylthiophene)
/Multiwalled Carbon Nanotubes Sensor for the Efficient Detection of Some
Pesticides. *Journal of the Brazilian Chemical Society*, Vol. 22, No. 4, (2011), pp. 746-755,
ISSN 0103-5053

The European Parliament and the Council. (December 2008). REGULATION (EC) No
1331/2008 OF THE EUROPEAN PARLIAMENT AND OF THE COUNCIL of 16
December 2008 establishing a common authorisation procedure for food additives,
food enzymes and food flavorings, In: *Official Journal of the European Communities*, 8

June 2011, Available from: <http://eur-lex.europa.eu/LexUriServ/LexUriServ.do?
uri=OJ:L:2008:354:0001:0006:EN:PDF>

The European Parliament and the Council. (December 2008). REGULATION (EC) No
1332/2008 OF THE EUROPEAN PARLIAMENT AND OF THE COUNCIL of 16
December 2008 on food enzymes and amending Council Directive 83/417/EEC,
Council Regulation (EC) No 1493/1999, Directive 2000/13/EC, Council Directive
2001/112/EC and Regulation (EC) No 258/97, In: *Official Journal of the European
Communities*, 8 June 2011, Available from: <http://eur-lex.europa.eu/LexUriServ
/LexUriServ.do?uri=OJ:L:2008:354:0007:0015:EN:PDF>

The European Parliament and the Council. (December 2008). REGULATION (EC) No
1333/2008 OF THE EUROPEAN PARLIAMENT AND OF THE COUNCIL of 16
December 2008 on food additives, In: *Official Journal of the European Communities*, 8
June 2011, Available from: <http://eur-lex.europa.eu/LexUriServ/LexUriServ.do?
uri=OJ:L:2008:354:0016:0033:en:PDF>

The European Parliament and the Council. (December 2008). REGULATION (EC) No
1334/2008 OF THE EUROPEAN PARLIAMENT AND OF THE COUNCIL of 16
December 2008 on flavorings and certain food ingredients with flavoring
properties for use in and on foods and amending Council Regulation (ECC) No
1601/91, Regulations (EC) No 2232/96 and (EC) No 110/2008 and Directive
2000/13/EC, In: *Official Journal of the European Communities*, 8 June 2011, Available
from:
<http://eur-lex.europa.eu /LexUriServ/ LexUriServ.do?uri=OJ:L:2008:354 :0034:
0050:en:PDF>

The European Parliament and the Council. (January 1997). REGULATION (EC) No 258/97
OF THE EUROPEAN PARLIAMENT AND OF THE COUNCIL of 27 January
1997 concerning novel foods and novel food ingredients, In: *Official Journal of the
European Communities*, 8 June 2011, Available from:
<http://eur-lex.europa.eu/LexUriServ/LexUriServ.do?uri=OJ:L:1997:043:0001:
0006:EN:PDF>

The European Parliament and the Council. (January 2002). REGULATION (EC) No
178/2002 OF THE EUROPEAN PARLIAMENT AND OF THE COUNCIL of 28
January 2002 laying down the general principles and requirements of food law,
establishing the European Food Safety Authority and laying down procedures in
matters of food safety, In: *Official Journal of the European Communities*, 8 June 2011,
Available from:
<http://eur-lex.europa.eu/LexUriServ/LexUriServ.do?uri=OJ:L:2002:031:0001
:0024:EN:PDF>

The European Parliament and the Council. (May 2009). COMMISSION REGULATION
(EC) No 450/2009 of 29 May 2009 on active and intelligent materials and articles
intended to come into contact with food, In: *Official Journal of the European Union*, 8
June 2011, Available from:
<http://eur-lex.europa.eu/LexUriServ/LexUriServ.do?uri=OJ:L:2009:135:0003
:0011:EN:PDF>

The European Parliament and the Council. *(October 2004).* *REGULATION (EC) No 1935/2004 OF THE EUROPEAN PARLIAMENT AND OF THE COUNCIL of 27 October 2004 on materials and articles intended to come into contact with food and repealing Directives 80/590/EEC and 89/109/EEC, In: Official Journal of the European Communities, 8 June 2011, Available from:* *<http://eur-lex.europa.eu/LexUriServ/LexUriServ.do?uri=OJ:L:2004:338:0004:0017:en :PDF>*

Torres-Giner, S., Gimenez, E., & Lagaron, J. M. (2008). Characterization of the morphology and thermal properties of zein prolamine nanostructures obtained by electrospinning. *Food Hydrocolloids*, Vol. 22, pp. 601-614

Travan, A.; Pelillo, C.; Donati, I.; Marsich, E.; Benincasa, M. & Scarpa, T. (2009). Non-cytotoxic silver nanoparticle-polysaccharide nanocomposites with antimicrobial activity. *Biomacromolecules*, Vol. 10, No. 6, pp. 1429-1435, ISSN 1525-7797

UNESCO. (2007). Boletín Semanal del Portal del Agua de la UNESCO n° 184: Las Aguas Residuales, Hechos y Cifras sobre Aguas Residuales, In: *UNESCO website*, 08 August 2011, Available from: <http://www.unesco.org/water/news/newsletter /184_es.shtml>

Ursache-Oprisan, M.; Focanici, E.; Creanga, D. & Caltun, O. (2011). Sunflower chlorophyll levels after magnetic nanoparticle supply. *African Journal of Biotechnology*, Vol. 10, No.36, (July 2011), pp. 7092-7098, ISSN 1684-5315

Utracki, L. A.; Broughton, B.; González-Rojano, N.; de Carvalho, L. H. & Achete, C. A. (2011). Clays for polymeric nanocomposites. *Polymer Engineering and Science*, Vol. 51, N° 3, pp. 559-572, ISSN 0032-3888

Vladimiriov, V., Betchev, C., Vassiliou, A., Papageorgiou, G., & Bikiaris, D. (2006). Dynamic mechanical and morphological studies of isotactic polypropylene /fumed silica nanocomposites with enhanced gas barrier properties. *Composites Science and Technology*, Vol. 66, pp. 2935–2944, ISSN 0266-3538

Walter, M.T., Luo, D., Regan, J.M., 2005. Using nanotechnology to identify and characterize hydrological flowpaths in agricultural landscapes. USDA-CRIS grant proposal 2004-04459

Wang, L. J.; Li, X. F.; Zhang, G. Y.; Dong, J. F. & Eastoe, J. (2007). Oil-in-water nanoemulsions for pesticide formulations. *Journal of Colloid and Interfacial Science*, Vol. 314, No. 1, (October 2007), pp. 230-235, ISSN 0021-9797

Wang, Q.; Chen, J.; Zhang, H.; Lu, M.; Qiu, D.; Wen, Y. & Kong, Q. (2011). Synthesis of Water Soluble Quantum Dots for Monitoring Carrier-DNA Nanoparticles in Plant Cells. *Journal of Nanoscience and Nanotechnology*, Vol. 11, No. 3, (March 2011), pp. 2208-2214, ISSN 1533-4880

Wolfe, J. (2006). Top Five Nanotech Breakthroughs of 2006, In: *Forbes/Wolfe Nanotech Report 2006*, 04 August 2011, Available from: http://www.forbes.com/2006/12/26/nanotech-breakthroughs-ibm-pf-guru-in_jw_1227soapbox_inl.html>

Yada, R. Y. (2011). Nanotechnology. How Science of the Small Could Have Large Implications for Your Food. The 18th Annual FFIGS Educational Workshop: Food Safety Today –A Micro and Macro Perspective!. 4 May, 2011, Available from:

<http://ffigs.org/2011%20R.%20Yada%20Food%20Safety%20Forum%20Ingersoll .
%20May%201%202011.pdf>

Yang, H. L.; Lin, J. C. & Huang, C. (2009). Application of nanosilver surface modification
    to RO membrane and spacer for mitigating biofouling in seawater desalination.
    Water Research, Vol. 43, pp. 3777-3786, ISSN 0043-1354

Yang, L.; Chakrabartty, S. & Alocilja, E. (2007). Fundamental building blocks for
    molecular biowire based forward error-correcting biosensors. *Nanotechnology*,
    Vol. 18, N°. 42, pp. 1-6, ISSN 0957-4484

Zeng, R.; Wang, J. G.; Cui, J. Y.; Hu, L. & Mu, K. G. (2010). Photocatalytic degradation of
    pesticide residues with RE(3+)-doped nano-TiO(2). *Journal of Rare Earths*, Vol. 28,
    No. 1, (December 2010), pp. 353-356, ISSN 1002-0721

Zhao, X.; Zhaoyang, L.; Wenguang, L.; Wingmoon, L.; Peng, S.; Richard, Y. T. K.; Keith, D.
    K. L. & William, W.L. (2011). Octaarginine-modified chitosan as a nonviral gene
    delivery vector: properties and in vitro transfection efficiency. *Journal of Naparticle
    Research*, Vol. 13, No. 2, (February 2011), pp. 693-702, ISSN 1572-896X

Zodrow, K.; Brunet, L.; Mahendra, S.; Li, D.; Zhang, A.; Li, Q. & Alvarez, P. J. J. (2009).
    Polysulfone ultrafiltration membranes impregnated with silver nanoparticles
    show improved biofouling resistance and virus removal. *Water Research*, Vol. 43,
    pp. 715-723, ISSN 0043-1354

8

Characteristics and Role of Feruloyl Esterase from *Aspergillus Awamori* in Japanese Spirits, *'Awamori'* Production

Makoto Kanauchi
Miyagi University
Japan

1. Introduction

Feruloyl esterases (EC 3.1.1.73), known as ferulic acid esterases, which are mainly from *Aspergillus* sp. (Faulds & Williamson, 1994), can specifically cleave the (1→5) ester bond between ferulic acid and arabinose. The esterases show high specificity of hydrolysis for synthetic methyl esters of phenyl alkanoic acids (Kroon and others, 1997). The reaction rate increases markedly when the substrates are small soluble feruloylated oligosaccharides derived from plant cell walls (Faulds and other, 1995; Ralet and others, 1994).

These enzymes have high potential for application in food production and other industries. Ferulic acid links hemicellulose and lignin. In addition, cross-linking of ferulic acids in cell wall components influences wall properties such as extensibility, plasticity, and digestibility, as well as limiting the access of polysaccharides to their substrates (Borneman et al., 1990).

Actually, feruloyl esterase is used for *Awamori* spirit production. *Awamori* spirits are Japanese spirits with a distinctive vanilla-like aroma. Feruloyl esterase is necessary to produce that vanilla aroma. Actually, lignocellulosic biomass is one means of resolving energy problems effectively. It is an important enzyme that produces bio-fuel from lignocellulosic biomass.

As explained in this paper, *Awamori* spirit production is described as an application of feruloyl esterase. The vanillin generating pathway extends from ferulic acid as precider, with isolation of *Aspergillus* producing feruloyl esterase, which is characteristic of the enzyme. Moreover, the application of feruloyl esterase for beer production and bio-fuel production is explained.

2. *Awamori* spirits

2.1 *Awamori* spirit characteristics

Awamori spirits have three important features. First, mash of *Awamori* spirit is fermented using *koji*, *Aspergillus* sp. are grown on steamed rice, which is the material and saccharifying agent used in *Awamori* spirit production. That fermentation is done in a pot still. Mash used in *Awamori* spirit processing is different from beer brewing, in which fermenting is done with saccharified mash by malt. Their fermentative form is call 'parallel fermentation' which progresses simultaneously with saccharification and fermentation. The resultant fermentative yeast can produce high concentrations of ethanol, approximately 16–18%, from mash of *Awamori* without osmotic injury.

Secondly, highly concentrated citric acid is produced in this process from *koji* made by black *Aspergillus* sp., classified as *Aspergillus awamori*. Because of this acid, the mash maintains low pH. It is usually made in the warm climate of Okinawa, with average temperatures of 25°C in all seasons. Koizumi (1996) describes that spoiling bacteria are able to grow in mash under pH 4.0 conditions. Moreover, although amylase from *Aspergillus oryzae* is inactivated at less than pH 3.5, that from *Aspergillus awamori* reacts stably at pH 3.0. Furthermore, the mash ferments soundly under those warm conditions.

Finally, aging is an important feature of *Awamori* spirits, which have a vanilla aroma that strengthens during aging. The *Awamori* spirit is aged in earthen pots for three years or more. Particularly, the spirit aged for more than three years, called '*Kusu*', is highly prized. The vanilla aroma in Scotch whisky, bourbon, or brandy is produced from lignin in barrel wood during aging. *Kusu* is not aged in barrels, but it does have a vanilla aroma resembling those of aged Scotch whisky, bourbon, and brandy.

Differences between *Awamori* spirit and other beverages are shown in the table. History and production methods of *Awamori* spirits are described below.

	Awamori	Sake	Whisky	Brandy
Type	Distilled beverage	Brewed beverage	Distilled beverage	Distilled beverage
Place	Okinawa	Mainly Japan	Worldwide	Worldwide
Production Temperature	All seasons average annual temperature (25°C)	Mainly winter 0–4°C	Room temperature (10–15°C)	Room temperature (15–20°C)
Material	*Indica* rice	*Japonica* rice	Barley, corn	Grapes
Mash	Parallel Fermentation Containing citric acid produced by	Parallel fermentation, Containing lactic acid produced by Lactic acid	Single Fermentation Not Containing Acid	Single Fermentation containing Malic acid from Material
Saccharifying agent	*Koji*	*Koji*	Malt	-
Microorganisms	*Aspergillus awamori* Awamori yeast	*Aspergillus oryzae* Sake yeast Lactic acid	Whisky yeast	Wine yeast

Fermentative Temperature	High temperature (27–30°C)	Low temperature (10–15°C)	Middle temperature (15–25°C)	Middle temperature (15–26°C)
Alcohol concentration	25–30%	15–16%	40–50%	40–50%
Aging period	Approximately 3 years or more	Very short term	More than 3 years	More than 3 years
Aging vessel	Mainly earthen pot	Mainly stainless tank	Barrel	Barrel
Taste and aroma	Vanilla like	Estery, fruit-like	Vanilla like	Vanilla like

Table 1. *Awamori* spirit and other alcoholic beverages

2.2 History of *Awamori*

Awamori spirits are traditionally produced in Okinawa, which has 47 production sites. *Awamori* spirits are produced from long-grain rice and rice imported from Thailand. Partly because it uses long grain rice imported from Thailand for production, it is believed that *Awamori* spirit production methods were brought from Thailand (Koizumi, 1996).

According to one account (Koizumi, 1996) of Okinawa's history, 'Ryukyu' was an independent country ruled by king Sho in 1420, which traded with the countries of Southeast Asia. At the time, the port of Naha bustled as a junction port between Japan and the South China Sea Islands, Indonesia, Cambodia, Vietnam, the Philippines, and Thailand. *Awamori* spirits were brought from there and also traded. In 1534, 'Chen Kan's Records', reported to his home country, China, noted that *Awamori* spirits have a clear aroma and were delicious; he noted also that *Awamori* spirits had been brought from Thailand.

Moreover, it was written that long-grain rice harvested in Thailand was used in *Awamori* spirit production, and the distilled spirits were aged in earthen pots. Their ancient technology of *Awamori* spirit production is followed by the present technology. Distillation technology was brought also via Thailand from China, as it was with *Awamori* spirits. Furthermore, they transported the technology eventually to the main islands of Japan.

The cradle of distillation technology is actually ancient Rome. In that era, distillation methods were used to produce essential oils in the following manner: plant resin was boiled in a pan on which a wool sheet had been placed. After boiling, the upper wool sheet was pressed to obtain the essential oil. That is a primitive distillation method.

The distillation method brought to Okinawa was superior to the Roman method, but the efficiency of distillation was low, according to Edo period accounts: 360 mL of distillate was obtained from 18 l of fermented alcohol beverages. Eventually, 72 ml of spirits were distilled from the first distillate (Koizumi, 1996). We can infer the alcohol concentration experimentally: fermentative alcoholic beverages (*Sake*) have approx. 10% alcohol concentration, the first distillate has approximately 20% alcohol concentration, and final spirits have approximately 30% alcohol concentration. The distillate yield by the original method was lower than that of the present method because the condenser was not a water-cooled system.

2.3 *Awamori* spirits production method

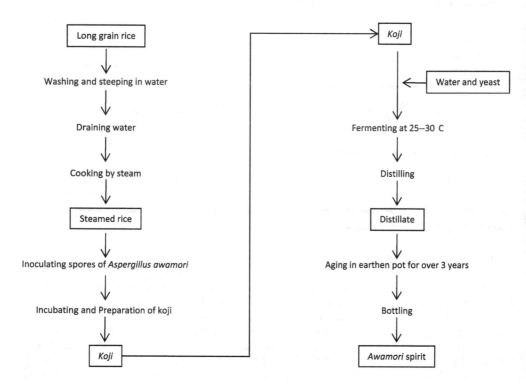

Fig. 1. Schemes of *Awamori* spirit production.

2.3.1 Rice
Awamori spirits are produced using long-grain rice imported from Thailand.
Oryza sativa is a perennial plant. Kato (1930) reported rice taxonomy. He reported some differences in *Indica* rice strains (imported from Thailand) and Japonica rice strains (grown in Japan): rice grains of *Indica* strain are longer than those of Japonica rice strain. Its leaves are light green. Moreover, *Indica* rice grains are longer than Japonica rice grains. The two strains sterilize in mating with each other. Furthermore, they relate to each other as subspecies, *Oryza sativa* subsp. *japonica* and *Oryza sativa* subsp. *indica*.
Some merits exist for the use of imported Thailand rice as the material for production.
1. The rice material is cheaper than Japanese rice. 2. Because the indica rice is not sticky, it is easy to work with during *koji* preparation. 3. Mash temperatures of mash using *Indica* rice are easy to control because this rice is hard and saccharifies slowly. 4. The alcohol yield from *Indica* rice is higher that from *Japonica* strains. 5. Indica rice has been used to produce *Awamori* spirit since it was brought from Chiame, Thailand (Nishiya, 1991).
The rice strains differ not only in grain size and shape but also in starch characteristics. Rice contains starches of two types: amylose and amylopectin. The structure is shown in the figure.

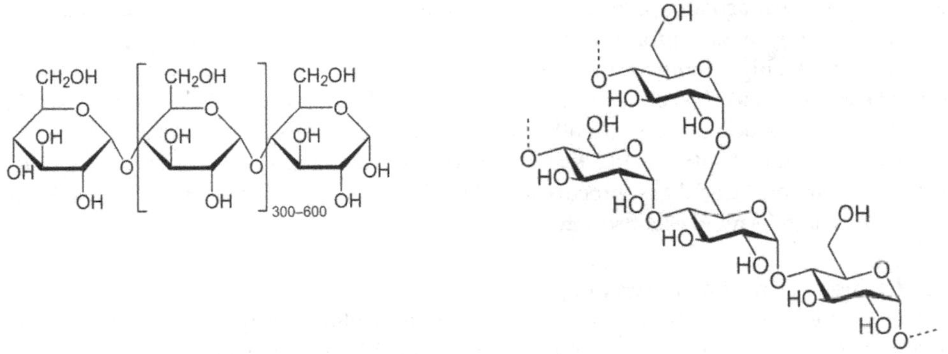

Fig. 2. Structures of amylose and amylopectin
Wikipedia (http://www.wikipedia.org)

Generally, *Japonica* rice strains contain 10–22% amylose, and amylase is not present in *Japonica* waxy strains. The starch is almost entirely composed of amylopectin (Juliano, 1985). In contrast, *Indica* rice contains about 18–32% amylase (Juliano, 1985). High concentrations of amylopectin make cooked rice sticky. Therefore, *Indica* rice is not sticky and is therefore suitable for preparation of *koji* for *Awamori* spirit production. Moreover, Horiuchi and Tani (1966) reported amylograms of nonwaxy starch prepared from some *Japonica* rice and *Indica* rice. The gelatinization temperature of *Indica* rice is 71.5°C (69.5–73.5°C), although the mean pasting temperature for *Japonica* varieties is 63.5°C (59–67%). No definite differences were found in the values of maximum viscosity and breakdown between *Japonica* and *Indica* varieties. Juliano et al. (1964a) obtained narrower pasting temperature ranges for *Japonica* rice flour (62–67°C) than for *Indica* (62–76.5°C). Gelatinization temperature of *Indica* rice was the highest among major cereals as waxy (*Japonica*) 62°C, 66°C for maize, and 62°C for wheat. (Dendy and Dobraszczyk, 2000).

2.3.2 Preparation of 'koji' growing *Aspergillus awamori* on steamed rice

A. awamori have a role of saccharification during fermentation. Furthermore, the strain has a black area (conidia), and it differs completely from *A. niger*. *A. awamori* was isolated by Inui from *koji* for *Awamori* spirit production and named *A. luchuensis*, Inui (Inui, 1991). The strain was later renamed *A. awamori*. The strain is an important strain for alcohol production in Japan. The mold strain prevents contamination during fermentation processing by produced citric acid and acid tolerance α-amylase supplied by the strain saccharified starch as substrate under acid conditions during fermentation processing. Raper and Fennell (1965) investigated Black *Aspergillus*. Murakami (1982) compared the size of conidia, lengths of conidiophores, and characteristics of physiology, and classified two groups as *Awamori* Group and *Niger* Group in Black *Aspergillus*. *A. awamori* can grow at 35°C. It assimilates nitrite, and it has a short conidiophore: less than 0.9–1.1 mm.

2.3.3 Preparation of *koji*

Koji is used for alcohol beverage production or production of Asian seasonings such as miso paste and soy sauce. However, *koji* for *Awamori* spirits is prepared with a unique method (Nishiya, 1991). After steaming the rice, the seed mash is inoculated to the steamed rice.

Furthermore, the inoculated rice is incubated at 40°C and reduced step-by-step. Generally *koji* for seasoning or *sake* production is prepared at 30–35°C or with temperatures raised step-by-step. At high temperatures, amylase is produced and other enzymes such as feruloyl esterase break the cell walls. At the same time, fungi grow in rice grains. After 30 hr, the temperature is decreased gradually. At low temperatures of less than 35°C, A. *awamori* produces very much citric acid. The characteristic vanilla aroma of *Awamori* spirits is produced from ferulic acid as a precursor. Fukuchi (1999) reported that A. *awamori* grown at 37–40°C has high feruloyl esterase activity.

2.3.4 Fermentation of *Awamori* mash

Awamori spirit yeast, '*Awamori* yeast', that had been isolated from *Awamori* brewery was used during fermentation process and it might be peopling in brewer's house. The yeast is tolerant of low pH and high alcohol contents, and it can grow and ferment at low pH and high citric acid concentrations, producing very high alcohol concentrations (Nishiya, 1991). Nishiya (1972) and Suzuki (1972) reported that yeast growth was inhibited in conditions of greater than 11% alcohol and 35°C temperatures. However, the yeast ferments up to approximately 17% alcohol.

Awamori mash is prepared with *koji* and water only, and the water as 170% of koji weight is added. Fermented *Awamori* mash has 3.8–4.8% citric acid. The range of mash temperatures is 23–28°C.The yeast grows rapidly and fermentation finally ceases at temperatures higher than 30°C. At temperatures less than 20°C, the yeast grows slowly, and the mash is contaminated by bacteria. Generally, after 3 days, the mash has more than 10% of alcohol concentration. After 4 days, it has approx. 14%. After 7 days, it has more than 17%.

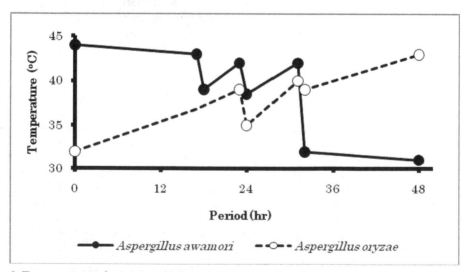

Fig. 3. Temperatures during two *Koji* preparations.

2.3.5 Mash distillation

After fermentation, the mash is distilled quickly because the mash aroma sours after yeast is digested by high alcohol and high temperatures. Two distillation systems exist as atmospheric distillation systems and reduced pressure distillation systems. *Awamori* spirits distilled with atmospheric distillation systems have three features (Nishiya, 1991).

1. Spirits have many components and a rich taste.
2. The spirit quality is good after aging.
3. Spirits have a distinctive aroma like scorching as furfural, which is produced by heat during distillation.

The distillation system with reduced pressure has no scorching aroma or higher alcohol, and it has a softer taste. It requires no long period of aging. During distillation, aldehydes and esters are distilled in the initial distillate. Then ethanol is distilled. The continuation of distillation decreases ethanol. The scorching aroma, that of furfural, is increased by heat during distillation. A direct heat distillation system is used traditionally: mash is heated in a kettle directly by fire. It has a scorched aroma. The compounds in distillate are shown during distillation of spirits.

2.3.6 Aging

Awamori spirits are aged in earthen pots using the *Shitsugi* method. The spirits show accelerated aging in earthen pots. Fatty acids or volatile acids as acetic acid in spirit are neutralized by calcium or magnesium released from earthen pots during aging. The acid or fatty smell in spirit is removed by aging. Recently, consumers favor a dry taste *Awamori* spirit aged in stainless steel tank after filtrate by ion exchange resin or activated charcoal.

The *Shitsugi* method is conducted by aging in earthen pots as follows. The distilled spirits are aged in earthen pots for 3–5 years. The spirits in the oldest earthen pots are consumed, and the consumed volume is supplied from the second oldest pot, and second oldest pot is supplied from third oldest pot is supplied from fourth oldest pot. Finally, the distilled spirit is poured in the newest pot. It resembles the solera system of sherry wine production. It is a specific aging method that has been used for *Kusu Awamori* spirits.

3. Mechanisms of making vanillin via ferulic acid in '*Awamori*' during aging within an earthen pot

3.1 Vanilla aroma of '*Awamori* spirit'

Awamori spirits have a vanilla aroma. Koseki (1998) reported mechanisms of vanilla aroma production in *Awamori* spirits. They found that not only vanillin but also phenol compounds as 4-vinyl guaiacol and ferulic acid from sufficient aged *Awamori* spirit. It does not age in barrels containing lignin. Therefore, they consider that the phenol compound is extracted from the material as rice.

3.2 Vanilla from ferulic acid in rice cell walls

Ferulic acid is contained in cell walls of rice. Then the ferulic acid is converted to vanilla during production. Koseki (1998) reported that ferulic acid dissolved in citric acid buffer was distilled, and that 4-vinyl guaiacol was detected in the distillate. Furthermore, the vanillin was detected from aged distillate containing 4-vinyl-guaiacol. Koseki mentions that the phenol compound was converted by heat. The ferulic acid is released from rice, and converts 4-vinyl guaiacol by heat when distilled. Furthermore, 4-vinyl guaiacol converts vanillin during aging. 4-Vinyl guaiacol is a well known compound as an off-flavor of beer and orange juice. Also, it is known as a characteristic flavor in weizen beer, or wheat beer. It is a distinct flavor.

Recently however, ferulic acid is converted to 4-vinyl guaiacol by microorganisms as *Saccharomyces cerevisiae* (Huang and others, 1993), *Pseudomonas fluorescens* (Huang and others, 1993), *Rhodotorula rubra* (Huang and others, 1994), *Candida famata* (Suezawa, 1995), *Bacillus pumilus* (Degrassi and others, 1995) and *Pseudomonas fluorescens* (Zhixian and others, 1994) via ferulic acid decarboxylase.

Furthermore, lactic acid bacteria converting ferulic acid to guaiacol were isolated from *Awamori* mash, and their ferulic acid decarboxylase was purified and compared (Watanabe, 2009). The lactic acid bacteria identified *Lactobacillus paracasei*, and their enzyme was the protein inferred as 47.9 kDa. The *p*-coumaric acid decarboxylase in *Lactobacillus plantarum* was 93 kDa; ferulic acid decarboxylase disagreed with those for *p*-coumaric acid decarboxylase (Cavin, 1997).

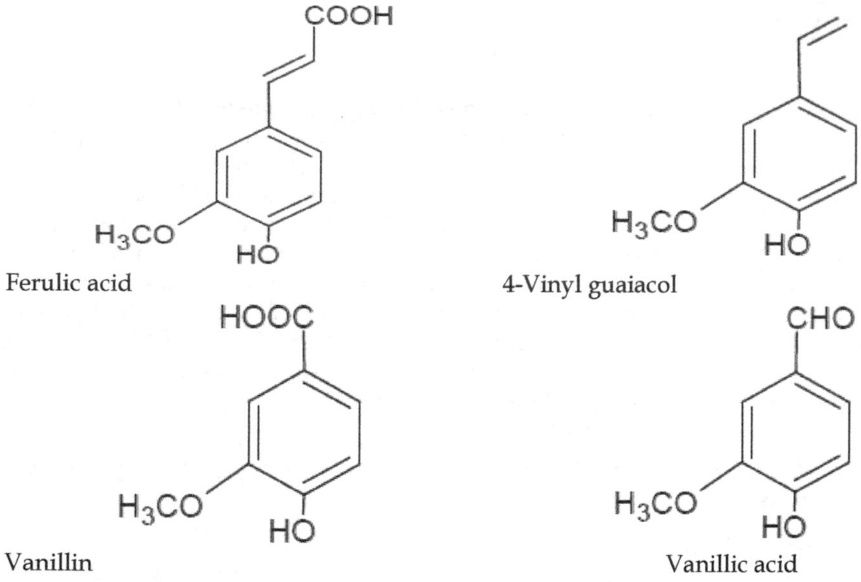

Ferulic acid   4-Vinyl guaiacol

Vanillin   Vanillic acid

Fig. 4. Structure of phenol compound.

The optimum temperature was 60°C, and the optimum pH was approximately pH 3.0. The *p*-coumaric acid decarboxylase from *Lactobacillus plantarum* disagreed with that reported by Cavin (1997).

The optimum temperature was 30°C; the optimum pH was 5.5–6.0. The *Awamori* mash has an acid condition that is maintained by citric acid produced using *A. awamori*. Consequently, the mash is not contaminated by other microorganisms. The enzyme was regarded as having been grown optimally at pH 3.0: high acidity and low pH conditions. The vanillin is produced not only by chemical conversion but also through bio-conversion in *Awamori* spirit or the mash during *Awamori* spirit production.

3.3 Feruloyl esterase is an important enzyme for high-quality *A. awamori*

Ferulic acid is supplied from combining xylan on rice cell walls. The ferulic acid is released by feruloyl esterase. In this awamori production, *A. awamori* on *koji* provides the enzyme.

According to their reports, microorganisms are important to produce vanillin in *Awamori* spirits. The released ferulic acid is converted to 4-vinyl-guaiacol by lactic acid bacteria. Then vanillin is produced from guaiacol.

The amount of ferulic acid is obtained in relation to feruloyl esterase activity: high concentration ferulic acid gives *Awamori* spirits a high vanilla aroma. Namely, the amount of vanillin in *Awamori* spirits is determined by the feruloyl esterase activity.

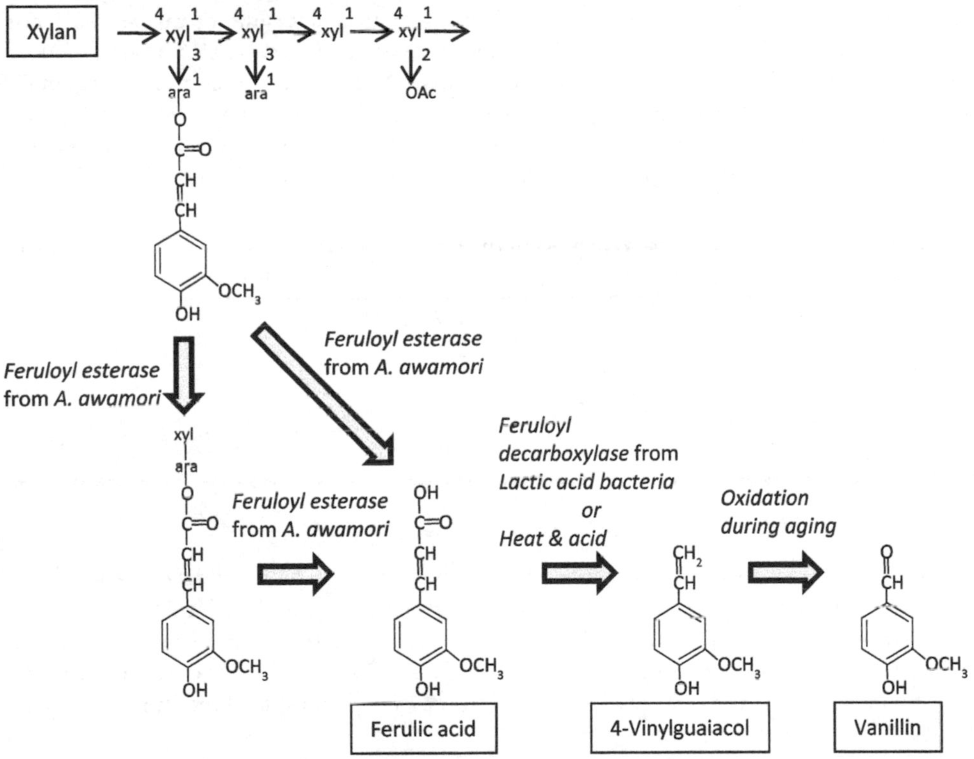

Xyl, xylose; ara, arabinose; Ac, acetate

Fig. 5. Mechanisms of production of vanillin from ferulic acid during *Awamori* spirit

4. Selection of *Aspergillus awamori* strain producing the highest feruloyl esterase and characteristics of their enzymes

4.1 Screening of *Aspergillus awamori* producing the highest feruloyl esterase

We next describe the importance of feruloyl esterase during *Awamori* spirit production. We isolated and screened A. *awamori* to give high feruloyl esterase activity for *Awamori* spirits production.

Mold strains were isolated and selected from some locations near *Awamori* spirits breweries in Okinawa. Consequently, black *Aspergillus* of many kinds was isolated. Then the strains were cultivated on xylan plate medium (1.5% xylan, 0.5% yeast extract, 0.5% polypeptone and 1% agar) for the first screening. The strain with a clear zone on the plate medium was screened and the strain was assayed for feruloyl esterase activity.

Results show that the largest clear zone was made by G-2. Its feruloyl esterase was 394 U/ml, which was 3–4 times higher than the others. As presented in Table 2, feruloyl esterase contents of G-2 were higher than those of either *Aspergillus awamori* NRIC1200 (118 U/ml) or *Aspergillus usami* IAM2210 (50 U/ml) as a standard strain.

The mold strains that produced feruloyl esterase formed a black colony. Many reports have described feruloyl esterase from the black *Aspergillus* group. Results of this study concur with those previously reported results. Conidiophores were observed using a microscope. The conidia were observed using scanning electric microscopy (SEM). Conidiophores of G-2 were similar to those of *Aspergillus* sp.: their conidia were black with a smooth surface. production.In addition, nitrite was not assimilated by G-2. Generally, black conidia were *A. niger* or *A. awamori*; other *Aspergillus* were brown, yellow, or green (Pitt and Kich, 1988).

Strains	Feruloyl esterase (unit/ml)
A. awamori NRIC1200	118
A. saitoi IAM2210	50
G-2 strain	394
Other selected strains	130–293

Table 2. Activity of feruloyl esterase from the strains

Moreover, *A. awamori* were not assimilated and the surfaces of their conidia were smooth. In contrast, *A. niger* was assimilated; its conidia surfaces were rough (Murakami, 1979). For those reasons, G-2 was identified as *Aspergillus awamori*, used to produce Awamori spirits as Japanese traditional spirits produced in Okinawa. The G-2 strain does not produce ochratoxin or aflatoxin as a mycotoxin. This *A. awamori* is applicable of food production and Awamori production.

4.2 Feruloyl esterase characteristics

The enzyme was purified using ion-exchange, size-exclusion, and HPLC chromatography. After purification, the specific activity of the enzyme was 20-fold higher and the yield was 16% higher than that of the crude enzyme solution. The enzyme solution was then analyzed using sodium dodecyl sulfate poly acrylamide gel electrophoresis (SDS-PAGE).

The molecular weight of feruloyl esterase in *A. awamori* was 35 kDa (Koseki, 1998b). The enzyme was 78 kDa, which was higher than that of *A. awamori*. The molecular weight of feruloyl esterase was measured using size-exclusion chromatography. The molecular weight was estimated as 80 kDa. The feruloyl esterase was inferred to be a monomer protein.

The optimum pH of the feruloyl esterase was pH 5.0 and the optimum temperature was 40°C. The activity stabilized at pH 3.0–5.0. The feruloyl esterase of *A. awamori* is reportedly (Koseki and others, 1998b) unstable at pH 3.0, which was 30–50% of non-treated enzyme activity. However, the enzyme was stable in acid. The enzyme was stable at 50°C. Feruloyl esterase of *A. niger* showed tolerance at 80°C (Sundberg and others, 1990). The enzyme is therefore acid-tolerant, but not heat-tolerant. Use of the enzyme is expected to be advantageous for food production under acid conditions, in which *A. awamori* produces citric acid very well.

Mercury ion (Hg^{2+}) inhibited feruloyl esterase activity completely; Fe^{2+} inhibited 80% of the activity. Both PMSF and DFP completely inhibited feruloyl esterase activity. In general, feruloyl esterase, produced by black *Aspergillus* group, was classified into the serin-esterase group enzyme, which was inhibited by Hg^{2+} and Fe^{2+} (Koseki and others, 1998b). They reported that the active center of feruloyl esterase from *A. awamori* had Ser-Asp-His. It was therefore considered that the feruloyl esterase had similar catalysis.

The feruloyl esterase activity was the highest among four substrates with 1-naphthyl acetate. The activity of the feruloyl esterase had 40% of the activity related to 1-naphthyl acetate, against 1-naphthyl propionate (composed of three carbons). In addition, 1-naphthyl butyrate (comprising four carbons) showed 5% activity related to 1-naphthyl acetate. The activity of feruloyl esterase was decreased against substrates containing more than three carbons; 2-naphthyl acetate did not react completely. The reported feruloyl esterase showed activity against 1-naphthyl propionate, as did 1-naphthyl acetate (Ishihara and others, 1999). The feruloyl esterase was inferred to be substrate-specific.

Adsorption of the feruloyl esterase to cell walls was not observed. The enzyme was not absorbed by xylan or starch that were present in the supernatant. However, it was absorbed by cellulose in the supernatant. It was also found in the precipitant with cellulose.

Strain	Mycelium color	Conidium color	Conidiophore (mm)	Conidial head diameter (μm)	Conidium diameter (μm)	Conidium surface	Assimilation of $NaNO_2$
A. awamori NRIC1200	White	Black	0.61	220	3.8	Smooth	-
A. niger NRIC1221	White	Black	0.86	170	3.4	Spinky	+
A. saitoi IAM2210	White	Black	0.91	190	4.1	Rough	-
G-2	White	Black	0.68	200	3.4	Smooth	-

+, positive; -, negative

Table 3. Morphological and physiological characteristics of black *Aspergillus* spp.

Feruloyl esterase from *A. niger* and *A. awamori* reportedly adsorbed specifically to cellulose (Ferreira and others, 1993), which agrees with the results shown for this study. The intensity of absorbing the enzyme with cellulose was conducted to wash cellulose by boiling for 10 min in SDS solution (1–11%). The enzyme had a cellulose binding domain, as reported also by Koseki and others (2006). The protein was unbound by 10% and 11% SDS solution. It is considered that they were bound strongly.

Lane 1   Lane 2

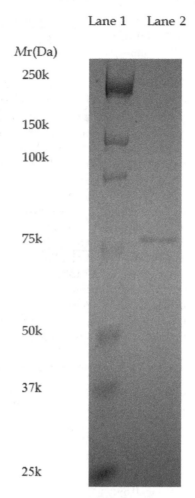

Fig. 6. SDS–PAGE of feruloyl esterase.

We investigated Michaelis constants of the feruloyl esterase. A Lineweaver–Burk plot is shown to calculate the Km of the feruloyl esterase: its Km was 0.0019% (0.01 mM). Koseki reported that Km was 0.26–0.66 mM (Koseki and others, 2006). The feruloyl esterase has higher affinity to 1-naphthyl acetate than that described by Koseki. The feruloyl esterase from A. *awamori* G-2 was more stable under the acid condition than the other black *Aspergillus* groups. In general, the esterase was as unstable under the acid condition as *Awamori* mash or shochu mash during fermentation. However, results suggest that the feruloyl esterase survived under the acidic conditions. The ferulic acid was related to the aroma of *Awamori* spirits or shochu spirits. The ferulic acid released from the cell wall was converted to vanilla via 4-vinyl guaiacol by yeast or lactic acid bacteria during aging (Gabe and Koseki, 2000; Watanabe and others, 2007). In addition, results showed that the enzyme-stabilized activity under acid and heat conditions of pH 3.0 and 50°C is applicable to food production.

Substrates	Relative activity (%)
1-naphthyl acetate	100
1-naphthyl propionate	40
1-naphthyl butyrate	5
2-naphthyl acetate	0

Table 4. Substrate specificity of feruloyl esterase

5. Application of feruloyl esterase to beer brewing

In beer brewing processes, the mash is used differently from the mash of *Awamori* spirits. Beer mash is saccharified and then fermented. In the mashing process, barley malt produces not only maltose as base material of ethanol for fermentation. The malt also supplies β-glucan in mash. β-glucan (called (1-3) (1-4)-β-D-glucans) are components of endosperm cell walls in barley, occupying 75% of the cell wall (MacGregor and Fincher, 1993). The amount of β-glucans is negatively correlated with yields of the amount of wort in the mashing process (Bourne and others,1982; Kato and others, 1995) . The amount of β-glucan in grain shows levels of decomposition of cell wall in barley endosperm. The malt contained a great amount of β-glucan, showing insufficiency in decomposing β-glucan in malt during germination. The malt insufficiently decomposes β-glucans and does not have sufficient starch or protein in endosperm. Therefore the yields fall. Furthermore, the β-glucans in malt used to make mash have high viscosity, which reduces the mash filtration speed (MacGregor and Fincher, 1993). Speed is a limiting factor of beer brewing. The slow filtration of beer presents problems for breweries. Moreover, glucan is a factor of invisible haze or pseudohaze (Jacson and Bamforth, 1983) (1-3), (1-4)−β-D-glucanase is usually added to elevate the activity (Bamforth, 1985).

Kanauchi and Bambort reported that β-glucan was decomposed not only by (1−3), (1−4)-β-D-glucanase but also by xylanase, arabinofuranosidase, and feruloyl esterase. At least, feruloyl esterase decomposed xylan or feruloylxylan layer covering β-glucan. The enzyme helps to access β-glucanase to b-glucan.

6. Application of feruloyl esterase for biomass processing

Global crude oil production is 25 billion barrels, but human societies are expected to reduce the use of fossil fuels after bio-fuel development (Benoit et al., 2006). Particularly in the US, these gasoline fuels contain up to 10% ethanol by volume. In the future, automobiles can use a blend of 85% ethanol and 15% gasoline by volume (Fazary and Ju, 2008).

Under this situation, lignocellulosic and other plant biomass processing methods have been developed recently. The material is not only a renewable resource but it is also the most abundant source of organic components in high amounts on the earth: the materials are cheap, with a huge potential availability.

Plant lignocellulosic biomass comprises cellulose, hemicellulose, and lignin. Its major component is cellulose (35–50%) with hemicellulose (20–35%) and lignin (25%) following, as shown in Table 5 (Noor and others, 2008). Proteins, oils, and ash make up the remaining fraction of lignocellulosic biomass (Wyman, 1994b). Cellulose is a high-molecular-weight linear polymer of b-1,4-linked D-glucose units which can appear as a highly crystalline material (Fun and others, 1982). Hemicelluloses are branched polysaccharides consisting of the pentoses D-xylose and L-arabinose, and the hexoses D-mannose, D-glucose, D-galactose and uronic acids (Saka, 1991). Lignin is an aromatic polymer synthesized from phenylpropanoid precursors (Adler, 1977).

It is noted throughout the world that ethanol is producible from biomass material. For their bioconversion, pretreatment is an important procedure for practical cellulose conversion processes (Bungay, 1992; Dale and Moreira, 1982; Weil and others, 1994; Wyman, 1994a). Ferulic acid is found in the cell walls of woods, grasses, and corn hulls (Rosazza and others, 1995). It is ester-linked to polysaccharide compounds, where it plays important roles in plant cell walls including protein protection against pathogen invasion and control of extensibility of cell walls and growth (Fry, 1982).

	Composition (% dry weight basis)		
	Cellulose	Hemicellulose	Lignin
Corn fiber	15	35	8
Corn cob	45	35	15
Rice straw	35	25	12
Wheat straw	30	50	20
Sugarcane bagasse	40	24	25

Table 5. Composition of some agricultural lignocellulosic biomass (Noor and other, 2008)

In bio-ethanol production, complete enzymatic hydrolysis of hemicelluloses as arabinoxylan requires both depolymerizing and sidegroup cleaving enzyme activities such as FAEs.

Any hemicellulose-containing lignocellulose generates a mixture of sugars upon pretreatment alone or in combination with enzymatic hydrolysis. Fermentable sugars from cellulose and hemicellulose will fundamentally be glucose and xylose, which can be released from lignocellulosics through single-stage or two-stage hydrolysis. In Europe, portable alcohol manufacturing plants are based on wheat endosperm processing, with the hemicellulosic by-product remaining after fermentation consisting of approximately 66% (W/W) arabinoxylan (Benoit et al., 2006). A synergistic action between cellulases, FAEs, and xylanases might prove to be more effective when applied at a critical concentration in the saccharification of steam-exploded wheat straw (Borneman and others, 1993; Kennedy and others, 1999; Tabka and others, 2006).

7. References

Adler, E. (1977). Lignin chemistry: Past present and future. *Wood Sci. Technol.*, Vol. 11, pp. 169-218

Bamforth, C.W. (1985). Biochemical approaches to beer quality. *J. Instit. Brew.*, Vol. 91, pp. 154-160

Benoit, I., Navarro, D., Marnet, N., Rakotomanomana, N., Lesage-Meessen, L., Sigoillot, J-C. & Asther, M. (2006). Feruloyl esterases as a tool for the release of phenolic compounds from agro-industrial byproducts. *Carbohydr. Res.*, Vol. 341, pp. 1820-1827

Borneman, W.S., Hartley, R.D., Morrison, W.H., Akin, D.E. & Ljungdahl, L.G. (1990). Feruloyl and *p*-coumaroyl esterase from anaerobic fungi in relation to plant cell wall degradation. *Appl. Micro. Biotech.*, Vol. 33, pp. 345-351

Bourne, D.T., Powlesland, T. & Wheeler, R.E. (1982). The relationship between total b-glucan of malt and malt quality. *J. Instit. Brew.*, Vol. 88, pp. 371-375

Bungay, H. (1992). Product opportunities for biomass refining. *Enzyme Microb. Technol.*, Vol. 14, pp. 501-507

Cavin, J.F., Barthelmebs, L.G., Van Beeumen, J., Samyn, J., Travers, B. & Divieés, J.-F. (1997). Charles Purification and characterization of an inducible *p*-coumaric acid decarboxylase from *Lactobacillus plantarum*. *FEMS Microbiology Letters*, Vol. 147, pp. 291-295

Dale, B.E. & Moreira, M.J. (1982). A freeze-explosion. *Biotechnol Bioeng Symp*, Vol. 12, pp. 31-43

Degrassi, G., De Laureto, P.P. & Bruschi, C.V. (1995). Purification and characterization of ferulate and *p*-coumarate decarboxylase from *Bacillus pumilus*. *Appl. Environ. Microbiology*, Vol. 61, No. 1, pp. 326-332

Dendy, D. A. V., Dobraszczyk, B. J. (2000). Cereals and cereal products: chemistry and technology, Aspen publishers, Inc., Gaithersburg, Maryland, USA, pp. 305-306

Faulds, C.B. & Williamson, G. (1994). Purification and characterization of a ferulic acid esterase (FAE-III). from *Aspergillus niger*: specificity for the phenolic moiety and binding to micro-crystalline cellulose. *Microbiol.*, Vol. 144, pp. 779-787

Faulds, C.B., Kroon, P.A., Saulnier, L., Thibault, J.F. & Williamson, G. (1995). Release of ferulic acid from maize bran and derived oligosaccharides by *Aspergillus niger* esterases. *Carbohydr. Polym.*, Vol. 27, pp.187-190

Fazary, A.E. & Ju, Y.H. (2008). The large-scale use of feruloyl esterases in industry. *Biotech. Molecular Biology Reviews*, Vol. 3, No.5, pp. 95-110

Fry, S.C. (1982). Phenolic components of the primary cell wall. *Biochemistry Journal*, Vol. 203, pp. 493-504

Fukuchi, K., Tamura, H. & Higa, K. (1999). The studies on preparation of koji for *Kusu Awamori* spirits. *Report Indust. Tech. Center in Okinawa.*, Vol. 2, pp. 77-83 (in Japanese)

Fun, L.T., Lee, Y.H. & Gharpuray, M.M. (1982). The nature of lignocellulosics and their pretreatment for enzymatic hydrolysis. *Adv. Biochem. Eng.*, Vol. 23, pp. 158-187

Horiuchi, H. & Tani, T. (1966). Studies of cereal starches. Part V. Rheological properties of the starch of rices imported into Japan. *J. Agr. Biol. Chem.*, Vol. 30, pp. 457-465

Huang, Z., Dostal, L. & Rosazza, J.P.N. (1993). Microbial transformations of ferulic acid by *Saccharomyces cerevisiae* and *Pseudomonas fluorescens. Appl. Environ. Microbiol.*, Vol. 59, pp. 2244-2250

Huang, Z., Dostal, L. & Rosazza, J.P.N. (1994). Mechanism of ferulic acid *Rhodotorula rubra. J. Biol. Chem.*, Vol. 268, pp. 23594-23598

Inui, T. (1901). Untersuchungen uber die der Zubereitung des alkoholischen Getrankes "Awamori Betheiligen. *J. Coll. Sci. Imp. Univer. Tokyo,* Vol. 15, pp. 465-476

Ishihara, M., Nakazato, N., Chibana, K., Tawata, S. & Toyama, S. (1999). Purification and characterization of feruloyl esterase from *Aspergillus awamori. Food Sci. Tech. Res.*, Vol. 5, No.3, pp. 251-254

Jacson, G. & Bamforth, C.W. (1983). Anomalous haze readings due to β-glucan. *J. Instit. Brew.*, Vol. 89, pp. 155-156

Juliano, B.O. (1985). Criteria and test for rice grain quality. Rice chemistry and technology. American Association of Cereal Chemists, Inc., St. Paul, Minnesota, USA, pp. 443-513

Juliano, B.O., Bautista, G.M., Lugay, J.C. & Reyes, A.C. (1964). Studies of the physicochemical properties of rice. *J. Agric. Food Chem.*, Vol. 12, pp. 131-138

Kato, S., Kosaka, H., Hara, S., Maruyama, Y. & Takigushi, Y. (1930). On the affinity of the cultivated varieties of rice plants, *Oryza sativa* L. *J. Depart. Agri., Kyushu Imperial University*, Vol. 2, pp. 241-276

Kato, T., Sasaki, A. & Takeda, G. (1995). Genetic Variation of β-glucan Contents and β-glucanase Activities in Barley, and their Relationships to Malting Quality. *Breeding Science,* Vol. 45, No. 4, pp. 471-477 (in Japanese)

Kennedy, J.F., Methacanon, P. & Lloyd, L.L. (1999). The identification and quantification of the hydroxycinnamic acid substituents of a polysaccharide extracted from maize bran. *J. Sci. Food Agric.*, Vol. 79, pp. 464-470

Koizumi, T. (1996). Birth of Quality Liquor Tokyo: Kodansha

Koseki, T. & Iwano, K. (1998). A mechanism for formation and meaning of vanillin in Awamori. *Soc. Brew. Japan,* Vol. 93, No. 7, pp. 510-517 (in Japanese)

Koseki, T., Furuse, S., Iwano, K. & Matsuzawa, H. (1998). Purification and characterization of a feruloyl esterase from *Aspergillus awamori. Biosci. Biotechnol. Biochem.*, Vol. 62, No.10, pp. 2032-2034

Koseki, T., Takahashi, K., Hanada, T., Yamane, Y., Fushinobu, S. & Hashizume, K. (2006). N-Linked oligosaccharides of *Aspergillus awamori* feruloyl esterase are important for thermostability and catalysis. *Biosci. Biotechnol. Biochem.*, Vol. 70, No. 10, pp. 2476-2480

Kroon, P.A., Faulds, C.B., Brezillon, C. & Williamson, G. (1997). Methyl phenylalkanoates as substrates to probe the active sites of esterases. *Eur. J. Biochem.*, Vol. 248, pp. 245-251

MacGregor, A.M. & Fincher, G.B. (1993). Carbohydrates of the barley grain, pp. 73-130, In: A.W. MacGregor, & R.S. Bhatty, (eds.), Barley: Chemistry and technology. The American Association of Cereal Chemists, Inc., St. Paul, Minnesota, USA

Murakami, H. (1979). The taxonomic studies of koji mold (26th), The methods of physiological examination and characteristics of cultivation on black *Aspergillus* sp. *J. Brew Soc. Japan,* Vol. 74, No.5, pp. 323-327 (in Japanese)

Murakami, H. (1982). Classification of the black *Aspergillus* sp. *J. Brew. Soc. Japan*, Vol. 77, No. 2, pp. 84 (in Japanese)

Nishiya, S., Suzuki, A. & Shigaki, K. (1972). Kinetic studies of yeast growth (part 4), *J. Soc. Brew. Japan*, Vol. 67, pp. 445-448 (in Japanese)

Nishiya, T. (1991). *Honkaku Syochu-seizo-gijyutsu*, Brewing Society of Japan, Tokyo, pp. 1-279 (in Japanese)

Noor, M.S., Zulkali, M.M.D. & Ku Syhudah, K.I. (2008). Proceedings of Malaysian University Conferences on Engineering and Technology, Ferulic acid from lignocellulosic biomass: Review, Malaysian Universities Conferences on Engineering and Technology (MUCET2008)

Pitt, J.Y. & Kich, M.A. (1988). A laboratory guide to common *Aspergillus* species and their teleomorphs. CSIRO, Division of Food Research Sydney. Academic Press Inc., Australia

Ralet, M.-C., Faulds, C.B., Williamson, G. & Thibault, J.F. (1994). Degradation of feruloylated oligosaccharides from sugar-beet pulp and wheat bran by ferulic acid esterases from *Aspergillus niger*. *Carbohydr. Res.*, Vol. 263, pp. 257-269

Raper, K.B. & Fennell, D.I. (1965). The Genus *Aspergillus*, pp. 1-686, Baltimore: Williams & Wilkins

Rosazza, J.P.N., Huang, Z., Dostal, L., Vom, T. & Rousseau, B. (1995). Biocatalytic transformations of ferulic acid: an abundant aromatic natural product. *J. Ind. Microbiol.*, Vol. 15, pp. 457-471

Saka, S. (1991). Chemical composition and Distribution. Dekker, New York pp. 3-58

Suezawa, Y. (1995). Bioconversions of Ferulic Acid and *p*-Coumaric Acid to Volatile Phenols by Halotolerant Yeasts.: Studies of Halotolerant Yeasts in Soy Sauce Making (Part I). *J. Agri. Chem. Soci. Japan*, Vol. 69, pp. 1587-1596 (in Japanese)

Sundberg, M., Poutanen, K., Makkanen, P. & Linko, M. (1990). An extracellular esterase of *Aspergillus awamori*. *Biotechnol. Appl. Biochem.*, Vol. 12, pp. 670-680

Suzuki, A., Nishiya, T. & Shigaki, K. (1972). Kinetic studies of yeast growth (part 5). *J. Soc. Brew. Japan*, Vol. 67, pp. 449-452 (in Japanese)

Tabka, M.G., Herpoël-Gimbert, I., Monod, F., Asther, M. & Sigoillot, J.C. (2006). Enzymatic saccharification of wheat straw for bioethanol production by a combined cellulase xylanase and feruloyl esterase treatment, *Enzyme and Microbial. Tech.*, Vol. 39, No.4, pp. 897-902

Watanabe, S., Kanauchi, M., Kakuta, T. & Koizumi, T. (2007). Isolation and characteristics of lactic acid bacteria in Japanese spirit awamori mash. *J. Am. Soc. Brew. Chem.*, Vol. 65, pp. 197-201

Watanabe, S., Kanauchi, M., Takahashi, K. & Koizumi, T. (2009). Purification and characteristics of feruloyl decarboxylase produced by lactic acid bacteria from *Awamori* mash, *J. Am. Soc. Brew. Chem.*, Vol. 65, pp. 229-234

Wei, J., Westgate, P., Koholmann, K. & Ladisch, M.R. (1994). Cellulose pretreatments of lignocellulosic substrate, *Enzyme Microb. Technol.*, Vol. 16, pp. 1002-1004

Wyman, C.E. (1994a). Ethanol from lignocellulosic biomass: technology, economics, and opportunities, *Bioresour. Technol.*, Vol. 50, pp. 3-16

Wyman, C.E. (1994b). Alternative fuels from biomass and their impact on carbon dioxide accumulation. *Appl. Biochem. Biotechnol.*, Vol. 45, No. 46, pp. 897-915

Zhixian, H., Dostal, L. & Rosazza, J.P.N. (1994). Purification and characterization of a ferulic acid decarboxylase from *Pseudomonas fluorescens*. *J. Bacteriol.*, Vol. 176, No. 19, pp. 5912-5918

Part 2

Social and Economic Issues

9

The Industrial Meat Processing Enterprises in the Adaptation Process of Marketing Management of the European Market

Ladislav Mura

Dubnica Institute of Technology, Department of Specialised Subjects
Slovak Republic

1. Introduction

Coming on globalization, international economical interdependency, generate new business chances and move into territorially and qualitatively new dimensions of enterprising. In the current sharp competitive struggle the successful and quick application of results of the research and technological improvement together with innovation of products and services to forecast the needs of consumers are on the first place. Adaptation to the new conditions of the market is expected in the field of marketing management selection of adequate marketing and competitive strategy for building a strong market position and for perspective additional development of business. [6]

Agro-food market conditions have dynamically changed in the last 20 years. The biggest influence on the market had the ownership transformation, later the penetrating of foreign investors into the several branches of food industry. [11]

Formation of business environment was markedly influenced by the accession of theSlovak Republic into the EU, together with the need of adaptation to new conditions of the united market. Enterprises have adjusted to competitive environment, they changed the intracompany management, and implemented innovations.

Especially for the branches of food industry it was important to modernize the technological accessories, to fulfill the challenging quality and hygienic standards asked by the European norms. Aforesaid aspect was obligatory to put into the marketing management adaptation of concrete businesses. [1] New conditions, which are characterized by a highly developed market - where the supply overtops the demand - relative consumers saturation of their basic wants, called for adaptation of supplying companies to consumers needs. The marketing management is one of the basic expectations of effective and successful company operations. The managers of Slovak companies are aware of the demanding market background. [7]

2. Materials and methodology

Based on exploration of marketing management adaptation and on analyzed sample of meat processing plants in the period of changing market conditions, in the horizon of the years 2002 - 2006, is the main aim of the article to define the determining factors of successful marketing management adaptation to the conditions of the common EU market. Partial aims are as follows: to analyze the marketing management of the examined enterprises before the enter of the Slovak Republic into the EU, to analyze the marketing management of the examined enterprises after the Slovakia´s entry into the EU, to characterize the business activities of meat processing plants on the domestic and foreign markets, to prepare the SWOT analysis and to specify the factors of the successful marketing management of analyzed meat processing plants.

Objects of research are the biggest and the most important plants of meat processing industry of Slovakia. The influential sample is created by the enterprises producing 74,8% of the Slovak meat processing companies production.

To fulfill the stated aim a primary research within the meat processing companies was needed. Basic information and data were obtained by a questionnaire, by managed interviews with top management and by panel discussion in order to define the key factors of successful adaptation of the marketing management. To keep the sensitive data we will use the following identifications: "comp.1", "comp.2", etc. in our article. Additional sources of information were secondary sources such as the Slovak agricultural and food industry reports (so-called "Green report"), analytical works of the Slovak payment agency, analytical and internal materials of the Ministry of Agriculture of the Slovak Republic, and outputs of research works in the given topic.

To process the primary and secondary information logical methods, selected mathematical and statistical methods, methods of descriptive statistics, SWOT analysis were used. Relations between particular characters were quantified and examined on the significance level of $\alpha=0,05$ by the chi quadrate test and the correlation coefficients. Interpretations of the outcomes are done by the "p-value".

3. Theoretical scopes

Conditions of business activities in the sector of agro-food complex are markedly influenced by turbulent background of the agro-market, changing rules of financial supports, social and economical spheres of the life in Slovakia. In the current period, just the smaller parts of the managements of agro-food industry apply marketing practices. [8] With a consecutive enter of international companies into the Slovak food processing enterprises, the increasing need of marketing management in relation to successfulness on the market can be observed. A marketing approach to the business management is the eminent condition for successful business activities.

The success of an enterprise depends on numbers of factors and conditions, among which the qualitative marketing management takes the first place in dynamically changing conditions. Marketing management is a systematic and goal-seeking activity, aimed at the maximum utilization of abilities and properties of the enterprise, with the goal of stable status on the market and competitive advantage besides meeting the customers´ needs. In today´s globalization conditions and marketing structures integration, the territorial expansion of marketing management becomes a scope on the target markets. [8]

Based on identified demands and requests, a company creates the most adequate marketing strategy for placement on the given market. A selected marketing strategy is declared by a marketing mix. Practically, the successful marketing mix application depends on the three conditions [4]:

- tools of marketing mix must chronologically form a constant and harmonic unit,
- facilities of marketing mix tools have to reflex of eventual market development and company situation
- intensity of the usage of several tools of marketing mix must be sufficient

Seriousness of the listed operations consists in the moving conditions in time and in their correlations.

The latest trends in marketing shows, that marketing is an integrated complex of actions focused on customers and the market. Marketing steps must be at the same time re-bounded with other processes in the company and to be an integral part of the management.

Marketing management is a continuous process of analysis, planning, implementation and control. Its sense is the creation and maintenance of long-standing relationships with target customers and consumers, which help companies to reach the given targets. [2] The aim of marketing management is to identify consumers´ needs and wishes , to create a vision of innovative products and to set up the company processes in the way to be able to present for competition a product of higher quality and efficiency.

Coming on with globalization of the world economics and the integration processes causes that enterprises operate their businesses more and more in the international environment, rank into the foreign markets to reach a better valorization of the company capital. [5] The companies are under an extreme competitive pressure.

4. Result and discussion

Integration of the Slovak Republic into the market structures of the EU besides the positive sides was also taken negatively by the business sector. This is valid for most of the agro-food companies. On one hand, for Slovak companies it was a chance to join the united market of the EU, on the other hand, it was a must to fulfill the demanding conditions of hygienic, qualitative and veterinary norms. The high level of these parameters caused the downfall of many companies, which were not able to adapt to the changed conditions. The fulfillment of particular norms and standards necessitated serious investments into the technological facilities and the reconstruction of existing producing companies.

Adaptation of meat processing companies to the new legal and market conditions was the first and basic premise of the successful business activities on the internal EU market. This form of adjustment, so called compulsory adaptation, allows the concrete enterprises to practice their business activities in the field of agro-food business.

The European Union represents an internationally, extra-nationally and nationally marked business environment. Some of the legal, political, economical and technological factors of a marketing mix of the EU members do not have clear international or national character, but rather a combination of national and extra-national norms, rules and politics. The objects of company marketing management of integrated economics are first of all the high quality production orientation, its ability to compete and the ability to achieve a place on the united market. Together with the strategic marketing management they are focused on the product differentiation and increasing the surplus value of products. [12] This field of marketing management can be denominated as a voluntary form of adaptation for the new conditions with the aim to achieve a competitive advance and a bigger market share. The choice which marketing strategy will used to reach the marked target is on the concrete company.

The objects of our analysis were nine biggest animal firms producing plants in Slovakia whose production covers 74,8% of the market demand. Data obtained by research are considered as a case study. The focus is on selected marketing management and economical indicators. The secondary data were obtained from Statistical office of the Slovak Republic. The table 1 presents the biggest meat processing factories in the years 2002 - 2006 in Slovakia based on data of the Statistical office of the Slovak Republic. [10]

Nr.	2002	2003	2004	2005	2006
1.	Tauris, a. s.	Tauris, a. s.	Tauris, a. s.	Tauris, a. s.	Tauris, a. s.
2.	Hrádok Mäsokombinát, s.r.o.	Hrádok Mäsokombinát, s.r.o.	Hrádok Mäsokombinát s.r.o.	Mecom, a. s.	Mecom, a. s.
3.	Mecom, a. s.	Mecom, a. s.	Mecom, a. s.	Hrádok Mäsokombinát, s.r.o.	THP, a. s.
4.	THP, a. s.	THP, a. s.	THP, a. s.	PM Zbrojníky, a. s.	Hrádok Mäsokombinát, s.r.o.
5.	Hyza, a. s.	PM Zbrojníky,a. s.	PM Zbrojníky, a. s.	THP, a. s.	PM Zbrojníky, a. s.
6.	Hydina ZK, a. s.	Tauris Danubius, a. s.	Tauris Danubius, a. s.	Tauris Danubius, a. s.	Hyza, a. s.
7.	Tauris Danubius, a. s.	Hyza, a. s.	Hyza, a. s.	Hyza, a. s.	Tauris Danubius, a. s.
8.	Hydina, a. s.	Hydina ZK, a. s.	Hydina ZK, a. s.	Hydina ZK, a. s.	Hydina ZK, a. s.

Table 1. The biggest meat-processing companies in Slovakia. Source: own processing

As the table shows, the biggest and at the same time the most successful business entity in the branch of meat-processing in 2002 - 2006 was the company Tauris, a.s. It holds its leading position for a long term thanks to business strategy, innovations and the effective marketing management. The second and third most important subjects – Hrádok Mäsokombinát, a. s. and Mecom, a. s. hold their position stable. The companies THP, a. s. and Hyza, a. s. started to fall behind the leading processing companies and as separate subjects they could not stand the competing struggle what culminated in a fusion to a single company.

For the identification of strengths and opportunities, the weaknesses and threats there was prepared a SWOT analysis in the sample of analyzed meat-processing companies. A summary review about the situation in the particular companies, overall information about the products, delivered certificates, turnover development, strengths and opportunities, the specification of weaknesses and the forecast of threats from the point of view of the company documents are given in the table 2.

The business entities which took part in the research reacted the changes in their macro-background and stepwise adapted to the new social and economical conditions. Examined enterprises not only implemented, but also permanently kept the production process in accordance to the strict conditions of critical control points of HACCP (Hazard Analysis Critical Control Point). Besides of this system they had implemented a system of quality management based on norms of ISO 9001, or ISO. 9002. The research showed the time slip in the implementation of systems of quality management at the particular companies which caused a competitive advance for some of them. Company 1 and company 3 which were flexible with the implementation of systems of the quality management based on norms of ISO became the leaders on the market with an adequate development of their turnover.

Adaptation to the strong technological norms asked for high investments to purchase and implement new technologies and technological procedures for meat processing to meat products. The investment activities were insured by companies by credit lines what on the other hand caused a high finance and capital indebtedness (Company 2 and Company 5). The opportunity to obtain the indebted companies motivated financial groups to enter the agro-food industry. Financial and investment groups as Penta Investments, a. s. and Eco Invest, a. s. are such examples.

	Company 1	Company 2	Company 3	Company 4
Product portfolio	132 types of products	252 types of products	224 types of products	110 types of products
System of quality management and the date of implementation	ISO 9001 – 2003 SK 15 – 1996 SK 5 – 2001 SK 618 – 2004	ISO 9001 – 2003 SK 6061 – 2004	ISO 9001 – 1998 ISO 9002 – 1998 BRC Food – 2003 IFS – 2003 SK 63 – 2003	ISO 9001 – 2002 ISO 14001 – 2006 SK 61 – 2003
HACCP	from the year 1997	from the year 1998	from the year 2004	from the year 2001
The turnover evolution during the analyzed period	107,5 mil. € - 107,9 mil. € ↑	75,6 mil. € - 63,0 mil. € ↓	54,8 mil. € - 81,3 mil. € ↑	70,3 mil. € - 38,6 mil. € ↓
Strengths	modern technology, product innovations, capable human resources, long term relationships with chains	modern technology, capable human resources, original recipes, good geographical location in relation to foreign markets, own abattoir	experienced management, strategic marketing, investments to the technologies, extension of producing capacities, acquisitions of small companies	capable human resources, own cannery
Weaknesses	Capital indebtedness, improper capital structure	tight range in chains, promotion	duplicity of management, improper organization and management structure	stray marketing, absence of management skills
Opportunities	penetration to the foreign markets, EU founds, increasing of assurance of consumers	penetration to the foreign markets, reinforcement of promotional activities, wider product portfolio in the chains	penetration to the foreign markets, innovation of product portfolio	penetration to the foreign markets, concentration of the branch, acquisition of companies out of EU norms
Threats	overhead costs, price pressure of chains - product quality fall down, competitive faith	overhead costs, the real income fall down of consumers	overhead costs, competitive faith	overhead costs, price pressure of chains - product quality fall down,
Promotion	active promotion by participation on home and foreign exhibitions	active promotion by participation on home and foreign exhibitions	active promotion by participation on home and foreign exhibitions	active promotion by participation on home and foreign exhibitions

Table 2. Some aspects of marketing mix and SWOT analysis. Source: own processing

	Company 5	Company 6	Company 7	Company 8	Company 9
Product portfolio	100 types of products	190 types of products	40 types of products	slaughter	40 types of products
System of quality management and the date of implementation	ISO 9001 – 2000 SK 64 – 2004	ISO 9001 – 2004 SK 3092 – 2004	ISO 9002 – 1996 SK 15 – 1996	SK 26 – 1997	ISO 9001 – 2004 SK 630 ES – 2004 SK 3031 – 2004
HACCP	From the year. 1998	From the year 2004	From the year 1995	From the year 1999	From the year 2000
The turnover evolution during the analyzed period	256 mil. € - 430 mil. € ↑	95 mil. € - 161 mil. € ↑	373 mil. € - 336 mil. € ↓	-	140 – 484 mil. € ↑
Strengths	quality of products,own retailing, capital interconnection whit the company 1	Flexibility, adaptability and giving on regional changes, available prices, tailor-made approach	product and packing innovations, capital interconnection whit the company 1	available prices, modern technology, bio-meat production	original recipes, flexibility, capable human resources, own abattoir,
Weaknesses	absence of marketing activities	only local entreprene-urship, weak marketing policy	high operating costs (special financial costs), absence of export	only jointing meat, fluctuation of human resources	absence of export, narrow range of goods, weak marketing policy
Opportunities	capital interconnection whit company 1, penetration to the foreign markets, new hypermarkets	penetration to the foreign markets,	penetration to the foreign markets, innovation of product portfolio	promotion, penetration to the foreign markets	penetration to the foreign markets, promotion
Threats	absolute submission of business policy to company 1, competitors fight	competitors struggle	operating costs, competitors fight, force of trade string	force of trade string, operating costs	operating costs, force of trade string, competitors fight
Propagation	Passive propagation	Passive propagation	active promotion by participation on home and foreign exhibitions	Passive propagation	Passive propagation

Table 3. Some aspects of marketing mix and SWOT analysis in other companies. Source: own processing

The tables 2 and 3 give a synthetic overview about the situation in particular businesses and give an integrated information about the products, quality certificates, sales dynamics, strengths and weaknesses of companies inclusive the identification of opportunities and threats from the point of view of a concrete company.

The synthesis of determinated facts of the SWOT analysis identifies the absence of strategic management elements in the selected group of companies. In one case there is an absence of the strategy of human resources development with limited motivation, two fifths of businesses do not use actively the possibilities of the communication mix and none of the companies entered an international market during the analyzed period. The potential threats are the epidemic diseases of animals (BSE, KMO), low net margin in the meat production, the enforcement of the law, limited defense of the home market. As substandard factors remain the increasing prices of energies and of raw materials and the decreased consumer´s acceptance. The managements of enterprises agree that under the impact of the entrance of Slovakia into the EU domestic and external competition sharpened and the companies do not feel an adequate defense of their home market against external suppliers. Discriminating practices of foreign commercial chains express themselves by abusing their dominant position and by increasing power against processers.

Strengths are positive factors influencing the future successfulness what concerns the analyzed sample of enterprises: the implemented system of quality management, modern technological equipment's, qualitative and innovated products, capital cohesion of companies with basic industry. With the implementation of quality systems we can see the time difference in achieving it, which is followed by a competitive advantage for companies with an earlier certification. On the other hand, the purchase of modern technologies caused indebtedness of some subjects and consequently a takeover by stronger subjects. Opportunities are new distribution channels (purchasing alliances, hypermarkets), shopping practices of consumers (packed meat, meat semi-products), internalization of business, penetrating into the foreign markets, and reinforcement of marketing activities.

Sharpened competitive struggle on the market expresses itself in the dynamics of sales of analyzed companies. We were interested in the development of total sales in the period 2002 - 2006. Based on our detections, the growth achieved six subjects (66,66%), stagnation two companies (22,22%) and decrease - one company (11,11%).

Through built-up questionnaire we detected changes in the amount of sales while the companies were selling goods under a private brand name. We tested the following hypotheses:

H_0: We expect that there is no dependence between the share of sales on the foreign market and sales under the private brand names.

H_1: We expect that there is a dependence between the share of sales on the foreign market and sales under the private brand names.

The hypothesis was examined with the chi-square test of independence. The strength of dependence was determined by the Persons contingency coefficient C. The results are shown in the table 4.

Test description	Test	Statistics	P-value
test of independence	chi-square	18,7682	0.0009
strength of dependence	contingency coefficient		0.7145

Table 4. Testing of independence. Source: SAS software

Calculated value of test criteria χ^2 is bigger than the critical value of χ_{tab}. We refuse the hypothesis H_0 and accept the hypothesis H_1 based on which there exists a dependency between listed qualitative attributes. According to the values of Persons contingency coefficient we can declare a very strong dependency. Production of goods under private brand names as well the expansion in the foreign markets show a strong adaptability for the intensive competitive environment of the company and show utilization of potential chances of achieving a position on the market.

In the next part of the research we focused on the store types which take the biggest part on the sales of goods during the analyzed period. The conclusions are in the graph 1.

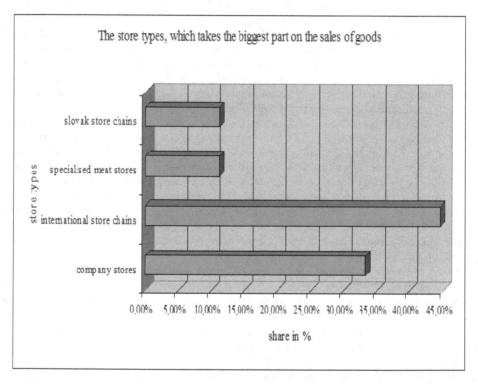

Fig. 1. The store types, which takes the biggest part on the sales of goods. Source: own processing

As the graph 1 shows, the biggest part of production is sold through international chains – four companies with share of 44,44%. Thanks to mentioned store types the production of Slovak meat processing companies is taken onto the foreign markets too. This fact helps enterprises to ensure the sales of bigger part of the production, to increase the profit and to build up the status on the target markets. By 11% less, the Slovak chains participate on sales of production - in case of three analyzed companies. The same share on sales - 11,11% have specialized meat stores and the company store. It is true in one case in company 1. In spite of the low share on the sales channels the specialized meat stores and the company stores have, they provide to customers the full range together with products with a higher added value.

The Slovak entry into the EU meant for Slovak companies a possibility to achieve the foreign markets of the EU members too, but on the other hand the competition accelerated. We have tested if there is a dependency between increased domestic competition after the Slovakia´s entry into the EU and if there are larger requests to adapt to the quality requirements and to the adherence of quality standards. We have verified the hypothesis:

H_0: We expect that there is no dependence between increased domestic competition after the EU entry and the larger requests to adapt to the quality requirements and to the adherence of quality standards.

H_1: We expect that there is a dependence between increased domestic competition after the EU entry and the larger requests to adapt to the quality requirements and to the adherence of quality standards.

The hypothesis was examined with chi-square test of independence. In regarding to achieved values the strength of dependence was not determined. The results are given in the table 5.

Test description	Test	Statistics	P-value
test of independence	chi-square	1,1688	0,2796
strength of dependence	contingency coefficient	-	-

Table 5. Testing of independence. Source: SAS software

Calculated value of test criteria χ^2 is lower than the critical value of χ_{tab}. We accept the hypothesis H_0 based on which there does not exist a dependency between listed qualitative attributes. We refuse the hypothesis H_1. It means that companies were not motivated to take a bigger attention to ensure the quality of their production after the entry to the EU. In the future perspectives we recommend companies to revaluate the possible impact of sharpened competition on their businesses and to actively participate on the building of their position on markets.

In the frame of marketing management analysis in the examined group of processing companies we asked the enterprises if there is an influence of various marketing steps to increase the sales. The scale of answers was from 1 to 10. The lowest influence has 1, the biggest 10. The businesses rated the influence of concrete marketing actions differently. To analyze the results we used the methods of descriptive statistics - median. The found status is illustrated in the graph 2.

The marketing step "price, price proceeding, price reduction" has clearly achieved the highest score on the ranking scale. Based on the opinion of analyzed companies, this has the biggest influence. The second most eminent marketing step is the B2B marketing which includes good relationships with consumers , contacts, business presentations. A close result was achieved within the factor "production under private brand names". It helps companies to rank into the different market segments. According to experience of managements within effects less influencing the sales increase there are factors such as promotion, in-store testing and product innovations. Different usage of marketing actions and tools in practice causes the differentiation of producers´ offers.

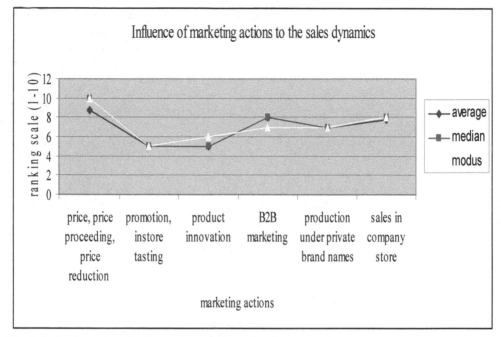

Fig. 2. Influence of marketing actions to the sales dynamics. Source: own processing

5. Conclusion

The branch of meat processing is a significant part of the food industry branch. It cannot be separated as it is an integrated part of food industry. The research revealed weak and limited connections between agricultural production and capacities of processing companies and sales to the stores network. Based on the achieved information and the carried out research we recommend the management of meat processing companies to focus on the following fields which we consider as a market expansion:

- production of meat products with pro-biotics,
- production of meat products with reduced content of fat,
- product portfolio diversification
- internalization of business activities, especially a penetration into the markets of V4 countries and the Russian federation,
- to administer the innovation strategies of products,
- active development of PR possibilities,
- to stabilize business relationships by informal meetings and common undertakings,
- to develop the B2B marketing,
- to renew human resources with language skills and business experience on foreign markets.

The target market identification and the identification of several market segments helps to fulfill not only the current aims but also potential expansive objectives. Selected aspects of marketing management in the business practice of meat processing companies and perspective fields for expansion determinate the winning factors of successful marketing management adaptation for the new EU conditions. Based on Agricultural payment agency [10] it is possible to expect in the future a sharp competitive struggle on the domestic markets of the EU and the countries with open economics such as in Slovakia which must adapt to the import of cheap meat and meat products for example from Poland or Romania.

6. Acknowledgment

This scientific work has been supported by the internal research project of Dubnica Institute of Technology: „Internationalization of small and medium enterprises in chosen region".

7. References

Horská, E.; Oremus, P. (2008). Territorial Approach to International Marketing Channel and Value Added: Case of Agribusiness. In: *Impacts of Globalization on Agribusiness: Trends and Policies*, IV. International Conference on Applied Business Research ICABR, ACCRA. Ghana, Publisher: Mendel University in Brno, 2008, ISBN 978-80-7375-154-8

Kotler, P.; Armstrong, G. (2004). *Marketing*. Praha: Grada Publishing, 2004, pp. 43 – 47, ISBN 80-247-0513-3

Kozelová, D.; Mura, L. et al. (2011). Organic products , consumer behavior on market and European organic product market situation. In: *Potravinárstvo* Vol. V, No. 3/2011, pp. 20-25, ISSN 1338-0230 02.08.2011 Available from: http://www.potravinarstvo.com/journal1/index.php/potravinarstvo/article/view/96/131

Kretter, A. et. al. (2010). *Marketing*. Nitra: SPU, 2010, 288 s., ISBN 978-80-552-0355-3

Malá, E. (2009). Language and intercultural dimension in the process of internationalisation of higher education. In: *Ianua ad Linguas Hominesque Reserata II*. Paris: INALCO 2009, p 100 – 123. ISBN 978-2-91525596-9

Mura, L. (2011). The network approach of internationalization – study case of SME segment. In: *Scientific Papers of the University of Pardubice – Series D*, No. 19, Vol. XVI č. 1/2011, pp. 155-161, ISSN 1211-555X 27.06.2011 Available from: http://www.upce.cz/fes/veda-vyzkum/fakultni-casopisy/scipap/posledni-obsah.pdf

Mura, L.; Gašparíková, V. (2010). Penetration of small and medium sized food companies on foreign markets. In: *Acta Universitatis Agriculturae et Silviculturae Mendelianae Brunensis*, MZLU Brno, LVIII, 2010, 3, pp. 157 – 164, ISSN 1211-8516 27.06.2011 Available from: http://www.mendelu.cz/dok_server/slozka.pl?id=45392;download=63053

Oremus, P. (2008). Adaptácia marketingového manažmentu podnikov potravinárskeho priemyslu na vnútornom trhu Európskej únie. Nitra: DDP, 2008, 150 p.

Šimo, D. (2006). Agrárny marketing. Nitra: SPU, 300 p., ISBN 80-8069-726-4

ŠÚ SR. Štatistiky. Poľnohospodárstvo. 14.06.2011 Available from http://portal.statistics.sk/showdoc.do?docid=11005

Šúbertová, E. (2010). The structural changes and results of agricultural companies in the Slovak republic. In: Megatrend Review, Vol. 7, 2/2010, pp. 53-62, ISSN 1820-4570

Zelená správa. (2011). Správa o poľnohospodárstve a potravinárstve. 12.06.2011 Availed from: http://www.land.gov.sk/sk/?start&navID=121

Zelené štatistiky. (2011). Pôdohospodárska platobná agentúra pri Ministerstve pôdohospodárstva, životného prostredia a regionálneho rozvoja SR. 12.06.2011 Availed

10

Facilitating Innovations in a Mature Industry-Learnings from the Skane Food Innovation Network

Håkan Jönsson[1], Hans Knutsson[1] and Carl-Otto Frykfors[2]
[1]Lund University,
[2]Linköping University
Sweden

1. Introduction

The food industry is today one of the most globalized industries and subjected to growing strains. Globalization and the shift towards innovation-driven knowledge-based competitiveness between firms and regions have rendered traditional national and regional policies and concerted efforts for enhancing industrial economy and prosperity less effective. Baldwin (2006) refers to globalization as "the great unbundling". It is not just a matter of slicing up of value chains and relocating various stages of production to more comparably advantageous regions. Innovation and intrinsic knowledge creation are important parts of the renewal of national and regional policies and efforts to compete internationally.

The food industry, though, is a mature industry. Such tend to be regarded less innovative than emerging Science & Technology-based industries. A possible reason is that they tend to be governed by social and technological regimes (Winter, 1983) related to experience and tacit knowledge built up during the long history of the sector. The food industry is not only mature, but is further characterized by complex and long value chains – shaped and formed over time.

This chapter was written from a regional perspective. In the region of Skane, located in the south of Sweden, a strong food industry has developed over centuries. The potential of Skane as a food region has been acknowledged and supported by changes in the national Swedish R&D and industrial growth policies. The concept of sectorial innovation system and a triple helix approach is now used for enhancing the knowledge-based transition in the Swedish system (Frykfors & Klofsten, 2011).

The major challenge in Skane was that the different actors in the food industry had different ultimate goals. Another challenge was that no single actor owned the overall strategic problem of enhancing the innovative dynamic capabilities (Teece 2007) of the local food industry forming an innovative cluster. It was necessary to reach an agreement about how to govern this joint strategic problem.

The Skane Food Innovation Network (SFIN) was established at the time of Sweden joining the European Community. As the international competition grew more apparent to companies in the Skane region, companies, universities and public authorities saw the need to establish a model for facilitating innovations in the food sector and to create a joint ownership of strategic questions. Since 1994, SFIN is gradually shaping the concept of innovation community. It creates dedicated sub-communities for specific areas and the innovative capability in this mature industry is gradually evolving.

This chapter describes and analyses the policy-making impact on the transition of a locally mature food industry of the Swedish province Skane into an innovative and ever more competitive international food region. The theoretical frame of reference is based upon the concept of social and technological regimes. These regimes, as will be shown, differed between major stakeholders in the region. This called for bridging activities. Several of these activities have been orchestrated by Skane Food Innovation Network (SFIN) and have helped forming new cognitive maps, in turn helping new innovation communities to form and navigate in new market spaces (Frykfors & Jönsson, 2010).

The aim of this chapter, thus, is to present and discuss the learnings of the development of the Skane Food Innovation Network. This relates directly to the ongoing discussion about how stakeholders, both private and public, may facilitate the development of a regional innovation system.

2. Background

2.1 The Skane food industry

Skane is located in Scandinavia, in the south of Sweden and has approximately 1.2 million inhabitants. It is the center of Sweden's food industry. The region has a high density in food-related activities: within the 11 000 square kilometers, all sectors of the food business area are found, covering the total chain from farm to fork. Primary production, the food processing industry, packaging, production machinery, distribution, warehousing and quality control can be found in the cluster. Competence in product and process development, both in industry and academia, is abundant and the marketing and R&D functions of large companies are well developed. Retailing and food distribution companies are also present in the cluster (Lagnevik 2006).

In 2007, the industries forming the core of the Skane food industry employed approximately 25,000 people (Henning et al 2010). Definitions of the boundaries of the food industry, which companies should be included etc, are debated. While discussing the dimensions of the food industry, it has been argued that it not only includes the traditional areas of the food industry, but also industries and disciplines that have a strong link to or act in symbiosis with the food industry (Oresund food 2011:40). If the Skane food industry is broadly defined, including related businesses, an employment at about 100,000 people can be accounted for. This is high in relation to the population in the region (Wastenson et al, 1999). Moreover, Skane accounted for about 21 per cent of the total number of employees in the Swedish food sector's core industries. As Skane has about 12 per cent of the total number of economically active individuals in Sweden, this indicates a strong regional position of the food industry (Henning et al 2010).

2.2 A changing foodscape

During most part of the 20th century, the Swedish food market was protected from international competition. The idea of national self-subsistence guided Swedish food policy. International competitiveness was not a major issue, since surplus production was limited.

During the last decades though, the Swedish food industry has been rapidly changing from a sheltered national industry into an industry exposed to strong international competition. This has occurred gradually in three steps (Lagnevik 2006).

In 1986 the Swedish government declared that the food sector gradually should be exposed to international competition. This induced a change in the Swedish agricultural sector and many Swedish agricultural companies began to adapt to the new working conditions. The second step and a major change in the competitive situation occurred when Sweden joined the European Community on January 1st, 1995. By the entry, trade barriers for finished food products were completely removed. The third step occurred on May 1st, 2004, as the new member states joined the European Community. The Swedish food industry and agriculture is now exposed to fierce international competition. The closest neighbor states, e.g. Lithuania and Poland, produce agricultural bulk products at a cost that cannot be met by Swedish farms and agricultural companies. In addition, the food industry has experienced revolutions in the IT and the Biotech sectors. These technological revolutions have radically changed the working environment for food companies and increased consumer interest for organic and local food, food safety and healthy eating. All in all, this means that the whole context in the food sector has changed dramatically the last 25 years.

It is not only the competitive situation that has changed, but also how the consumers view food quality and food safety. A term developed to capture the multidimensional aspects of food, trying to give equal value to the material, social and mental aspects of food is "foodscape". The term has been used by, among others, Rick Dolphijn (2005) and Pauline Adema (2009). Building on Arjun Appadurai's (1990) influential use of different "scapes" in order to understand the processes of deterritorialization in a global world economy, we use the term 'foodscape' to try to capture the complex intertwining of people, food products, places, emotions etc. that happens in food-related situations, resulting in communal identities as well as economic, physical and social structures linked to food. This broader view on the changing conditions for the food actors is necessary in order to understand the context within which the activities of SFIN have taken place.

2.3 The triple helix setting

The companies in the Skane food industry are a mixture of larger national and multinational companies (such as IKEA Food services, Nestle, Findus and Atria), and SMEs, from specialized food manufacturers to small innovative research-based companies, such as Oatly. Food-related education and research is to be found at all four universities in the region, especially at Lund University (Scandinavia's largest establishment for higher education and research with over 40,000 students, founded in 1666), but also at the Swedish University of Agricultural Science, Alnarp, at the Malmö University, and at Kristianstad University College. Research in the region covers all kinds of scientific knowledge in the food chain from farming to consumer studies as well as scientific knowledge in industries related to and supporting the food chain. The Skane region also has a strong position in research and development in the companies of the food branch. Many Swedish companies have located their R&D centers in this region, as have packaging, processing and distribution plants.). Among the food related industries, a special note should be made about the packaging industry as it represents a large sector in the region with internationally successful actors such as Tetra Pak.

State, regional and municipal authorities support development in the food industry. In particular, the Region of Skane invests actively in the development of food-related initiatives and pursues a vision of becoming the food center of northern Europe in 2025. The state-governed county council supports local companies and initiatives consciously and forcefully.

2.4 Features of the Swedish innovation system and innovation & research policies

The arena for the Swedish food industry has changed quickly and radically in just a few years. Skane, being the center of Sweden's food industry and closest to the international market, has faced an urgent need for increasing the innovative capabilities of the food industry. A brief look at the Swedish innovation policy is needed in order to describe the framework in which SFIN operates.

Understanding the rationalities of the current Swedish use of the triple helix approach requires recognition of the main structure of the Swedish economy. It reveals several distinctive features grounded in *la longue durée* of niche management, e.g., industrial specialization and creative use of innovative technology procurement. This has orchestrated the main features of the innovation and research policy system and its growth model for industrial development and prosperity (Freeman, 1987, Lundvall, 1992).

The concept of "development pairs" (Fridlund, 1999) describes the long-term development cooperation and close partner relationship formed between specific state client and private firms in order to exploit and develop new technology. Such relationships often played the role of driving force and a catalyst for growth of industrial development blocks (Dahmén, 1987), which later matured into defined innovation system or what Carlsson & Stankiewicz (1995) called "technology systems". The evolution of the Swedish innovation system offers a number of examples of niche management, e.g the state power authority which historically enjoyed great freedom from political interventions and the Swedish telecom industry, and different key companies or group of key firms (Fridlund, 1999, Laestadius & Berggren, 2000).

However, the application of development pairs has exposed some negative structural features of the concept. The most striking feature is the bifocal policy structure that is split between two subsystems – one that supports innovation and business development, and another for academic education and research. Another shortcoming in the approach is a diminutive structure of intermediaries and research institutes. A third characteristic feature is that the regional governance level is by-passed and weakened (Pierre & Peters, 2005). The regional politico-administrative level has primarily been pre-occupied with financial support and the governance of common healthcare and communication infrastructure, while policies for research, technology and innovation (RTI) always have been a state affair.

As Sweden became member of the EU in 1995, the country faced new regulations for compliance with EG legislation of procurement and trade. As a result, development pairs being the significant part of the Swedish innovation system were no longer viable. This called for a change in national innovation and research policies.

2.5 The dawn of the Skane model of food innovation

Influenced by the EU, the regional level in Sweden has to be more influential in issues of innovation and development policies. The shift in command, underlined by ongoing globalization of markets and of growing multinational companies, has made national innovation and growth policies less effective and more difficult to manage solely from a national level. In 2003, VINNOVA, the state agency for innovation and systems-oriented research, launched a regionally oriented program for research, technology and deployment/demonstration (RTD) in line with this shift. The program is based on the concept of sectoral innovation systems and the triple helix approach. The aim is to promote upgrading and renewal of local innovation and R&D capabilities and skill building in certain important growth areas with strong regional profiles. The focus on enhancing the strength of regional constituents and supporting structures of the sector innovation system involving regional authorities has materialized in the Winn-Growth program. This program, though, calls for the development of new types of governance and practices.

One initiative within the Winn-Growth program was granted to the Skane Food Innovation Network, an innovation intermediary, a regional "facilitator" composed of companies, universities and regional authorities. The idea was to increase the return on investment and added value in the regional food industry by strategically enhancing the R&D and innovation capacity in the food sector. A previous innovation strategy for the food sector of Skane, based on increasing internationalization and more knowledge-driven innovations, had been outlined prior to the Winn-Growth program. This had been done together with the establishment of an academic functional food research program (Nilsson 2008). Each of the regional Winn-Growth programs suggested by Vinnova run for a 10-year period and are subdivided in approximately 3-year stages.

3. The Skane food innovation network

The changing competitive conditions that occurred during the 1990's had spurred thoughts among leading industrialists, in the regional government and among researchers that measures were needed to be taken if the food industry in the region should stay competitive. In 1994, the three parties consequently created a joint network organization, Triple Helix style, and named it 'The Skane Food Innovation Network'. The objective was to facilitate the exchange of research and development between universities and the food industry. In the organization, leading actors from the region did investigations and started joint actions in areas related to competence and competitiveness. The insight that major changes were ahead made leading actors work hard together to improve the future for the food industry. In this context, it is worth noting that the original initiative came from the food industry. This has created a triple helix organization in which the industry has had a leading role for more than one decade (Asheim & Coenen 2005, Coenen 2006). From the outset, it was decided that the network CEO should have an industrial background and that the chairman of the board should be the County Governor.

The urgent need for a transition away from traditional bulk production was the major incentive for creating the organization. New knowledge and collaborations with other actors were needed in order to increase the production of specialized and highly processed products. This made some of the major players in the food industry dedicated to the organization. At the time, functional foods and convenience foods seemed to have a very promising future. Both segments demanded academic research and, although to a lesser extent, regional collaboration.

The Skane Food Innovation Network was organized as a flexible network organization, with a small, operative executive board. Through the co-operation over the years, it came to be seen as a constructive force in the regional development from both the industry and the regional authorities. In spite of the small size of the organization, when competing for the

"Winn-Growth programme" it was seen as logical that SFIN should represent the joint innovation efforts in the region.

3.1 From project management to innovation communities

The Winn-Growth program multiplied SFIN's resources in a single blow. As any organization experiencing sudden growth, measures needed to be taken in order to fulfill the new goals. Over the years, continuous reconsiderations have been made. These can be divided into four phases: the starting phase, the foresight phase, the governance phase, and the community phase. The different phases were all induced by discussions about strategic issues and are therefore described in some detail. The story is based on interviews with key actors, both during the foresight process in 2006-2007 (Jönsson & Sarv 2008) and on follow-up interviews in 2009-2010. Two of the authors (Jönsson and Knutsson) became members of the management group in 2009. The material is thus complemented by inside information, in an anthropological tradition described as participant observation (Davies 2008).

Phase 1 – Making things start

The SFIN Winn-Growth initiative's point of departure, outlined by professor Magnus Lagnevik at Lund International Food Studies, was that many innovations are born at interfaces. The primary aim was therefore to encourage interaction at interfaces between different areas of scientific knowledge, between different technologies, between academic research and commercial enterprises and between private and public organizations. Knowledge integration (Grant, 1996) between different knowledge areas should produce new products, services and concepts. The Triple Helix dimension was the second cornerstone of the SFIN organization, reflected in the board already from the start, with the county governor as chairman and representatives from the food business and academia in the board and management group.

In order to get things started quickly, without having to start by building a central organization, the vast majority of the funding was canalized through existing organizations. Four key development areas were defined; Functional foods, International Marketing, Food Service, and Innovation. The former three were canalized through The Functional Food research centre, Lund International Food studies and Food Centre Lund, all academic research centers within Lund University. The latter was assigned to Ideon Agro Food, a foundation working mainly on strengthening the links between academia and business, aiming to support innovation projects and technological development in the food industry.

The key development area during the first years was Functional Foods. The area was at the time believed to have an enormous growth potential, and the need for knowledge integration between all sides of the triple helix was obvious. On the research side, nutrition, medicine, food technology, food engineering, marketing and consumer behavior were all relevant but shared neither traditions nor agendas. Food companies, food ingredient companies and packaging companies could join from the corporate side, all with a more congruent agenda. The county government, responsible for the healthcare in the region, could enter the cooperation with a lot of knowledge, but also as a major receiver and user of the innovation results.

During the first three-year period, the program was primarily engaged in stimulating the innovation system by providing new competence and deepening knowledge about the needs, prerequisites and opportunities for renewal of the food industry. Several PhD projects were initiated, primarily in the functional foods and marketing areas. The research projects were quite successful, as were many of the smaller innovation projects. Yet still, doubts were raised when the first three years of the program was about to be concluded.

Within the literature on clusters and cluster management, questions regarding the governance of clusters emerge (Nilsson 2008). Is it at all possible to govern a cluster, based upon intricate networks of actors, imprinted with different ways of thinking and acting? The Skane food innovation network had expectations from both its members and its funders (primarily Vinnova) to govern the innovation efforts in the Skane food industry. These expectations created an urgent need for rethinking the initial approach, based on the idea of outsourcing dedicated projects to organizations beyond the control of the SFIN organization. There were good reasons for the reconsideration: there was a wish to keep administration at a minimum and a clear need to position SFIN distinctly in relation to other facilitating organizations active in the industry. After the first three years, the problem of organizing the innovation efforts became obvious. It was probably efficient to get research and R&D project conducted by delegating them to specialized organizations and institutions. However, with a central organization without the resources to fully coordinate the outcomes of the separate projects, the Winn-Growth project appeared to be any other funder of research and innovation projects, with the strategic challenge and direction being left out. Another problem was the decreasing engagement from the business. The first years saw massive investment in research, which takes time to deliver useful results, especially since many of them were four-year PhD projects. The fact that the hype around functional foods cooled down these years made the connections between business and academia weaker. Concerns were raised about the dedication of resources from the industry. A long-term strategic discussion was initiated, both from Vinnova and from SFIN. What was the role of SFIN, really?

Phase 2 – Foresight as pit stop

Each of the regional Winn-Growth programs granted by VINNOVA ran for a 10-year period and was subdivided in approximately 3-year stages. Prior to each stage, an update of the program content was undertaken, involving the reconsideration of a relevant set of regional stakeholders and measures. In the intersection between the first and second stage of the SFIN program, a foresight process was suggested in order to navigate in a changing "foodscape" since both key stakeholders and markets were in transition. The foresight process should also strengthen the Innovation Network's governance of the "regional food cluster" of the sectoral innovation system, in accordance with the experiences mentioned above. The foresight exercise was part of an EU-funded four-party foresight development project titled Foresight Lab (www.innovating-regions.org). The main goal for the "systemic" foresight was not to do a full-scale foresight, but to strengthen the innovation community in order to make the food cluster more competitive when future opportunities emerged.

The foresight identified three major challenges for the governance of the food cluster. The first and foremost was to make actors in the triple helix–perspective more profoundly committed to the idea of building a regional-based food innovation community. The regional perspective was not self-evident to many of the actors – they were regularly working in organizations where either a larger geographical area (national and/or global for the major companies) or a smaller area (sub-regional or local for the micro businesses) was being in focus for the business strategy and development activities.

The second was the heterogeneous character of the food innovation system. There were many small businesses throughout the value chain, from farmers over food producers to grocery stores, restaurants and public eating-places (retail and wholesale being somewhat more centralized). Some actors did not feel connected to other levels in the food chain, while others felt a need to establish collaborations, but uncomfortable with the process. At this point in the process it became obvious that the Swedish food system (as many other mature industrial sectors) had established strong regimes on each level of the food chain, with little or no understanding of the actors on other levels.

The third challenge for the foresight activities rose from the different social and technological regimes of the actors. This called for bridge-building activities that could translate and transform the disparate cognitive maps in order to build an innovation community that could navigate together in the new market. All actors were imprinted with values emergent from a time when the Swedish food system was based on national self-subsistence, with limited market competition. This was a major impediment for cross-sectoral innovation, since the new situation demanded new ways of collaboration and consistent interpretation of the foodscape. Although difficult, the timing was right. Since the regimes already were in transition, all actors were aware of the need for new working methods and collaborations.

The foresight confirmed several strengths of the Skane food industry, such as the presence of all vital parts of the supply chain, a strong, academic research tradition, and a well developed, common sense approach for food pleasure at the regional community levels. But findings also pointed especially to some weaknesses and threats in terms of "gaps". Hence, different needs and measures for strengthening the "bridges" across knowledge areas, long and short term plans and actions, big and small businesses, parties in the supply chain, research and business and among different levels of the innovation system (Jönsson & Sarv 2008).

Four potential sub-communities were identified from the mapping as being of particular interest to SFIN. In line with the systemic idea of developing the communities from a learning partition perspective, with knowledge and governance partition support, four systemic meetings were launched, with open invitations to all relevant actors. After the meetings, an ex post analysis of the foresight exercise and its outcomes was made. The focus was upon governance and how the organization could be developed in order to take a strategic leadership of a heterogeneous food innovation and production system. Governance-related services were hereby identified as particularly interesting services in the upcoming foresight work. Consequently, four focus areas were proposed: 1) *launch new services* (such as systemic meetings for sub communities in the food innovation system and dedicated knowledge partition services for hot areas), 2) *develop innovation community and foresight governance*, 3) *increase communication and bridge building* by continuously publish innovation community progress on an interactive basis, 4) *investigate in innovation system research* (research that not only formulates bases for food innovations, but also contributes to the development of the innovation system as a whole).

For all focus areas, specific activities were proposed. The analysis finally concluded that the different suggestions reinforced one another, and should therefore be developed concurrently, on an experimental and learning-oriented basis. Both "direct innovations" and cross discipline research projects were noted in the results in addition to the new knowledge and experiences regarding foresight work. In particular, the linking of the food innovation system with the health innovation system and the improved integration of retailers into the food innovation network was very promising (Jönsson & Sarv 2008). The foresight became a catalyst in the development process that Skane Food Innovation Network initiated in 2006. In 2010, most of the proposals of the Foresight's concluding report have been implemented.

Phase 3 – Establishing governance and credibility

Parallel and in continuous dialogue with the foresight process, organizational changes within SFIN took place. A new CEO, Lotta Törner, was recruited in 2006, and a re-organization of the board with the new county governor as chairman took place. In the dialogue with the major players from the business side, it was obvious that, from their perspective, there had been too much emphasis on long term academic research. Given the question what SFIN as an organization could do, the answer was primarily to enhance the attractiveness of the food business. For a long time, the industry had faced problems with recruiting highly educated younger people. A stronger focus on supporting the latter phases of the innovation process, to get new innovative products, concepts and services on the market, was further proposed.

The period led to a greater focus on meeting places. A number of meeting places were developed. Here, representatives of various interests and competencies could exchange and discuss ideas and develop creative solutions and business ideas. The existing meeting places were upgraded. The annual "Network day" were turned into a meeting place for most of the food sector in the region by inviting internationally renowned speakers and awarding research prizes and scholarships of a combined value of almost 100 000 euros in collaboration with a large foundation in the food and health area.

Furthermore, dedicated sub networks were established. A network of CEO's was formed, in which the most prominent CEOs of the food business now meet on a regular basis to exchange ideas and to discuss present and future challenges. This was a direct result of a foresight activity, where it was identified that the CEO's needed a special network in order to be committed to development projects outside their own companies. A Research Network was created as a meeting place for interdisciplinary contacts for food scientists from all faculties and universities in the region. The network is organizing seminars and workshops on "hot areas". A Retailer Network was organized in order to get the retailers more involved in the Innovation community. This was arranged as a part of a platform for innovative market places in order to integrate the retail side in the food innovation community and to promote innovative market solutions. The platform Future Meal Service, focusing on meals in the public sector was another initiative during this period. An Entrepreneur Council, where entrepreneurs can present their ideas and get professional advice and seed money in order to make their business ideas successful was further established.

The constant need for upgrading the creative capability of the innovation system was recognised as an important factor for the future, not least since the Winn-Growth program ends in 2013. The establishment of foresight as a continuous process was decided as a strategy in this work. Systemic meetings are being used whenever interesting projects materialize. A new arena for developing contacts between the packaging industry and the food industry in order to develop "Innovative food in innovative packages" is one example of the results generated from the systemic meetings. As part of the foresight process, there has been an increased focus on international bench learning.

A lot of focus was given to the Student Recruitment Program, reflecting the business representatives wish for activities dedicated at enhancing the attractiveness for the food industry among younger people. A starting point was to establish an Advisory Board with students. Ten students from different educational programs gathered together with the assignment to give the Network and the individual companies valuable input on how the best brains can be recruited to the food sector in the future. In order to improve contacts between companies and educational programs, the students on the Advisory Board arranged field trips to companies and lectures by CEOs at universities and colleges. In collaboration with Skane Food Innovation Network's communication manager they have designed a web-based "career" site. The advisory board's activity led to the establishment of a joint trainee program, where five companies assigned in total eight trainees for a period of sixteen months. The trainee program offered a mix of company-based work and joint activities. The establishment of a joint trainee program showed the high degree of credibility as a neutral arena as the SFIN had established from the companies perspective.

Finally, attempts were made to use diversity as a tool for adding new perspectives to the innovation climate. Traditionally, the food industry has been dominated by male, ethnic Swedes. This means that both women, and the fast growing immigrant population, have been under-represented in key positions in the food industry, even though there is a lot of competence and entrepreneurship in both groups. The research project Power over Food, that started in 2009 is addressing the question "How can better knowledge of gender and equal opportunity issues in organizations and companies create more and better food innovations?" There are suspicions that many capable female researchers and entrepreneurs terminate their careers in the food field. That implies a risk that many ideas for potential innovations come to nothing. The project wants to find out if this loss is related to attitudes about gender and equal opportunity, and then come up with possible changes that can encourage more people to remain in the business. The development of "Etnos", a project devoted to producers of ethnic foods in Malmö funded by the Skane Food Innovation Network and the Skane County Administrative Board considerably increased the diversity in the program. This ethnic network consists of food entrepreneurs with background in other countries than Sweden.

Phase 4 – Expanding the innovation community

After reassuring commitment from the main initiators of the network, SFIN in 2009 took several steps in order to strengthen the governing capacity and expanding the innovation community. The management group was expanded and re-organized into strategic areas. These were "Strategy & Cooperation", "Jobs & Careers", "Tomorrow's Meal Services", "Innovation & Entrepreneurship", and "Taste of Skane" (formed in 2010).

The distinct areas of responsibility demand a deliberate cross-fertilization and commitment from the different areas. The heterogeneous food innovation system requires "multilingual" skills. This was explicitly searched for by the CEO when recruiting the members of a new, larger management group. Apart from having knowledge and authority in the various areas, the members were constantly asked to make sure that the integration and cross-fertilization between the different areas was functioning.

Strategy & Cooperation was dealing with the overarching questions for the whole SFIN network. *Jobs & Careers* were targeted at the attractiveness of the food business, with the advisory board, trainee program and establishing connections between gifted students at universities and university colleges and the food industry. *Tomorrow's Meal Services* is focusing on meals in the public sector. Special emphasis is given to education and innovative purchasing procedures. The existing procedures don't promote new meal solutions or food products and the average education level in this part of the industry doesn't suffice in the contacts with private suppliers and competitors. In this platform the collaboration between all parts of the triple helix is most prominent, based on the idea of getting better meal solutions. Food for elderly people and hospital meals are top priorities within the area. *Innovation & Entrepreneurship* is the area in which the development and marketing of new products and processes are supported, either by dedicated small funding or by competent coaching from experienced business people connected to SFIN.

The initiative for SFIN had come from the traditional large-scale food industry, with the regional authorities and academia as main supporters. However, the changes in the foodscape, especially at the demand side of the food system, had created new opportunities and challenges for the food sector. While the food industry faced decreasing margins and fierce competition, the gastronomic side of the food business started to boom during the 1990s. From being a nation in the culinary outskirt, Sweden and its neighbouring Scandinavian nations had a rapid development in the fine dining sector. Swedish chefs started to win prestigious culinary awards and the restaurants with Michelin stars had an impressive development. Sourdough bakeries made people lining up for buying crispy bread at prices that must have seemed like a wet dream for the large bakeries delivering bread to discount stores and super markets. The media's infotainment programs made new generations discover the pleasures of cooking, starting to demand high-class products, new vegetables and spices, and rare cuts of meat from the local supermarket. And the on-going debate on food safety and food quality led to a growing appreciation for food experiences based on authentic, traditional and local values. Small-scale food manufacturers got a revival, often integrating their production with culinary tourism.

Restaurants and small-scale food manufacturers were already connected in networks, based on mutual commercial interests. But the connections with the large-scale food and retail systems were poor. The region hosted a separate development organisation for culinary tourism, restaurants and food manufacturers. However, a mutual interest had been starting to grow between the separated food domains. The food industry saw that large numbers of consumers were willing to pay considerably higher prices for authentic high-class taste experiences, an added value that not even the hi-tech products of convenience and functional foods had been able to accomplish. The small food manufacturers had limited access to necessary retail and distribution channels in order to expand and/or make their sales profitable. Here they saw a potential in collaborating with the larger players.

In 2010, the regional authorities decided to let their separate development organisation for culinary tourism and small-scale food manufacturers be handed over to SFIN. The former platform Innovative Market Places was integrated as a part of the platform *Taste of Skane*, directed at the concrete local development of on one side the small-scale food production and distribution, and on the other side the local culinary tourism. From now on, the whole food business, from fine dining to bulk production, was hosted by one single organisation in order to facilitate cross-fertilization between small-scale and large-scale food activities.

If the different regimes of the food industry, the academia and the regional authorities had been somewhat difficult to bridge, this was nothing compared to the separate traditions of food service and restaurants, and small-scale and large-scale food producers. Some of the differences were hidden, but some were clearly outspoken. Many of the small food manufacturers had started their business in direct opposition to the food industry's way of growing and processing food. Although in essence being committed to collaboration, the mistrust is not easy to overcome. But some development projects have been mutually attractive. A successful pilot project has concerned the development of a joint retail brand for local food. The retailers' network had already been discussing the need for an easy way for the consumer to buy local premium food in ordinary supermarkets. Despite consumers' positive attitudes, local food (i.e. local provenance) had not been commonly promoted in the Swedish supermarkets. Local food with a strong local profile had mainly reached the consumers through alternative channels, such as the Farmers' Market (Ekelund & Fernqvist 2007). Since Swedish food retailing is highly concentrated to four actors, accounting for 76 per cent of all food retail sales (Market Magasin, 2008), it seemed necessary for food products promoted as local, to be sold through the supermarkets in order to grow beyond being pure niche products. By relocating the food products in the store, and developing information material and a joint brand, the small-scale food products are now easy to locate within the stores of the participating retailers. The brand "Taste of Skane – carefully selected local food" remedies the producers' concern that their products were insufficiently exposed in ordinary supermarkets. So was their concern about low profitability, since the retailers agreed upon having somewhat lower margins on the selected products. It shows the potential of collaboration if there is both a short-term gain for the users, and a larger good (regional food) that could engage the actors apart from strict business considerations.

A parallel development is to be found in the SFIN "Chefs' network". It turned out that some of the more prominent chefs in the region were interested to contribute with the experience and knowledge to the public sector. Since this particular part of the food industry, municipal and county foodservices, is a low-status, low-wage part of an already meagre sector, the initiative was welcomed. One of the most renowned chefs in the region started up a small chefs' network with only two chefs from the public sector and two from the private sector. These four people had the assignment of identifying urgent development areas where the cross-fertilization between sectors could be meaningful. A new dimension was introduced to the SFIN agenda, industry development hand in hand with corporate social responsibility. Gradually, the chefs' network has turned into a "meal network", where a much wider array of competencies and backgrounds now join forces in order to revitalise the entire meal situations in various areas within the public sector. It is behavioural consultants, architects, chefs, nutritionists, etc. Different projects have up until today covered elderly care, education in food and meal knowledge, health care and disabled people. They all share the characteristics of not being commercially "hot" and the innovative solutions are all needed in the entire sector. The pilot projects are high-risk and designed with replication criteria and business potential, in case the projects turn out well. The common denominator for the meal network is to redefine the meal situations for those who do not choose for themselves what and when to eat. Although the agenda is clear, the bridging of different areas is not easy. Being public operations, these are all relying on political decisions. The public foodservice has been neglected for a long time and is devoid of any national, regional and even local co-ordination and suffers from poor funding and low status. However, the regime of public foodservice is increasingly scrutinized, media interest grows and points out the social as well as nutritional importance of public meals. The sector has a potential to raise the bar for product and process innovation within the entire Swedish food industry. Thus, the natural extension of the meal network within the area of Tomorrow's Meal Services is the establishment of a politicians' network, which is currently underway. The public authorities may turn out to be the most vital part of the triple helix structure underpinning the SFIN operations – not only as a primary funding source, but also as an increasingly demanding end-user in the Swedish food industry.

Both the retailers' network and the meal network initiate and run risky pilot projects with a business potential for involved entrepreneurs and other incumbents. Pilot projects are carefully selected and consciously and strategically granted. In this way, relatively small initial pilot project fundings may reproduce themselves in the industry, on sound commercial conditions.

4. Summary on findings

The story of Skane Food Innovation Network and its efforts in stimulating innovation in the regional food industry boils down to four main topics: the organization has been developed, managerial crossroads have been designed, stakeholder interests have been aligned, and SFIN has found a viable modus operandi in its risky pilot projects with reproduction and diffusion potential.

4.1 Organizational development
The SFIN organization has evolved, in seven years, from strict and arms-length R&D funding into a multilevel and multidisciplinary innovation community.
An important insight was the need for dedicated sub-communities, i.e. "focal networks", with hands-on activities, in turn co-ordinated in the overarching food innovation network. This combination creates opportunities for direct innovation in the sub-communities as well as creating a breeding ground for cross-fertilization between the sub-communities.

4.2 Managerial crossroads
The SFIN story reveals how the Triple Helix approach has been gradually operationalized through board representation and a "multilingual" management group. The role of the management team has been sharpened and requires both depth and breadth from the area managers. The managers need to be skilled intermediaries with an ability to identify and translate differences across differing regimes. Cultural differences between large and small, public and private, primary production and retailing, etc, influence community building activities. At the same time, cultural impediments may well hold the potential innovation (Jönsson, 2008).

4.3 Alignment of stakeholder interests
SFIN has gone through a drastic change from being preoccupied with the funding of R&D projects without any joint strategy, into engaging in cross-sectorial collaborations. A recent study of the Skane food industry concluded that SFIN have had a significant impact on reducing the fragmentation in the industry (Henning et al, 2010). Gradually, joint arenas become legitimate and the strategic and long-term nature of the SFIN operations spread among different actors. This makes it less threatening to engage in open innovation-kind of collaboration, as long as it resides under the "neutral" SFIN label.

4.4 Emergence of a modus operandi
In seven years, SFIN has turned into a network of cross-functional networks defining and funding risky pilot projects, commercially viable and with diffusion potential. The diffusion process is backed up by support activities such as marketing and communication, but the business potential is for the entrepreneurs to realize. In this way, funding exploits the

incentives of entrepreneurs, at the same time receiving a "reality-check" from the level of interest expressed by entrepreneurs in the first place. Furthermore, small funding may be leveraged into, at best, increased economic activity and societal improvement.

5. Discussion and outlook

The Skane Food Innovation Network operates on many fronts to boost the impact of innovation in the food industry. The Network also tries to create the conditions for innovation by enhancing the attractiveness of the food business. It has been a tricky road, with successful paths as well as some dead ends. Although promising, the work of SFIN is only at the beginning of the transition processes in the Skane food sector. The changing foodscape, with completely new situations for both farmers, industries and retailers, calls upon new and innovative solutions. Most of the mature industries tend to have the same problems, such as well-established regimes, created in very different markets. These regimes are de facto impediments to innovation. In this section, we will discuss the learnings of the work of SFIN with facilitating innovations in a mature industry, focusing on the use of a multilevel triple helix approach, the importance of analyzing and bridging regimes and the importance of an end user perspective as a guiding principle.

5.1 The multilevel triple helix network approach
The food innovation system is heterogeneous, with a mix of large national and inter-nordic producers and retailers and many small local businesses all through the value chains. It spans from farmers over food producers to grocery stores, restaurants and public eating-places. This characteristic was initially a major problem when trying to build an overarching innovation community and common ground for concerted efforts. Some actors did not feel connected to other levels of a presumptive cluster in the food chain, while others felt a need to establish collaborations but felt insecure about the process. This mirrored the fact that the Swedish food system, like many other mature industrial sectors, had established strong regimes on each level of the food producer-consumer chains, with little or no understanding of the actors on other levels. All actors were imprinted with values emergent from a time when the Swedish food system was based on national self-subsistence and strong public regulations, resulting in limited market competition (Beckemann 2011). This was a major impediment for cross-sectoral innovation, since the new situation demanded new ways of collaboration.

The triple helix approach is concerned with joint efforts and coordination, in our case between universities, business firms, local production units of large multinational companies and regional authorities. Different stakeholders have their individual rationales and logic. In order to be successful, the triple helix approach should rest on mutual trust and goal congruence between incumbents. Bridging regimes is like an evolution, an emerging understanding of the common good of a change of strategy and behavior.

5.2 The importance of analysing and bridging regimes
The fact that processes underlying innovation and industrial and economic transformation are governed by social and technological regimes have been acknowledged by, among others, Cooke, (2005), Geels, (2008), Bergek et al, (2007), Klein Wollthuis et al, (2005) and Malerba and Orsenigo, (1997). Winter (1983) defines regimes in a sector as a specific set of

not only regulative institutions and norms but regimes also regulate codified formal as well as tacit informal habits and routines related to common collective and individual practices and beliefs. These practices and beliefs shape and coordinate actions between various groups, individuals, and organizations in the sector. The notion that technological regimes and their production of knowledge are shaped by historical and cultural factors have, from a different starting point, been repeatedly argued in the tradition of cultural analysis, which have become used more often in both product development and marketing in the latest decades (Pink, 2005. Sunderland & Denny, 2005. Kedia & Van Willigen, 2005). The tradition of cultural analysis further stresses that technological systems do not function independently from the human actors within the system. The actors are seen as embedded in social groups with cultural requisites, such as traditions, norms and beliefs (Law, 1999).

Breschi and Malerba (1997) concluded in their studies of sector characteristics of national innovation systems that technological regimes are defined by the level and type of opportunity and appropriate condition to innovate. This is bonded to the history, nature and the cumulativeness of knowledge as well as to the means of communication and transmission of knowledge within the sectoral systems of production. Following Levinthal (1991) and Scott (1995) regimes have three dimensions: i) cognitive rules, related to belief systems, ii) normative rules expressed in missions, goals, and identity, and iii) strategies and strategic orientations towards the surrounding external socio- technical and politico-economic environment. Regimes are closely related to the concept of institution. An institution could be defined as "patterns of routinized behaviour" (Hodgson, 1988) and may be analysed on a number of different levels.

In the networks of SFIN, it is a number of sub-systems who engage and challenge current regimes. The result is ideas, tested in "risky pilot projects" with diffusion and profitability potential. The networks may be understood as "liquid environments", where different knowledge, experience and values meet (Johnson, 2010). Such liquid environments define the so-called "adjacent possible" (ibid.). This means that the configuration of single networks and the links between the individual networks constitute the limits to what the network may produce. i.e., configurations are imperative. The regimes are influenced by the experience, thoughts and ideas, values and objectives of each and every individual within the different networks. SFIN organizes the network of networks and initiate pilot projects. On this level, cognitive rules are tested and different belief-systems are bridged in the work of individuals.

Leaving the individuals' level, the next level may be approached from a business angle. Business firms are normally run with a profit incentive. Innovations aim at creating new or better value to customers, leading to sales and profits. Following Christensen (1997), the average company inherently faces difficulties innovating a thriving business. Organizational routines and activities are shaped for efficient use of resources. Business innovation implies the change of product offerings, markets or resource use and the re-shaping of the "theory of the business" (Drucker, 1994). In terms of regimes, innovation by definition alters the business regime in one way or another, disruptively or incrementally (Christensen, ibid.). Govindarajan and Trimble (2010) are pre-occupied with "solving the execution challenge", focusing the way that an on-going business may handle challenging ideas and taking them to market. A new idea could form a spin-off initiative and be the start of an entirely new company. However, firms also need to innovate their current businesses, why it is necessary to establish a formalised co-operation between the existing business and the innovation initiative. What Govindarajan and Trimble (ibid.) suggest is a gradual and well-managed integration of old and new regimes in terms of both social and technological challenges. SFIN relieves the established firms from the direct disturbances of challenging ideas, working as an outside test-lab without worrying the on-going business. Still, the CEO network and the Entrepreneur Council bridge the gap between established practices and innovations developed in the pilot projects. Both social and technological regimes are bridged by way of the SFIN networks.

Different regimes that have their own specific cultures have been developed during the long history of the national food system in Sweden. By opening up arenas for individuals to meet, to identify and test new ideas, SFIN helps established firms to engage in innovation without compromising their running business. At the same time, entrepreneurs are invited to a vibrant group of people all joined in the common interest of developing the food industry by "open innovation".

5.3 The importance of an end user perspective

The end-user perspective is notable in the literature on open innovation (Chesbrough, 2006. von Hippel, 2005. Wallin, 2006), which stresses the importance of integrating the demand side in the study with the development of innovation systems. The action taken in the extended network all followed one of the major recommendation from the foresight process: take the end user perspective as a starting point. The user side was also used as a start for community building activities, whether it was the student recruitment program, the local food or the food for elderly activities.

Bringing the end-users helped to synchronize the agendas of the different actors in a multilevel food triple helix space and a multilevel foodscape. The end user perspective made the participating actors really feel that although they couldn't solve the problem by themselves, they all had important contributions to make. We conclude that the user side cannot be reduced to the result of the innovation system or the triple helix actors' achievements in a conventional way, since the user side interacts with every level and affects the outcomes from an early stage. The end-users have of course always been the important landing point for innovation work, the place where the success of the attempts to innovate is determined. But our point is that they may also be the best starting point, since it is the only level to which all actors of the triple helix can relate.

5.4 A new innovation systems model

The experience this far shows that SFIN engage a wide range of stakeholders in its different networks. Small-scale food producers, public servants, small service businesses, large retailers, politicians, entrepreneurs, large international food-related companies, researchers etc join the different networks of SFIN. Individuals meet in focal networks, form pilot projects which drive economic development in the industry from within. The former Swedish innovation system using "development pairs" in order to direct – "top-down" – the formation of an entire industry through a single company is gradually supplanted by the bottom-up network model strengthening the inherent innovative capabilities of a wide range of small and medium-sized firms, as well as larger corporations and public organizations. We suggest that the SFIN triple helix-based network form of organization holds several strengths. It is dynamic in its formation, it is resilient to temporary failure and it is cost-efficient in its selection and execution phases – it uses entrepreneurial incentives and helps isolating innovation initiatives from on-going business in established firms. Although it remains to be tested, this could be considered an efficient way of bridging strong regimes of a stifled and mature industry. It could be the Swedish food industry, but the mechanisms controlling the network of networks may well be transferred to other industries sharing these characteristics.

6. Concluding remarks

We would argue that the related work methodology of SFIN may be part of a transition from the prevalent Swedish innovation and development mode and work as a model for facilitating innovations in mature industries. The combination of an overarching innovation network responsible for issues of governance, in combination with dedicated sub-communities implementing hands-on activities and projects was a major step forward from the original SFIN organization, which was based on a traditional way or organizing innovation facilitators in Sweden. We would like to call the refined methodology a *Régime-bridging strategy*, with a *multilevel triple helix approach* and an *end user perspective* as fundamental cornerstones.

7. Notes on contributors

Håkan Jönsson, associate professor in European Ethnology is researcher and lecturer at the Department of Arts And Cultural Sciences at Lund University, where he teaches at the Master of Applied Cultural Analysis program (www.maca.ac). He is also head of operations in the Skane Food Innovation Network, responsible for the area of small scale food manufacturers and culinary tourism.

Hans Knutsson is assistant professor at the School of Economics and Management, Lund University. He teaches accounting, management control, and strategy and focuses his research on public management and cluster development. He is head of operations in the Skane Food Innovation Network, running the area Foodservice of Tomorrow.

Carl-Otto Frykfors is affiliated to the Department of Management and Engineering at Linköping University and prior senior program manager at VINNOVA, The principal governmental agency for knowledge driven industrial renewal and innovation. He was further director of The Dahmén Institute in charge for evaluation of Foresight activities related to an European Foresight project between regions in Sweden, Italy, Germany and Poland.

8. References

Adema, P. 2009. Garlic Capital of the World: Gilroy, Garlic, and the Making of a Festive Foodscape. Jackson: University Press of Mississippi.

Appadurai, A. 1990. 'Disjuncture and Difference in the Global Cultural Economy' Public Culture 2: 1990, pp 1-24.

Asheim, B. T. & Coenen, L 2005 Knowledge bases and regional innovation systems: Comparing Nordic Clusters in Research Policy 34 pp 1173-1190

Baldwin,R., 2006. Globalisation: the great unbundling(s). Report Prime Minster's Office, Economic Council of Finland.

Beckemann, M. 2011. The Potential for Innovation in the Swedish Food Sector. Lund: Lund University Press (diss.)

Bergek, A., S. Jacobsson, M. Hekkert, and K. Smith. 2007. Functionality of innovation systems as a rationale for, and guide to innovation policy. In: Innovation policy, theory and practice: An International handbook, ed. R. Smits, S. Kuhlmann and S. Shapira. Cheltenham, UK: Edward Elgar.

Breschi, F., and F. Malerba. 1997. Sectoral innovation system: Technological regimes, Schumpeterian dynamics and spatial boundary. In System of innovation, technologies, institutions and organizations, ed. C. Edqvist. London: Pinter.

Carlsson, B. and Stankiewicz, R. (1995): 'On the nature, function and composition of technological systems'. In: Carlsson, B. (ed.): *Technological Systems and Economic Performance: The Case of Factory Automation*. Dordrecht: Kluwer Academic Publishers. pp. 21-56.

Chesbrough, H.W, 2006. in Open Innovation: Researching a new Paradigm, eds Chesbrough; Vanhwerbeke; West, Oxford University Press.

Christensen, C.M. 1997 The innovator's dilemma. When new technologies cause great firms to fail. Harvard Business School Press, Boston.

Coenen, L 2006 Faraway, so close! The changing geographies of regional innovation. CIRCLE. Lund University, Lund.

Cooke, P. 2005 Regionally asymmetric knowledge capabilities and open innovation: Exploring Globalisation 2 -A new model of industry organization. Research policy 34(8):1128-49

Davies, C.A 2008: Reflexive Ethnography. New York: Routledge

Dahmén, E. 1987. 'Development Blocks' in Industrial Economics'. *Scandinavian Economic History Review*.

Dolphijn, R., 2005. Foodscapes. Towards a Deleuzian Ethics of Consumption. Delft. Eburon Publishers.

Drucker, P.F. 1994, The Theory of the Business, Harvard Business Review, September, pp. 95-105.

Ekelund, L., Fernqvist, F 2007 'Organic Apple Culture in Sweden'. The European Journal of Plant Science and Biotechnology

Freeman, C. 1987 Technology Policy and Economic Performance: lesson learned from Japan. London, Printer Publisher.

Fridlund, M, 1999. Switching Relations and Trajectories: The Development Procurement of the Swedish AXE Switching Technology, in: Charles Edquist, Leif Hommen & Lena Tsipouri, eds., Public Technology Procurement and Innovation, Economics of Science, Technology and Innovation v. 16. Norwell, Mass.: Kluwer Academic Publishers, 143–165.

Frykfors, C-O., Klofsten, M. 2011. Emergence of the Swedish Innovation System and the Support for Regional Entrepreneurship: A Socio-Economic Perspective, in Science and Technology Based Regional Entrepreneurship: Global Experience in Policy and Program Development, ed. Mian, S.A. Edward Elgar.

Frykfors, C-O, Jönsson, H; 2010 'Reframing the multilevel triple helix in a regional innovation system: a case of systemic foresight and regimes in renewal of Skåne's food industry' Technology Analysis & Strategic Management 2010, 22, 819-829

Geels, F. W. 2004. From sectoral systems of innovation to socio-technical system. Insights about dynamics and change from sociology and institutional theory. Research Policy 33 (6-7) pp 897-92.

Govindarajan, V., & Trimble, C. 2010 The other side of innovation. Solving the execution challenge. Harvard Business Press, Boston.

Grant, R. M 1996, 'Toward a knowledge based theory of the firm', Strategic Management Journal, 17, Special Issue Winter, 93–109.

Henning, M, Moodysson, J, Nilsson, M 2010. Innovation and regional transformation From clusters to new combinations. Malmö: Region Skåne

Von Hippel, E 2005 Democratizing Innovation. MIT Press Chambridge

Hodgson, G. 1988 Economics and Institutions, Polity Press.

Johnson, S. 2010 Where good ideas come from. The natural history of innovation, Penguin Group, New York.

Jönsson, H 2008. The Cultural Analyst – an Innovative Intermediary?' // ETN 2/2008. Lund: Etnologiska Institutionen.

Jönsson, H & Sarv, H 2008. On the Art of Shaping Futures. Lund, Skånes Livsmedelsakademi.

Kedia, S.; Van Willigen, J. 2005. eds. Applied Anthropology. Domains of Application. Westport, CT: Greenwood.

Klein Woolthuis, R.; Lankhuizenb, M.; Gilsing, V, 2005. A system failure framework for innovation policy design in Technovation 25 (6), pp 609-19.

Laestadius, S., Berggren, C., 2000. The embeddedness of industrial clusters : the strength of the path in the Nordic telecom system. Stockholm: Royal Institute of Technology.

Lagnevik, M 2006. 'Food innovation at interfaces: experience from the Öresund region', in Hulsink, W.; Dons, H. (Eds.) Pathways to High-Tech Valleys and Research Triangles. Wageningen: Wageningen UR Frontis Series, Volume 24

Law, J. 1999, ed. Actor Network Theory and After. Oxford, Blackwell.

Levinthal, D.A. 1991. Organizational adaption and environmental selections-interrelated processes of change. Organization Science 2: 140–45.

Lundvall, B-Å. 1992, National Innovation System : Towards a Theory of Innovation and Interactive Learing, London: Printer Publisher.

Malerba,F.; Orsenigo, L. 1997 Technological Regimes and Sectoral Patterns of Innovative Activities. Industrial and Corporate Change 6: 83-119.

Market Magasin 2008

Nilsson, M. 2008. A tale of two clusters. Sharing resources to compete. Lund: Lund Business Press.

Oresund Food. 2011 Redefining the Food sector. Copenhagen: Oresund Food

Pierre, J.; Peters, B.G. 2005. Governing Complex Societies: Trajectories and Scenarios. Basingstoke, Palgrave.

Pink, Sarah, 2005. ed. Applications of Anthropology. Professional Anthropology in the Twenty-first Century. Oxford, Berghahn Books.

Scott, W.R. 1995 Institutions and Organizations, Sage, Sunderland, P. L.; Denny, R. M., 2007. Doing Anthropology in Consumer Research. Walnut Creek: Left Coast Press.

Teece, D.J. 2007 Explicating Dynamic Capabilities: The Nature and Microfoundations of (Sustainable) Enterprise Performance, Strategic Management Journal 28, 1319- 1350.

Wallin, J. 2006. Business Orchestration: Strategic leadership in the Era of Digital Convergence. John Wiley & Sons Ltd.

Wastenson, L., T. Germundsson, P. Schlyter, the Swedish Society for Anthropology and Geography, Statistics Sweden, Lund University. Department of Social and Economic Geography, National Atlas of Sweden and National Land Survey of Sweden 1999. Sveriges nationalatlas (National Atlas of Sweden). Vällingby, Gävle, National Atlas of Sweden (SNA); Publisher and distributor of maps.

11

Functional Foods in Europe: A Focus on Health Claims

Igor Pravst
Nutrition Institute, Ljubljana
Slovenia

1. Introduction

The functional foods concept started in Japan in the early 1980s with the launch of three large-scale government-funded research programs on *systematic analyses and development of functional foods, analyses of physiological regulation of the functional food* and *analyses of functional foods and molecular design* (Ashwell 2002; Pravst et al. 2010). In 1991, in an effort to reduce the escalating cost of health care, a category of foods with potential benefits was established (Foods for Specified Health Use – FOSHU) (Ashwell 2002). In the USA, evidence-based health or disease prevention claims have been allowed since 1990, when the *Nutrition Labelling and Education Act* was adopted; claims have to be approved by the Food and Drug Administration (FDA) (Arvanitoyannis and Houwelingen-Koukaliaroglou 2005). *Codex Alimentarius Guidelines for the use of nutrition and health claims* were accepted in 2004, and amended in 2008 and 2009, followed by *Recommendations on the scientific basis of health claims* (Grossklaus 2009). In the European Union, harmonisation was achieved in 2006 with *Regulation (EC) No 1924/2006 on nutrition and health claims made on foods*, which requires authorization of all health claims before entering the market.

The definition of functional foods is an ongoing issue and many variations have been suggested (Arvanitoyannis and Houwelingen-Koukaliaroglou 2005). A consensus on the functional foods concept was reached in the European Union in 1999, when a working definition was established whereby a food can be regarded as functional if it is satisfactorily demonstrated to beneficially affect one or more target functions in the body beyond adequate nutritional effects in a way that is relevant to either an improved state of health and well-being or a reduction of disease risk. Functional foods must remain foods and demonstrate their effects when consumed in daily amounts that can be normally expected (Ashwell 2002). In practice, a functional food can be: an unmodified natural food; a food in which a component has been enhanced through special growing conditions, breeding or biotechnological means; a food to which a component has been added to provide benefits; a food from which a component has been removed by technological or biotechnological means so that the food provides benefits not otherwise available; a food in which a component has been replaced by an alternative component with favourable properties; a food in which a component has been modified by enzymatic, chemical or technological means to provide a benefit; a food in which the bioavailability of a component has been modified; or a combination of any of the above (Ashwell 2002). Regardless of the various definitions, the main purpose of functional food should be clear – to improve human health and well-being. However, health claims are a very convenient tool when it comes to marketing functional foods. The latest analyses show that in some European countries health claims are now used in over 15% of commonly eaten packaged foods (Lalor et al. 2010). Consumers are very sensitive to health-related communications and the use of health claims is unfortunately often connected with intentions to mislead them (Pravst 2011b). While this is clearly prohibited by law (Colombo 2010), companies are employing innovative strategies to avoid these rules so as to make a greater profit. Sophisticated regulation of this area is needed to provide consumers with high quality foods and non-misleading information. In this review the assessment of functional foods and their use in Europe will be critically discussed.

2. Consumer acceptance of functional foods and nutrition labelling

Health-related information on food labels may play a crucial role in informing consumers and influence their purchase decisions (Pothoulaki and Chryssochoidis 2009). However, consumer acceptance is far from unconditional and depends on beliefs about functional foods being viewed as a marketing stunt, as well as on familiarity and perceptions related to the perceived fit of ingredients and carrier or base products rather than on classical demographic characteristics (Verbeke 2010). In addition, it is unlikely one can count on consumer willingness to compromise on the taste of functional foods for health (Verbeke 2006). Recent studies show that the consumer might have a stronger preference for simple health statements compared to the implied benefits that result from consuming a functional food product (Bitzios et al. 2011). Indeed, consumers' attitudes to functional foods and the use of nutrition information or claims on food labels have been studied extensively in recent years (Bitzios et al. 2011; Hoefkens et al. 2011; Krutulyte et al. 2011; Pothoulaki and Chryssochoidis 2009; Urala and Lahteenmaki 2004; Verbeke 2005; Verbeke 2006; Verbeke et al. 2009) and some useful reviews of these topics are available (Bech-Larsen and Scholderer 2007; Pothoulaki and Chryssochoidis 2009; Verbeke 2010; Wills et al. 2009).
A recent multi-country study confirmed that consumers perceive some nutrients as qualifying (fibre, vitamins and minerals) or disqualifying (energy, fat, saturated fat, salt, sugars) (Hoefkens et al. 2011) and only small differences are observed between countries. Overall, consumers perceive the nutritional value of foods as important when selecting foods, particularly when it comes to qualifying nutrients (Hoefkens et al. 2011). Such a finding is quite interesting since public health nutrition awareness campaigns across Europe mainly target disqualifying nutrients, i.e. salt (Celemin et al. 2011). The marketing campaigns of the food industry can obviously have very strong effects, particularly if consumers perceive them as trustworthy. It must be noted that the labelling of nutrition information on foods in the EU is currently not mandatory, except for foods with nutrition and health claims. This is about to change under the accepted *Regulation on the provision of food information to consumers*. However, most food labels already contain at least back-of-pack nutrition labelling or related information, although there are significant differences between various food categories and countries (Bonsmann et al. 2010).

3. Assessment of functionality

The effects of particular functional foods on human health must be based on scientific evidence of the highest possible standard. However, because a direct measurement of the effect a food has on health is not always possible, the key issue is to identify critical biomarkers that can be used to monitor how biological processes are being influenced (Howlett 2008). If appropriate biomarkers are chosen it is possible to study the effect of the food on the final endpoints by measuring the biomarkers. The markers could be chosen to reflect either some key biological function or a key stage in development that is clearly linked to the study endpoint (markers of an intermediate endpoint) (Fig. 1) (Howlett 2008). In such a way we can study the long-term effects in the shorter term. Nonetheless, the biomarkers must be very carefully selected.

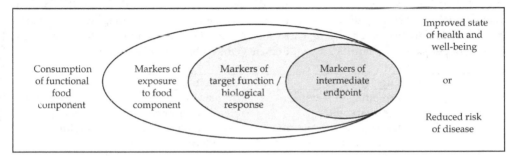

Fig. 1. The use of markers to link food consumption to health outcomes. Reproduced with permission of ILSI Europe (Howlett 2008).

In Europe the consensus on the criteria for the scientific support for claims on foods was reached in 2005 within PASSCLAIM – the Process for the Assessment of Scientific Support for Claims on Foods (Aggett et al. 2005). It delivered criteria to assess the scientific support for claims on foods. The set of criteria developed were derived through an iterative continuous improvement process that involved developing and evaluating criteria against key health claim areas of importance in Europe. These areas included diet-related cardiovascular disease, bone health, physical performance and fitness, body-weight regulation, insulin sensitivity and diabetes risk, diet-related cancer, mental state and performance and gut health and immunity. Expert groups reviewed the availability of indicators of health and disease states within their respective areas of expertise. They have demonstrated the limitations of existing biomarkers and have identified the need for better biomarkers (Aggett et al. 2005).

3.1 Food vs. medicine
In a time when the role of a healthy diet in preventing non-communicable diseases is well accepted, the borderline between foods and medicine is becoming very thin. The concept of foods should obviously go beyond providing basic nutritional needs. While products intended to cure diseases are classified as medicine, a healthy diet consisting of foods with functional properties can help promote well-being and even reduce the risk of developing certain diseases (Howlett 2008). However, a diet can only be healthy if the combination of individual foods is good. Limiting certain components (e.g. salt and saturated or trans fatty acids) or simply delivering an intake of nutrients cannot be regarded as a healthy diet. While it is easy to give advice on a *healthy diet*, it is much harder to define it for a particular need.

The borderline between food and medicine is shown in Table 1. As with traditional foods, functional foods should also not have serious side effects when consumed in reasonable amounts, although some functional foods can be intended for specific populations. This is contrary to medicine, which is intended for curing a disease or treating its symptoms, and used in exactly defined quantities to minimize side effects and assure therapeutic action. Food legislation also defines *food supplements* and *foods for special medical purposes*. *Food supplements* are concentrated sources of nutrients or other substances with a nutritional or physiological effect and are meant for supplementing a normal diet. Supplements are usually marketed as capsules, pastilles, tablets, pills and other similar forms of liquids and powders designed to be taken in measured small unit quantities. In contrast, *foods for special medical purposes* are specifically formulated, processed and intended for the dietary management of diseases, disorders or medical conditions of individuals who are being treated under medical supervision. These foods are intended for the exclusive or partial feeding of people whose nutritional requirements cannot be met by normal foods.

	Functional food	**Medicinal food**	**Medicine**
Usage	Providing nutritional needs, weight management, supporting health of digestive system, skeleton, heart, reducing risk of development of certain diseases.	Food management of some diseases or certain medical conditions, in which there are specific nutritional needs (i.e. problems with swallowing, lost appetite, postoperative nutrition)	Curing disease or treating its symptoms
To be consumed:	As wanted	As needed	As prescribed

Table 1. The borderline between food and medicine (Raspor 2011)

4. Functional foods in the context of regulation

To promote the use of any particular functional food its beneficial effects must be communicated to consumer. This is usually done through the use of nutrition and health claims in the labelling and advertising of foods. In this context, functional foods in Europe are probably most critically affected by the *Regulation (EC) No 1924/2006 on nutrition and health claims made on foods* (EC 2006). The regulation applies to all nutrition and health claims made in commercial communications, including trademarks and other brand names which could be construed as nutrition or health claims. The general principle is that claims should be substantiated by generally accepted scientific data, non-misleading and pre-approved on the EU level. Additionally, claims shall not give rise to doubt about the safety of other foods or encourage excessive consumption of a food.

4.1 Nutrition claims
The regulation defines a *nutrition claim* as any claim stating, suggesting or implying that a food has particular beneficial nutritional properties due to the calorific value or composition with respect to the presence (or absence) of specific nutrients or other substances. Only nutrition claims which are listed in the regulation are allowed, providing they are in accordance with the set conditions of use. Some examples of authorised nutrition claims and the conditions applying to them are listed in Table 2.

Nutrition claim	Conditions of use
low energy	less than 170 kJ/100 g for solids or 80 kJ/100 ml for liquids
low sodium	less than 0.12 g of sodium per 100 g or per 100 ml (different limits for waters)
source of fibre	at least 3 g of fibre per 100 g or at least 1.5 g of fibre per 100 kcal
high fibre	at least 6 g of fibre per 100 g or at least 3 g of fibre per 100 kcal
source of vitamins or minerals	a significant amount[1] of vitamins or minerals
high in vitamins or minerals	at least twice the amount of 'source of vitamins or minerals'
source of omega-3 fatty acids	at least 0.3g alpha-linolenic acid per 100g and per 100kcal, or at least 40mg of the sum of eicosapentaenoic acid and docosahexaenoic acid per 100g and per 100kcal.
high omega-3 fatty acids	at least 0.6g alpha-linolenic acid per 100g and per 100kcal, or at least 80mg of the sum of eicosapentaenoic acid and docosahexaenoic acid per 100g and per 100kcal.

Note: [1]As a rule, 15 % of the recommended allowance specified in the Annex of Directive 90/496/EEC on nutrition labelling for foodstuffs supplied by 100 g or 100 ml (or per package if the package contains only a single portion) should be taken into consideration in deciding what constitutes a significant amount.

Table 2. Examples of nutrition claims and the conditions applying to them (EC 2006)

Labelling of the content of vitamins and minerals must also comply with the *Council Directive 90/496/EEC on nutrition labelling for foodstuffs*, which was amended in 2008. In this directive vitamins and minerals which may be declared on labels and their recommended daily allowances (RDAs) are set (Table 3).

Vitamins	RDA	D-A-CH RI/AI[1]	Minerals and trace elements	RDA	D-A-CH RI/AI[1]
Vitamin A (μg)	800	1,000	Potassium (mg)	2,000	2,000
Vitamin D (μg)	5	5	Chloride (mg)	800	830
Vitamin E (mg)	12	14	Calcium (mg)	800	1,000
Vitamin K (μg)	75	70	Phosphorus (mg)	700	700
Vitamin C (mg)	80	100	Magnesium (mg)	375	350
Thiamin (mg)	1.1	1.2	Iron (mg)	14	10
Riboflavin (mg)	1.4	1.4	Zinc (mg)	10	10
Niacin (mg)	16	16	Copper (mg)	1	2.0-5.0
Vitamin B6 mg)	1.4	1.5	Manganese (mg)	2	1.0-1.5
Folic acid (μg)	200	400	Fluoride (mg)	3.5	3.8
Vitamin B12 μg)	2.5	3	Selenium (μg)	55	30-70
Biotin (μg)	50	30-60	Chromium (μg)	40	30-100
Pantothenic acid (mg)	6	6	Molybdenum (μg)	50	50-100
			Iodine (μg)	150	150-200

Notes: [1]RI- Recommended intake, AI- Estimated value for adequate intake (provided value for adult men, 25-51 years).

Table 3. Vitamins and minerals which may be declared and their recommended daily allowances (RDAs) as set in Council Directive 90/496/EEC on nutrition labelling for foodstuffs (amended in 2008) in comparison to D-A-CH reference values (D-A-CH 2002)

4.2 Health claims

Regulation (EC) 1924/2006 defines *health claim* as any claim which states, suggests or implies that a relationship exists between a food category, a food or one of its constituents and health. All health claims should be authorised and included in the list of authorised claims. The quantity of the food and pattern of consumption required to obtain the claimed beneficial effect should be included in the labelling and reasonably expected to be consumed in context of a varied and balanced diet. The claim must be specific, based on generally accepted scientific data and well understood by the average consumer. Reference to general, non-specific benefits of the nutrient or food for overall good health or health-related well-being may only be made if accompanied by a specific health claim (EC 2006). All authorised health claims are listed in a public *Community Register of nutrition and health claims made on food* which includes the wordings of claims and the conditions applying to them, together with any restrictions. Basically, the regulation distinguishes three categories of health claims (Table 4). General function claims describe the role of a food in body functions, including the sense of hunger or satiety and not referring to children, while disease risk reduction claims state that the consumption of a food or food constituent significantly reduces a risk factor in the development of a human disease. The regulation also mentions claims referring to children's development and health, for which no further definition is given.

General function health claims	Claims referring to children's development and health	Reduction of disease risk claims
Claims not referring to children and describing the role of a nutrient or other substance in growth, development and the functions of the body; or psychological and behavioural functions; or slimming or weight control or a reduction in the sense of hunger or an increase in the sense of satiety or to the reduction of the available energy from the diet.	No further definition provided.	Claims that state, suggest or imply that the consumption of a food category, a food or one of its constituents significantly reduces a risk factor in the development of a human disease.

Table 4. Types of health claims according to EC regulations (EC 2006)

5. Evaluation and authorisation of health claims

The regulation requires authorization of all health claims by the European Commission through the Comitology procedure, following scientific assessment and verification of a claim by the European Food Safety Authority (EFSA) (Pravst 2010). Claims referring to children's development and health, reduction of disease risk claims and general function claims which are based on newly developed scientific evidence are submitted directly by companies. In cases where health claim substantiation is based on (unpublished) proprietary data and if a claim cannot be substantiated without such data, the applicant can request 5 years of protection of such data.

In addition to the described procedure of health claims submission by applicants, all health claims which were on the market prior to 2006 were included in the so-called *Article 13 evaluation process*. Lists of general function claims on the market were provided by EU member states in collaboration with the industry and included in a consolidated list which formed the basis for the EFSA evaluation. In 2009 it became clear that the process of evaluating existing general function health claims would be much more demanding than expected. After examining over 44 000 claims supplied by the EU member states, the EFSA has received a consolidated list of over 4600 general function claims which entered the evaluation process. The evaluation of most claims was finished in 2011 with 341 published scientific opinions, and providing scientific advice for 2758 general function health claims, about 20% being favourable. While some claims were withdrawn, about 1500 claims on so-called *botanicals* have been placed on hold by the European Commission pending further consideration on how to proceed with these.

Scientific and technical guidance for the preparation and presentation of an application for authorisation of a health claim (EFSA 2011c) and *General guidance for stakeholders on the evaluation of health claims* (EFSA 2011a) were prepared by the EFSA. Scientific substantiations of claims are performed by taking into account the totality of the available pertinent scientific data and by weighing up the evidence, in particular whether:

- the effect is relevant for human health;
- there is an established cause-and-effect relationship between the consumption of the food and the claimed effect in humans;
- the effect has been shown on a study group which is representative of the target population; and
- the quantity of the food and pattern of consumption required to obtain the claimed effect could reasonably be achieved as part of a balanced diet.

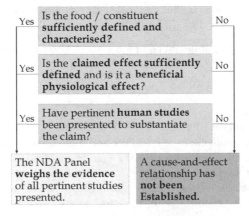

Fig. 2. Key questions addressed by the EFSA in the scientific evaluation of health claims (EFSA 2011a)

The process of the implementation of health claims legislation has involved a steep learning curve which is far from complete. While the EFSA has had to cope with an unprecedented and unforeseen workload, coupled with very short deadlines, the industry is financing expensive trials which are often still not being performed using standards that would enable successful substantiation. More specific guidance is therefore being released to help industry to substantiate health claims, starting with *Guidance for health claims related to gut and immune function* (EFSA 2011b). Guidance for health claims related to (1) antioxidants, oxidative damage and cardiovascular health, (2) bone, joints and oral health, (3) appetite ratings, weight management and blood glucose concentrations, (4) physical performance, and (5) cognitive function is also being developed. The key questions addressed by the EFSA in the scientific evaluation of health claims are presented in Fig. 2.

5.1 The wording of a health claim

The wording of a health claim must be specific, it must reflect the scientific evidence and it must be well understood by the average consumer. In the process of the scientific evaluation the EFSA can propose changing the wording to reflect the scientific evidence, but such wording is sometimes hard to understand by the general population. An example of such a procedure can be shown with a health claim for specific water-soluble tomato concentrate and maintenance of a healthy blood flow (Table 5). The EFSA considered that the wording which was proposed by the applicant did not reflect the scientific evidence because only measures of platelet aggregation have been used, whereas blood flow and circulation also depend on many other factors. It changed the wording to reflect the scientific evidence, although the Commission later reworded it to be understood by consumers (Pravst 2010).

	Claim wording
Applicant proposal:	Helps to maintain a healthy blood flow and benefits circulation
EFSA's proposal:	Helps maintain normal platelet aggregation
Authorised by the EC:	Helps maintain normal platelet aggregation, which contributes to healthy blood flow

Table 5. The wording of a health claim for specific water-soluble tomato concentrate and maintenance of a healthy blood flow (EFSA 2009; Pravst 2010).

When discussing the wording of health claims it should also be noted that, for reduction of disease risk claims, the wording should refer to the specific risk factor for disease and not to disease alone.

5.2 Food characterisation

The sufficient characterisation of foods or food constituents is important for the proper scientific evaluation of a health claim application and must enable authorities to control the market. In the scientific evaluation process the characterisation is needed to identify the food or food constituent, to define the appropriate conditions of use, and to connect it with the provided scientific studies. These studies should be performed using the same food or food constituent (which should also be sufficiently characterised). The lack of characterisation was one of the most common reasons for the EFSA's non-favourable opinions regarding general function claims (Pravst 2010).

In relation to characterisation it is necessary to distinguish between a specific formulation, specific constituent and a combination of constituents. All combinations must be characterised in detail, particularly in relation to the active constituents. Beside the physical and chemical properties and the composition, the analytical methods must also be provided (EFSA 2011a). In cases where variations in composition could occur, variability from batch to batch should be addressed together with stability with respect to storage conditions during shelf life (EFSA 2011a). Where applicable it is useful to show that a constituent is bioavailable and provide a rationale of how the constituent reaches the target site.

For microorganisms genetic typing should be performed at the strain level with internationally accepted molecular methods and the naming of strains according to the International Code of Nomenclature. Depositing samples in an internationally recognised culture collection for control purposes has been suggested (EFSA 2011a). The stability of the microorganisms and the influence of the food matrix on their activity should be studied.

For plant products the scientific name of the plant should be specified, together with the part of the plant used and details of the preparation used, including details of the extraction, drying etc. It is beneficial when the applicant can show that the composition of the plant-derived product can be controlled by analyses of specific chemical ingredients. For example, the above mentioned specific tomato concentrate was characterised on the basis of a clearly described production manufacturing process of tomatoes (*Lycopersicum esculentum*) together with a detailed chemical specification of the most important components and demonstrated batch-to-batch reproducibility (EFSA 2009). Chemical compounds which have been shown to have a beneficial effect *in vitro* were identified and quantified using the HPLC-MS technique, and the presence of unspecified constituents was limited. Several chemical and physical characteristics were assessed during stability testing, including breakdown products and the microbiological status. The bioactive components were shown to survive and to retain their activity *in vitro* over typical product shelf lives and when the product was included in specified matrixes.

Study population	Function
Subjects with untreated hypertension	maintaining normal blood pressure (for general population)
Overweight and obese subjects	weight reduction (for general population)
Patients with functional constipation	bowel function (for general population)
Irritable Bowel Syndrome (IBS) patients	bowel function (for general population)
Irritable Bowel Syndrome (IBS) patients	gastro-intestinal discomfort (for general population)
Subjects with immunosuppression	immune function (for population groups considered to be at risk of immunosuppression)

Table 6. Some specific study populations which were found as appropriate for scientific substantiation of health claims (EFSA 2011b; Pravst 2010)

5.3 Specific conditions of use

The quantity of the food or food constituent and pattern of consumption must be specified together with possible warnings, restrictions on use and directions for use. It is important that the consumer can consume enough food as part of a balanced diet to obtain the claimed effect (EFSA 2011a). The target population must be specified and in relation to this, it is critical that the specific study group in which the evidence was obtained is also representative of the target population for which the claim is intended (Pravst 2010). A key question is whether the results of studies on patients can be extrapolated to the target population. A judgement on this is made on a case-by-case basis, but in most cases patients were not found to be an appropriate study group. Usually, such studies are not considered as pertinent. In cases where studies are not performed on a representative of the target population, evidence must be provided that the extrapolation can be performed. No scientific conclusions can be drawn from studies on patients with genetically and functionally different cells and tissues. Some examples of specific study populations which were found as appropriate for scientific substantiation of health claims are listed in Table 6.

5.4 Relevance of the claimed effect

The claimed effect should be clearly defined and relevant to human health (Pravst 2010). This can be demonstrated with the example of the *Caralluma fimbriata* extract, and its effect on one's waist circumference (EFSA 2010c). Studies showed a statistically significant reduction in waist circumferences, but a reduction in one's waist circumference is not a beneficial physiological effect if it is not accompanied by an improvement in the adverse health effects of excess abdominal fat. On the contrary, only a slight improvement in parameters which are widely accepted as important to human health can be recognised as key evidence, such as in the case of a general functional claim for the role of omega-3 in maintaining normal blood pressure (EFSA 2009i). The EFSA concluded that high doses (3 g per day) of docosahexaenoic (DHA) and eicosapentaenoic acid (EPA) may have smaller, but statistically significant, effects in normotensives of about 1 mmHg; better results were observed in subjects with untreated hypertension.

5.5 Scientific substantiation of the claimed effect

Human data are critical for substantiating a claim and particular attention is paid to whether such studies are pertinent to the claim. Studies need to be carried out with the subject product with similar conditions of use in a study group representative of the population group and using an appropriate outcome measure of the claimed effect (EFSA 2011a). Using appropriate outcome measures can be a challenge because a limited number of validated biomarkers is available (Cazaubiel and Bard 2008). Biomarkers are characteristics that are objectively measured and evaluated as indicators of normal biological processes, pathogenic processes or pharmacologic responses to therapeutic intervention. Well-performed human intervention trials are particularly important for successful substantiation. Double-blind, randomised, placebo-controlled trials are considered the *gold standard* not only for the substantiation of disease risk reduction claims but also for general function claims (Pravst 2010). During the scientific evaluation such trials are assessed critically to assure there are no weaknesses. A good study design, proper performance, well-defined statistics and appropriate statistical power (enough subjects) are key issues in this context. In some cases, non-blind studies are also acceptable, particularly in the case of non-processed foods where blinding is not possible. This was confirmed recently in the case of a general function claim application for dried plums and their laxative effect (EFSA 2010h) – a study in which subjects free of gastrointestinal and eating disorders were randomised to consume either dried plums or grape juice was found to be pertinent. Nevertheless, taking the results of other studies into account there was insufficient evidence to establish a cause-and-effect relationship.

Human observational studies and data from studies in animals or model systems are considered only as supporting evidence.

6. Components of functional foods

As already mentioned, limiting certain food components or simply delivering nutrient intake cannot be regarded as a healthy diet. However, the functional effects of foods are usually studied in relation to their composition and bioactive components (in some cases also with isolated components). In this chapter over 300 scientific opinions on general function health claims (for the ones which are currently present on the European market) were reviewed. On the basis of these opinions and discussion within member states the European Commission will prepare a list of substantiated health claims and their conditions of use.

6.1 Vitamins, minerals and trace elements

When talking about essential nutrients we need to consider that there is a well-established consensus among scientists on many functions of such nutrients. In the evaluation of health claims we may rely on such a consensus and in such cases it may not be necessary to review the primary scientific studies on the claimed effect of the food (EFSA 2011a). On such bases many general function health claims for vitamins, minerals and trace elements received favourable opinions in the assessment process (Table 7). In most cases the proposed condition of use of such a claim is that the food is at least a source of the nutrient (15% of the RDA specified in Table 3 per 100 g or 100 ml, or per package if the package contains only a single portion).

Several essential nutrients were recognised to possess antioxidant activity, which is commonly communicated on functional foods. While the current evidence does not support the use of antioxidant supplements in the general population or in patients with certain diseases (Bjelakovic et al. 2008), some food components are indeed included in the antioxidant defence system of the human body, which is a complex network including endogenous antioxidants and dietary antioxidants, antioxidant enzymes, and repair mechanisms, with mutual interactions and synergetic effects among the various components (EFSA 2009f). For example, vitamin C functions physiologically as a water-soluble antioxidant and plays a major role as a free radical scavenger. On such a basis a cause-and-effect relationship has been established between the dietary intake of vitamin C and the protection of DNA, proteins and lipids from oxidative damage. Other antioxidants include vitamin E, riboflavin, copper, manganese and selenium (Table 7).

A series of vitamins and trace elements is involved in the functioning of the human immune system, but their ability to promote immunity function is questionable, especially in populations with adequate intake. For example, zinc deficiency is associated with a decline in most aspects of immune function; lymphopaenia and thymic atrophy are observed, cell mediated and antibody mediated responses are reduced (EFSA 2009~). Additionally, zinc deficiency appears to induce apoptosis, resulting in a loss of B-cell and T-cell precursors within the bone marrow. Adequate zinc status is necessary for natural killer cell function and zinc deficiency renders people more susceptible to infections. A cause-and-effect relationship has been established between the dietary intake of zinc and the normal function of the immune system but it was noted that there is evidence for inadequate intake of zinc in the general EU population (EFSA 2009~). A function in the immune system was also recognised for copper, selenium and various vitamins (vitamins A, D, B6, B12, C and folate) (Table 7). In the case of vitamin D it was also concluded that it contributes to healthy inflammatory response (EFSA 2010*f*).

Health realtionship[1]	Vitamins	Minerals and trace elements
Function as antioxidant	Vitamin E (EFSA 2010,,) Riboflavin (EFSA 2010}) Vitamin C (EFSA 2009{)	Cu (EFSA 2009h) Mn (EFSA 2009r) Se (EFSA 2009v)
Function in immune system	Vitamin A (EFSA 2009x) Vitamin D (EFSA 2010*f*) Vitamin B6 (EFSA 2009z) Folate (EFSA 2009k) Vitamin B12 (EFSA 2009y) Vitamin C (EFSA 2009{)	Cu (EFSA 2009h) Fe (EFSA 2009n) Se (EFSA 2009v) Zn (EFSA 2009~)
Function in brain or nervous system	Thiamine (EFSA 2009c; EFSA 2010) Riboflavin (EFSA 2010}) Niacin (EFSA 2009s; EFSA 2010w) Pantothenate (EFSA 2009t; EFSA 2010x) Vitamin B6 (EFSA 2009z; EFSA 2010) Folate (EFSA 2010j) Vitamin B12 (EFSA 2010€) Biotin (EFSA 2010d) Vitamin C (EFSA 2009{; EFSA 2010,)	Cu (EFSA 2009h) Fe (EFSA 2009n; EFSA 2010n) I (EFSA 2010m) K (EFSA 2010{) Mg (EFSA 2009q; EFSA 2010r) Zn (EFSA 2009~)
Function in bone or teeth	Vitamin D (EFSA 2009g; EFSA 2009 |) Vitamin K (EFSA 2009})	Ca (EFSA 2009f; EFSA 2009g) Mg (EFSA 2009q) P (EFSA 2009u) Cu (EFSA 2009h) K (EFSA 2010{) Mn (EFSA 2009r) Zn (EFSA 2009~) F- (EFSA 2009j)
Function in skin, hair or connective tissues	Vitamin A (EFSA 2009x) Riboflavin (EFSA 2010}) Niacin (EFSA 2009s) Biotin (EFSA 2009e)	Cu (EFSA 2009h) I (EFSA 2009m) Se (EFSA 2009v; EFSA 2010~) Zn (EFSA 2010†)
Function in vision	Vitamin A (EFSA 2009x) Riboflavin (EFSA 2010})	Zn (EFSA 2009~)
Function in muscle	Vitamin D (EFSA 2010*f*)	Ca (EFSA 2009f) Cu (EFSA 2009h) K (EFSA 2010{) Mg (EFSA 2009q)

Health realtionship[1]	Vitamins	Minerals and trace elements
Function in blood, haemoglobin and oxygen transport	Vitamin A (EFSA 2009x) Vitamin K (EFSA 2009}) Riboflavin (EFSA 2010}) Vitamin B6 (EFSA 2009z) Folate (EFSA 2009k) Vitamin B12 (EFSA 2009y) Vitamin C (EFSA 2009{)	Ca (EFSA 2009f) Cu (EFSA 2009h) Fe (EFSA 2009n)
Function in cell division & differentiation	Vitamin A (EFSA 2009x) Vitamin D (EFSA 2009|) Folate (EFSA 2009k) Vitamin B12 (EFSA 2009y)	Ca (EFSA 2010e) Fe (EFSA 2009n) Mg (EFSA 2009q) Zn (EFSA 2009~)
Function in regulation of hormones	Pantothenic acid (EFSA 2009t) Vitamin B6 (EFSA 2010)	I (EFSA 2009m) Se (EFSA 2009v) Zn (EFSA 2010†)
Function in metabolism of nutrients	Thiamine (EFSA 2009c) Riboflavin (EFSA 2010}) Niacin (EFSA 2009s) Pantothenate (EFSA 2009t) Vitamin B6 (EFSA 2009z; EFSA 2010) Vitamin B12 (EFSA 2009y) Biotin (EFSA 2009e) Vitamin C (EFSA 2009{)	Ca (EFSA 2009f) Cr (EFSA 2010f) Cu (EFSA 2009h) Fe (EFSA 2009n) I (EFSA 2009m) Mg (EFSA 2009q) Mn (EFSA 2009r) Mo (EFSA 2010v) P (EFSA 2009u) Zn (EFSA 2009~; EFSA 2010†)

Notes: [1] See references for details on specific health claims. A reference to general, non-specific benefits of the nutrient for overall good health or health-related well-being may only be made if accompanied by a specific health claim.

Table 7. Selection of general function health claims for vitamins, minerals and trace elements as assessed by the EFSA

According to a recent Irish study, brain function is most commonly communicated on soft drinks. In fact, all known water-soluble vitamins, and several minerals and essential elements (copper, iron, iodine, potassium, magnesium and zinc) are recognised as important in this context (Table 7). Functions which were favourably assessed by the EFSA include maintenance of normal function of the nervous system, cognitive function, psychological functions, neurological functions, and reduction of tiredness and fatigue (depending on the particular nutrient). The role of vitamin B6 in the maintenance of mental performance was also evaluated. Human studies have shown the effect of vitamin B6 on symptoms of depression, cognition, ageing, premenstrual syndrome and memory performance, however the daily doses for supplementation ranged from 40 - 600 mg (EFSA 2009z), well above the Tolerable Upper Intake Level (UL) (25 mg). Such a claim is not appropriate because it would encourage excess consumption of vitamin B6.

The effect of nutrition on bone health is well established. The maximum attainment of peak bone mass achieved during growth and the rate of bone loss with advancing age are the two principal factors affecting adult bone health. With increasing life expectancy the epidemiology of bone-health related conditions is changing drastically and represents a major public health threat in the Western world. The careful formulation of functional foods represents an important step in the promotion of bone health and consequently on the quality of one's life, but to minimise health risks for consumers both the positive and the negative effects of active ingredients should be considered when developing such products. Calcium and vitamin D are considered the most important constituents of functional foods for the support of bone health (Earl et al. 2010; Palacios 2006) and received favourable opinions for maintaining normal bone and teeth. Additionally, the role of vitamin D in maintaining normal blood calcium concentrations and absorption and utilisation of Ca and P was confirmed (EFSA 2009 |). In the last decade, the role of vitamin K in γ-carboxylation of osteocalcin has also been recognised (Ikeda et al. 2006; Katsuyama et al. 2004) as about 10-30% of osteocalcin in the healthy adult population is in an under-carboxylated (and therefore inactive) state (Vermeer et al. 2004). Other nutrients recognised in maintaining normal bone health are magnesium, zinc, manganese, potassium, copper and phosphorus (Table 7). However, the enrichment of (functional) foods or drinks with phosphorus is controversial as its intake can easily exceed the recommendations and a bigger intake might have adverse effects on bone health (Pravst 2011b). Therefore, both health and ethical concerns arise as to whether such claims should be allowed, even though science is not yet clear on this issue. A useful solution in such cases would be to authorise the claim with more specific conditions of use.

Most of the nutrients with a function in bone health were also recognised as important in the maintenance of normal teeth (vitamin D, calcium, magnesium and phosphorus). Additionally, the beneficial role of fluoride for tooth health (by counteracting hydroxyapatite demineralisation and supporting remineralisation) is widely accepted as fluoride can replace hydroxyl ions in the hydroxyapatite crystal lattice of a tooth's tissues and make it more resistant to acid exposure (EFSA 2009j).

Biotin is the only vitamin that has received a favourable opinion for its role in the maintenance of normal hair (EFSA 2009e). While there were no studies available in which improvement in hair loss and hair quality were studied using objective methods, it is known that the symptoms of biotin deficiency include thinning hair and progression to a loss of all hair, including eyebrows and lashes (EFSA 2009e). Copper, selenium and zinc were also recognised as important in the maintenance of normal hair (Table 7). Other functions of these nutrients include the maintenance of normal nails, skin or mucous membranes, while a role in the normal formation of connective tissue was determined only for manganese (EFSA 2010s) and copper (EFSA 2009h).

Vitamin A and compounds with pro-vitamin A activity (i.e. beta-carotene) are recognised to have a function in maintaining normal vision. Retinal is required by the eye for the transduction of light into neural signals which are necessary for vision and without an adequate level of vitamin A in the retina night blindness occurs (EFSA 2009x). In a different way vitamin A deficiency leads to a reduction in mucus production by the goblet cells of the conjunctival membranes and the cornea becomes dry. Riboflavin (EFSA 2010}) and zinc (EFSA 2009~) also received favourable opinions.

With muscle weakness being a major clinical syndrome of vitamin D deficiency, vitamin D is the only vitamin with a favourable opinion about its role in the maintenance of normal muscle function (EFSA 2010*f*). Clinical symptoms of the deficiency include proximal muscle weakness, diffuse muscle pain and gait impairments such as a waddling way of walking.

Dietary intake of calcium, magnesium, potassium and copper was also connected with muscle function (Table 7). In a similar manner the importance of sodium was confirmed; however, current sodium intake in the general EU population is more than adequate and directly associated with a greater likelihood of increased blood pressure, which in turn has been directly related to the development of cardiovascular and renal diseases (EFSA 2011y). Promoting sodium intake with the use of health claims is in obvious conflict with current targets for a reduction in dietary sodium intakes (Cappuccio and Pravst 2011).

Several nutrients have received favourable opinions for their function in blood, formation of haemoglobin or oxygen transport (Table 7). In haemoglobin, iron allows reversible binding of oxygen and its transport in the erythrocytes to the tissues. The most common consequence of iron deficiency is microcytic anaemia. The role of iron in oxygen transport, normal formation of red blood cells and haemoglobin is obviously well established (EFSA 2009n). Interestingly, there is still a significant prevalence of iron deficiency in Europe, especially among children and women of reproductive age and during pregnancy. Because dietary iron is absorbed as Fe^{2+} (and not as Fe^{3+}), reducing agents, including vitamin C, promote non-haem iron absorption by keeping it reduced. The function of vitamin C and non-haem iron absorption therefore received a favourable opinion (EFSA 2009f). Other vitamins and minerals with a confirmed cause-and-effect relationship were vitamin A in the metabolism of iron, vitamin K and calcium in normal blood coagulation, folate in normal blood formation, copper in normal iron transport, and riboflavin, vitamin B6 and B12 in the normal formation of red blood cells.

Other specific functions of vitamins, minerals and trace elements which are rarely communicated on food labelling were also included in the evaluation process. The ones with favourable outcomes include a function in cell division & differentiation, regulation of hormones and metabolism of nutrients (Table 7). In addition, some of the following very specific health claims were assessed: selenium and the maintenance of normal spermatogenesis (EFSA 2009v); zinc and normal fertility and reproduction, normal DNA synthesis; normal vitamin A metabolism and normal acid-base metabolism (EFSA 2009~); vitamin C and normal collagen formation (EFSA 2009f) and regeneration of the reduced form of vitamin E (EFSA 2010,); vitamin B6 and normal glycogen metabolism (EFSA 2009z) and normal cysteine synthesis (EFSA 2010); folate and normal maternal tissue growth during pregnancy (EFSA 2009k); phosphorus and normal functioning of cell membranes (EFSA 2009u); magnesium and electrolyte balance (EFSA 2009q); and, calcium and normal functioning of digestive enzymes (EFSA 2009f).

Vitamin B6, B12 and folic acid were recognised as important for their contribution to normal homocysteine metabolism (EFSA 2009k; EFSA 2010€; EFSA 2010). Deficiencies in these vitamins lead to impaired homocysteine metabolism causing mild, moderate, or severe elevations in plasma homocysteine (depending on the severity of the deficiency) as well as the coexistence of genetic or other factors that interfere with homocysteine metabolism (EFSA 2010). In addition to these vitamins, a contribution to normal homocysteine metabolism has also been established for betaine (EFSA 2011g) and choline (EFSA 2011l). Choline can function as a precursor for the formation of betaine, which acts as a methyl donor in the remethylation of homocysteine in the liver by the enzyme betaine-homocysteine methyltransferase (EFSA 2011g). Choline can be biosynthesised in our liver and is therefore not a vitamin; however, most men and postmenopausal women need to consume it in their diets (Zeisel and Caudill 2010). The nutritional need for choline greatly depends on gender, age and genetic polymorphisms. For this reason the EFSA was unable to propose conditions of use; however, it was noted that a nutrient content claim has been authorised in the US, based on the adequate intake for adult males (550 mg of choline per day). Furthermore, choline has also received a favourable opinion for the maintenance of normal liver function and its contribution to normal lipid metabolism (EFSA 2011l). In relation to betaine it must be mentioned that the application for the use of betaine as a novel food ingredient in the EU was rejected, mainly due to concerns over the safety of betaine with long term use (EFSA 2005). Despite this, betaine as a constituent of traditional foods is not considered a novel food.

In practice, a functional food can also be a food from which a component has been removed to provide benefits not otherwise available (Ashwell 2002). Good examples of such foods are foods low or very low in sodium, whose consumption helps to maintain normal blood pressure (EFSA 2011p). In fact, sodium intake and blood pressure demonstrate a close and consistent direct relationship (Cappuccio and Pravst 2011) and there is extensive evidence to support this (He and MacGregor 2010). A 4.6g reduction in daily dietary salt intake decreases BP by about 5.0/2.7 mmHg in hypertensive individuals and by 2.0/1.0 mmHg in normotensive people. Dose-response effects have been consistently demonstrated in adults and children (He and MacGregor 2003). A 5g higher salt intake is associated with a 17% greater risk of total cardiovascular disease and a 23% greater risk of stroke (Strazzullo et al. 2009).

6.2 Fats, fatty acids and fatty acid composition

Fats are a major contributor to total energy intakes in most Western diets, supplying 35-40% of food energy through the consumption of 80-100g of fat per day (Geissler and Powers 2005). All fat sources contain mixtures of saturated, monounsaturated and polyunsaturated fatty acids, but the proportions vary significantly according to the source. Besides being a source of energy, fats also play other diverse roles in the human body. Two fatty acids, linoleic acid and α-linolenic acid are essential in the human diet as they cannot be synthesised in mammalian tissues. The human body can synthesise eicosapentaenoic acid (EPA) and docosahexaenoic acid (DHA) from α-linolenic acid, but the synthesis is not very efficient.

The balance between saturated and unsaturated fats in the diet is known to influence circulating cholesterol concentrations (WHO 2003). Current dietary recommendations for adults in the EU are to limit intake of fats to 20-35 energy % (E%) and to keep the intake of saturated fatty acid (SFA) and trans fatty acid as low as possible (EFSA 2010b). The proposed adequate intakes are: 4 E% for linolenic acid; 0.5 E% for α-linolenic acid (ALA); and 250 mg for EPA+DHA.

In Table 8 the proposed conditions of use of health claims related to fats, fatty acids and fatty acid composition are listed. Cholesterol related claims are one of the most common categories of health claims on the market (Lalor et al. 2010). The role of fatty acid composition in cholesterol management has been confirmed and the proposed conditions of use should stimulate a reduced intake of saturated fatty acids also through product reformulation (EFSA 2011o; EFSA 2011). Such a product should contain at least 30% less saturated fatty acids compared to a similar product. The role of linolenic and α-linolenic acid in cholesterol management also received favourable opinions, where a food contains at least 15% of the proposed labelling reference intake.

Food component[1]	Function	Conditions of use	Reference
Fat	Normal absorption of fat-soluble vitamins	no conditions of use can be defined	(EFSA 2011n)
Saturated fatty acids	Management of cholesterol	reduced amounts of saturated fatty acids by at least 30% compared to a similar product	(EFSA 2011f) (EFSA 2011o)
Linoleic acid	Management of cholesterol	15% of the proposed labelling reference intake values of 10g linoleic acid per day	(EFSA 2009p)
ALA	Management of cholesterol	15% of the proposed labelling reference intake value of 2g ALA per day	(EFSA 2009b)
Plant sterols	Management of cholesterol	0.8g per day	(EFSA 2010z)
DHA/EPA	Management of blood triglycerides	2-4g per day	(EFSA 2009i)
	Management of blood pressure	3g per day	(EFSA 2009i)
	Heart health	250mg per day	(EFSA 2010i)
DHA	Vision	250mg per day	(EFSA 2010g)
	Brain function	250mg per day	(EFSA 2010g)

Notes: [1]ALA - α-linolenic acid; DHA - docosahexaenoic acid; EPA - eicosapentaenoic acid.

Table 8. Conditions of use health claims related to fats, fatty acids and fatty acid composition as assessed by the EFSA

Plant sterols have been known for several decades to cause reductions in plasma cholesterol concentrations (St Onge and Jones 2003). These are also well known by consumers and are one of the most well-known functional ingredients. Plant sterols are structurally very similar to cholesterol, except that they always contain a substituent at the C24 position on the sterol side chain (Ling and Jones 1995). It is generally assumed that cholesterol reduction results directly from the inhibition of cholesterol absorption through the displacement of cholesterol from micelles. Their concentrations in mammalian tissue are normally very low - primarily due to poor absorption from the intestine and faster excretion from the liver compared to cholesterol (Ling and Jones 1995). The LDL-cholesterol lowering effects of plant sterols and stanols have been reviewed several times, lately by Demonty and co-workers (Demonty et al. 2008). In the general health claim evaluation process plant sterols and stanols were confirmed to be helpful in maintaining normal blood cholesterol levels when the food provides at least 0.8g of plant sterols or stanols daily in one or more servings (EFSA 2010z). Such products may not be nutritionally appropriate for pregnant and breastfeeding women, or for children under the age of five (EFSA 2010z). It must be mentioned that the related reduction of disease risk claims were already authorised for both plant sterols and stanol esters. Lowering blood cholesterol and reference to cholesterol as a risk factor in the development of coronary heart disease can be communicated on foods which provide a daily intake of 1.5-2.4g plant sterols/stanols. Most studies of cholesterol lowering effects were conducted with plant sterols or stanols added to foods such as margarine-type spreads, mayonnaise, and dairy products. Studies are showing an increase in the consumption of plant sterols in Europe. The consumption of foods enriched with plant sterols was recently studied in Belgium; the results indicate that plant sterol-enriched food products are also consumed by the non-target group and that efficient communication tools are needed to better inform consumers about the target group of enriched products, the advised dose per day and alternative dietary strategies to lower blood cholesterol level (Sioen et al. 2011). Although intakes of plant sterols have not produced significant adverse effects, it is not known whether constant consumption at high levels would have any toxic effects (St Onge and Jones 2003). It is therefore logical to advise the food industry to also formulate cholesterol-management foods with other effective functional ingredients, particularly with specific dietary fibres, being a more common constituent of the human diet (see chapter *6.3 Carbohydrates and dietary fibre*).

Intervention studies have demonstrated the beneficial effects of preformed n-3 long-chain polyunsaturated fatty acids (DHA, EPA) on recognised cardiovascular risk factors, such as a reduction in plasma triacylglycerol concentrations and blood pressure, albeit in quite high daily dosages (EFSA 2010b). The proposed daily dose for the management of blood triglycerides and the management of blood pressure is 2-4g and 3g of DHA and EPA respectively (Table 8). The question remains, however, as to whether such claims will be authorised. Lower daily dosages of DHA and EPA are required to maintain normal cardiac function (250mg DHA/EPA), normal vision and brain function (250 mg DHA).

6.3 Carbohydrates and dietary fibre

According to the degree of polymerisation we categorise carbohydrates into sugars (monosaccharides and disaccharides), oligosaccharides (3-9 residues), and polysaccharides with over 9 monomeric residues.

6.3.1 Glycaemic carbohydrates

Glycaemic carbohydrates are digested and absorbed in the human small intestine and provide glucose to body cells as a source of energy. According to their degree of polymerisation, these are sugars and some oligosaccharides and polysaccharides. The main glycaemic carbohydrates in the diet are glucose and fructose (monosaccharides), sucrose and lactose (disaccharides), malto-oligosaccharides and starch (polysaccharides) (EFSA 2010a). Glycaemic carbohydrates provide about 40% of energy intake in average Western diets, with a desirable level at around 55% (Geissler and Powers 2005). Far more glycaemic carbohydrates are consumed in developing countries.

Glucose is the preferred energy source for most body cells and the brain requires glucose for its energy needs. An intake of 130g of dietary glycaemic carbohydrates per day for adults is estimated to cover the glucose requirement of the brain (EFSA 2010a). In the health claim evaluation process a cause-and-effect relationship has been established between the consumption of glycaemic carbohydrates and the maintenance of normal brain function (EFSA 2011r). However, when talking about carbohydrates as constituents of functional foods we are usually discussing either lowering their amount or the functional properties of indigestible polysaccharides (dietary fibre).

Sugar replacement in foods or drinks with xylitol, sorbitol, mannitol, maltitol, lactitol, isomalt, erythritol, D-tagatose, isomaltulose, sucralose or polydextrose was found to be efficient in the reduction of post-prandial blood glucose responses as compared to sugar-containing products (EFSA 2011 l). Lowering sugar levels is also considered beneficial in maintaining the mineralisation of teeth (EFSA 2011s; EFSA 2011|). However, excessive amounts of polyols may result in a laxative effect and this should be communicated to the consumer with an appropriate (and mandatory) advisory statement.

6.3.2 Dietary fibre

Dietary fibre is mostly derived from plants and is composed of complex, non-starch carbohydrates and lignin that are not digestible within the small intestine – since mammals do not produce enzymes capable of their hydrolyses into constituent monomers (Turner and Lupton 2011). Dietary fibre is considered a non-nutrient and contributes no calories to our diet as it reaches the colon intact. However, in the colon dietary fibres are available for fermentation by the resident bacteria, and the metabolites released can be used to meet some of the energy requirements.

Regulatory fibre is defined in *Council Directive 90/496/EEC on nutrition labelling for foodstuffs* as carbohydrate polymers with three or more monomeric units, which are neither digested nor absorbed in the human small intestine and belong to the following categories: (1) edible carbohydrate polymers naturally occurring in the food as consumed; (2) edible carbohydrate polymers which have been obtained from raw food material by physical, enzymatic or chemical means and which have a beneficial physiological effect demonstrated by generally accepted scientific evidence; (3) edible synthetic carbohydrate polymers which have a beneficial physiological effect demonstrated by generally accepted scientific evidence. The EFSA has evaluated several specific dietary fibres for their role in the management of cholesterol, glycaemic response and gut health (Table 9).

The beneficial health effects of water-soluble dietary fibre mainly relate to their ability to improve viscosity of the meal bolus in the small intestine and thus to delay the absorption of nutrients. A lowering of blood cholesterol was established for beta-glucans, chitosan, glucomannan, hydroxypropyl methylcellulose (HPMC), pectins and guar gum when at least 3-10g of fibre was consumed daily (Table 9). Among these, the lowest daily dosage (3g in one or more servings) is required for non-processed or minimally processed beta-glucans from specific sources (EFSA 2009d). The structural features of beta-glucans greatly influence their molecular shape (conformation) and the behaviour of the polysaccharide in a solution, including viscosity. The primary source of beta-glucans and the production processes therefore have a great impact on their functionality. Beta-glucans are widely distributed as non-cellulosic matrix phase polysaccharides in cell walls of the *Poaceae*, which consist of the grasses and commercially important cereal species (Burton and Fincher 2009). Chemically, these are (1,3;1,4)-β-D-Glucans - as they consist of unbranched and unsubstituted chains of (1,3)- and (1,4)-β-glucosyl residues. Their physicochemical and functional properties in cell walls are influenced by the ratio of (1,4)-β-D-glucosyl residues to (1,3)-β-D-glucosyl residues. An example of beta-glucans with a recognised cholesterol effect includes beta-glucans from barley (AbuMweis et al. 2010; Talati et al. 2009), a cereal grain derived from the *Hordeum vulgare*. The proportion between β-(1,3) and β(1,4) linkages is 30 and 70%, respectively (Jadhav et al. 1998). In the polymer chain the blocks of 2 or 3 contiguous (1,4) linkages are separated by single (1,3) linkages; however, blocks of 2 or more adjacent (1,3) linkages are absent (Jadhav et al. 1998). Because the (1,3) linkages occur at irregular intervals the overall shape of the polysaccharide is irregular, which reduces its tendency to pack into stable, regular molecular aggregates and enable the formation of stable viscous solutions. It must be noted that beta-glucans are useful particularly in the production of hard functional foods such as bread, toasts, pasta, extruded flakes, crisps etc. On the market there are also some beta-glucan-enriched liquid functional foods (i.e. yoghurts) and drinks with a labelled cholesterol management effect; however, the length of the polymeric chain of beta-glucans contained in such products is shortened with chemical or enzymatic procedures. Such an ingredient has a lower capacity to form viscous solutions, and consequently a lower efficiency. Thus, the use of a health claim is not allowed in these cases (EFSA 2009d).

Dietary fibre	Management of cholesterol	Management of glycaemic response	Gut health	Reference
Beta glucans	3g per day (non-processed beta-glucans)	4g for each 30g of available carbohydrate (oats and barley beta-glucans)	high in fibre[2]	(EFSA 2009d) (EFSA 2011f) (EFSA 2011u)
Chitosan	3g per day			(EFSA 2011k)
Glucomannan	4g per day			(EFSA 2009l)
HPMC[1]	5g per day	4g per meal		(EFSA 2010l)
Pectins	6g per day	10g per meal		(EFSA 2010y)
Guar gum	10g per day			(EFSA 2010k)
Arabinoxylan		8g of arabinoxylan-rich fibre per 100g of available carbohydrates		(EFSA 2011e)
Resistant starch		14% of total starch as resistant starch (high carbohydrate baked foods)		(EFSA 2011w)
Rye fibre			high in fibre[4]	(EFSA 2011x)
Wheat bran fibre			high in fibre[2] 10g per day[3]	(EFSA 2010…)

Notes: [1]HPMC - hydroxypropyl methylcellulose; [2]Increase in faecal bulk; [3]Reduction in intestinal transit time; [4]Changes in bowel function; Reference to general, non-specific benefits of the nutrient for overall good health or health-related well-being may only be made if accompanied by a specific health claim.

Table 9. Conditions of use for general function health claims of various dietary fibres as proposed by the EFSA

Decreasing the magnitude of elevated blood glucose concentrations after consuming carbohydrate-rich food is a critical target in the production of low glycaemic index (GI) foods. It is well established that the management of glycaemic responses is beneficial to human health, particularly for people with impaired glucose tolerance (which is common among the general population). Various specific dietary fibres were recognised as beneficial in the reduction of post-prandial glycaemic responses. In most cases, the rationale for such a function is, similar to cholesterol lowering, related to their ability to achieve improved viscosity of the meal bolus in the small intestine. This enables a delay in the absorption of sugars - which is considered beneficial as long as post-prandial insulinaemic responses are not disproportionally increased. In the evaluation process the EFSA found a cause-and-effect relationship between the mentioned effect and the consumption of beta-glucans from oats and barley, HPMC, pectins, arabinoxylan produced from wheat endosperm and resistant starch. Various conditions of use were proposed (Table 9).

Both water-soluble and water-insoluble dietary fibres are also known to support gut health through changes in bowel function. Reduced transit time, more frequent bowel movements, increased faecal bulk or softer stools may be a beneficial physiological effect, provided these changes do not result in diarrhoea (EFSA 2011x). On the basis of such changes the functional role of rye fibre was confirmed to contribute to normal bowel function in foods providing at least 6g of such fibre per 100g (or at least 3g per 100kcal), being *high in fibre* (EFSA 2011x). The ability to increase faecal bulk has also been confirmed for wheat, oat and barley grain fibre (Table 9). Similar to the previous case, the proposed conditions of use are that the food is *high in fibre*. It is well established that the bulking effect of dietary fibre is closely related to the physico-chemical properties of the fibre, and in that way to the degree of fermentation by the gut microbiota in the large intestine (EFSA 2011u). The insoluble components of fibre are minimally degraded by colonic bacteria and thus remain to trap water, thereby increasing faecal bulk. In contrast, the bulking effects of soluble dietary fibre are determined by the higher extent of fermentation, and thus an increase in the bacterial mass in faeces (EFSA 2011u). A somewhat different support of gut health can be based on the ability of dietary fibre to reduce intestinal transit time. Such a function was found for wheat bran fibre, which increases the water holding capacity of the contents of the intestine, increases intestinal and pancreatic fluid secretion and thus increases the velocity of chyme displacement through the intestine if at least 10g per day is consumed (EFSA 2010...). A similar effect, but with a different mechanism of action, was also confirmed for lactulose – a synthetic sugar used in the treatment of constipation (EFSA 2010p). In the colon, lactulose is broken down to lactic acid and to small amounts of acetic and formic acids by the action of beta-galactosidases from colonic bacteria. This process leads to an increase in osmotic pressure and a slight acidification of the colonic content, causing an increase in stool water content and a softening of stools (EFSA 2010p). However, due to the medicinal use of lactulose in some EU countries the authorisation of such a health claim on foods is questionable.

6.3.3 Prebiotics

Prebiotics were defined as non-digestible food ingredients that beneficially affect the host by selectively stimulating the growth or activity of one or a limited number of bacterial species already resident in the colon, and thus attempt to improve host health (Gibson and Roberfroid 1995). An intake of prebiotics can modulate the colonic microbiota by increasing the number of specific bacteria and thus changing the composition of the microbiota.

Consumers perceive prebiotics as having health benefits. The *Guidance on the implementation of regulation No 1924/2006 on nutrition and health claims made on foods* specifies that a claim is a health claim if, in the naming of the substance or category of substances, there is a description or indication of functionality or an implied effect on health. The examples provided include the claim *contains prebiotic fibres*. Such claims should therefore only be used if the food contains prebiotic fibres with a scientifically proven effect and when such a claim is accompanied by a specific health claim. In practice, this is not yet the case and such claims are still very common on the market. Only a few (prebiotic) fibres have received favourable opinions from the EFSA. One example is the already-mentioned oat and barley grain fibres - whose bulking effects are determined by the higher extent of fermentation and thus an increase in the bacterial mass in faeces. On such a basis their ability to increase faecal bulk has been confirmed (EFSA 2011u) (Table 9). Several fibres were evaluated for other functions, i.e. for their role in maintaining healthy gastro-intestinal function by increasing the number of bifidobacteria in the gut. In some cases the studies clearly demonstrated a significantly increased number of bifidobacteria in the gut; however, there was no direct evidence provided that changes in the number of bifidobacteria in the gut are beneficial for gut function (EFSA 2009a). For such claims the beneficial effect needs to be shown.

6.4 Other food or food constituents
6.4.1 Antioxidants
Numerous food constituents possess antioxidant activity, yet, in the health claim evaluation process, only a few of them were assessed with a favourable outcome. As in the prebiotics case, the claim *contains antioxidants* is considered a health benefit and should be accompanied by a specific health claim. However, apart from vitamins and trace elements (Table 7), until now only olive oil polyphenols have been recognised to possess antioxidant activity which is beneficial to human health, specifically in the protection of LDL particles from oxidative damage (Covas et al. 2006a; Covas et al. 2006b; EFSA 2011v; Marrugat et al. 2004; Weinbrenner et al. 2004). Hydroxytyrosol and its derivatives (e.g. oleuropein complex and tyrosol) are the key compounds with such activity, and to bear the claim olive oil it should contain enough of them to provide 5mg of these compounds daily. It was noted that the concentrations in some olive oils does not allow consumption of such an amount of polyphenols in the context of a balanced diet (EFSA 2011v). Many other known antioxidants (including flavonoids and flavonols, lycopene, lutein etc.) received unfavourable opinions, mostly because of poor characterisation, non-specific health effects or poor evidence for such effects. In a recent draft of the *Guidance on the scientific requirements for health claims related to antioxidants, oxidative damage and cardiovascular health* the EFSA noted that it is not established that changes in the overall antioxidant capacity of plasma exert a beneficial physiological effect in humans. A beneficial physiological effect will therefore need to be proven for any specific antioxidant for a successful substantiation.

6.4.2 Probiotics
The claim *contains probiotics* is also considered a health benefit and should be accompanied by a specific health claim. However, the EFSA has not released a favourable opinion in relation to live organisms other than for live yoghurt cultures in yoghurt, which were shown to improve the digestion of lactose in yoghurt in individuals with lactose maldigestion (discussed below). The main reasons for the unfavourable opinion were that the microorganisms were not properly characterised (in either the health claim application or the supporting study of the claimed effect), or that there was poor evidence of the beneficial effect. However, many probiotics health claims were returned for re-evaluation and there has been a call to provide additional data for scientific evaluation. Discussion about the results of the re-evaluation has been very speculative, but it is clear that further research will be needed to support a beneficial physiological effect in humans in most (if not all) cases. The specific functions will need to be properly addressed.

6.4.3 Sport nutrition

A series of food constituents have been evaluated for sport related general function health claims. Caffeine was shown to contribute to increase in endurance performance, reduction in the rated perceived exertion during exercise and increased alertness (Table 10). However, it is noted that for children consumption of a dose of 5mg/kg body weight could result in transient behavioural changes, such as increased arousal, irritability, nervousness or anxiety. In relation to pregnancy and lactation, moderation of caffeine intake, from whatever source, is advisable (EFSA 2011i). The role of vitamin C in the maintenance of normal function of the immune system during and after intense physical exercise was also confirmed, but due to the high daily dosage (200mg vitamin C per day in addition to the usual diet) it remains a question if such a claim will be authorised. In some EU countries such a dosage of vitamin C is considered a medicinal use. Other health claims which received favourable opinions include protein and the maintenance of muscle mass, water and the maintenance of normal thermoregulation, and creatine and an increase in physical performance during short-term, high intensity, or repeated bouts of exercise. The proposed condition for use of these claims is presented in Table 10. The role of carbohydrate-electrolyte solutions in the enhancement of water absorption during exercise and in the maintenance of endurance performance was also evaluated. It was proposed that in order to bear the claim a carbohydrate-electrolyte solution should contain 80-350 kcal/L from carbohydrates, and at least 75 % of the energy should be derived from carbohydrates which induce a high glycaemic response, such as glucose, glucose polymers and sucrose. In addition, these beverages should contain between 460mg/L and 1150mg/L of sodium, and have an osmolality between 200-330mOsm/kg water (EFSA 2011j).

Food or ingredient	Function	Conditions of use	Reference
Vitamin C	Maintenance of normal function of the immune system during and after intense physical exercise	200mg vitamin C per day[1] (in addition to the usual diet)	(EFSA 2009f)
Protein	Maintenance of muscle mass	Source of protein	(EFSA 2010|)
Caffeine	Increase in endurance performance	3mg/kg body weight (one hour prior to exercise)	(EFSA 2011h)
	Increase in endurance capacity		(EFSA 2011h)
	Reduction in the rated perceived exertion during exercise	4mg/kg body weight (one hour prior to exercise)	(EFSA 2011h)
	Increased alertness	75mg caffeine per serving (adults)	(EFSA 2011i)
Creatine	Increase in physical performance during short-term exercise	3g per day	(EFSA 2011m)
Water	Maintenance of normal thermoregulation	2.0 L per day	(EFSA 2011~)

Food or ingredient	Function	Conditions of use	Reference
Carbohydrate-electrolyte solutions	Enhancement of water absorption during exercise	Energy: 80-350kcal/L from carbohydrates and at least 75 % from sugars; Sodium: 460-1,150mg /L Osmolality: 200-330mOsm/kg	(EFSA 2011j)
	Maintenance of endurance performance		(EFSA 2011j)

Notes: [1]In some countries 200mg of vitamin C daily is considered a medicinal use.

Table 10. Conditions of use for sport related general function health claims as proposed by the EFSA

6.4.4 Weight management

For food producers weight-management products are appealing as these can be marketed successfully. Weight loss can be interpreted as the achievement of a normal body weight in previously overweight subjects. In this context, weight loss in overweight subjects without the achievement of a normal body weight is also considered beneficial to health (EFSA 2010t). However, only a few favourable opinions of the EFSA were published (EFSA 2010o; EFSA 2010t; EFSA 2011}). It was established that substituting two daily meals with meal replacements helps to lose weight in the context of energy restricted diets. In order to bear the claims, a food should contain a maximum of 250 kcal/serving and comply with specifications laid in legislation covering foods intended for use in energy-restricted diets for weight reduction (EFSA 2010t). Similarly, it was established that replacing the usual diet with a very low calorie diet (VLCD) helps to lose weight (EFSA 2011}). VLCDs are diets which contain energy levels between 450 and 800 kcal per day, and 100% of the recommended daily intakes for vitamins and minerals. They should contain not less than 50g of high-quality protein, should provide not less than 3g of linoleic acid and not less than 0.5g alpha-linolenic acid, with a linoleic acid/alpha-linolenic acid ratio between 5 and 15, and should provide not less than 50g of available carbohydrates (CODEX STAN 203-19956) (EFSA 2011}). VLCDs are typically used for 8-16 weeks.

6.4.5 Foods for individuals with symptomatic lactose maldigestion

Lactose maldigestion is a common condition characterised by intestinal lactase deficiency. It is most prevalent in Asian, African, Hispanic and Indian populations, but is also common in Europe. Most people with primary lactose maldigestion are usually able to tolerate small amounts of lactose (EFSA 2009o). Ingested lactose is hydrolysed by an enzyme of the microvillus membrane of the enterocytes, called lactase. It is split into the monosaccharides glucose and galactose, which are rapidly and completely absorbed within the small intestine. In persons with lactose maldigestion, undigested lactose reaches the colon where it is degraded to lactic acid, acetic acid, water and carbon dioxide by intestinal bacteria (EFSA 2011q). In some lactose maldigesters this can elicit symptoms of lactose intolerance, which may develop one to three hours after the ingestion of lactose. These symptoms include abdominal pain, bloating, flatulence and diarrhoea. It was established that consumption of foods with reduced amounts of lactose may help to decrease gastro-intestinal discomfort caused by lactose intake in lactose intolerant individuals (EFSA 2011q). However, it was not possible to propose a single condition of use because of the great variation in the individual tolerances of lactose intolerant individuals. Additionally, an improvement in lactose digestion may be of interest to lactose intolerant subjects (EFSA 2009o). A cause-and-effect relationship has been established between the externally administered lactase enzymes and breaking down lactose in individuals with symptomatic lactose maldigestion, which can alleviate lactose intolerance symptoms. The recommended dose was 4500 FCC (Food Chemicals Codex) units with each lactose-containing meal. It was noted that the dose may have to be adjusted to individual needs for lactase supplementation and consumption of lactose-containing products (EFSA 2009o). Live yoghurt cultures in yoghurt were also shown to improve the digestion of lactose in yoghurt in individuals with lactose maldigestion (EFSA 2010q). The effect has been confirmed in a number of human studies and is based on the ability of specific bacteria to produce active β-galactosidase enzymes in the digestive tract. It was proposed that in order to bear the claim, the yoghurt should contain at least 10^8 colony-forming units (CFU) live starter microorganisms (*Lactobacillus delbrueckii* subsp. *bulgaricus* and *Streptococcus thermophilus*) per gram (EFSA 2010q).

6.4.6 Chewing gums

It was established that there is a cause-and-effect relationship between the consumption of sugar-free chewing gum and plaque acid neutralisation, the maintenance of tooth mineralisation, and a reduction of oral dryness (EFSA 2009w). The solubility of tooth hydroxyapatite crystals drops with the lowering of pH and buffering of acids, and limiting the duration of periods of pH drop can prevent demineralisation and promote remineralisation of the hydroxyapatite crystals. Acid is produced in plaque through the fermentation of carbohydrates by acid-producing bacteria and studies have shown that chewing a sugar-free chewing gum enhances saliva flow and counteracts pH drops upon sugar-induced acid production. Chewing for at least 20 minutes after meals may be needed to obtain the beneficial effect. In the absence of fermentable carbohydrates, no clinically relevant reduction on plaque pH may be expected by the consumption of sugar-free chewing gum (EFSA 2009w). In a separate opinion it was established that sugar-free chewing gum with carbamide contributes to plaque acid neutralisation over and above the effect achieved with sugar-free chewing gums without carbamide (EFSA 2011z). At least 20mg carbamide should be added per piece to communicate such claims.

6.4.7 Borderline substances

In some cases the functionality of possible food ingredients enters the borderline between food and medicine. A series of such cases has also been revealed in the evaluation of general function health claims. The most obvious example is the evaluation of monacolin K from rice fermented with the red yeast *Monascus purpureus* (EFSA 2011t). Red yeast rice is a traditional Chinese food product which is still a dietary staple in many Asian countries. Depending on the *Monascus* strains used and the fermentation conditions, it may contain Monacolin K, in the form of hydroxy acid and lactone (also known as lovastatin), which received a favourable opinion for the maintenance of normal blood LDL-cholesterol concentrations (EFSA 2011t). On the EU market there are a number of lovastatin-containing medicinal products and it remains controversial if it is possible to set appropriate conditions of use that would put this product into a food category. In any case, red yeast rice has not been traditionally consumed as a food in Europe and must go through a novel food authorisation procedure. However, its use is already possible in some specific food categories (i.e. in food supplements) in some EU countries. Somewhat similar is the case of melatonin. While in a number of EU countries this hormone is considered a medicinal product, in some countries it is sold as a food supplement. In the health claim evaluation process it was confirmed that melatonin helps to reduce the time taken to fall asleep (EFSA 2011d) and contributes to the alleviation of subjective feelings of jet lag (EFSA 2010u). It is proposed that 1mg of melatonin is consumed close to bedtime (EFSA 2011d).

7. Nutrient profiles

To exclude the use of nutrition and health claims on foods with overall poor nutritional status nutrient profiles should be established, including the exemptions which food or certain categories of food must comply with in order to bear the claims (EC 2006). Unfortunately, this part of the legislation has not yet been implemented, even though the scientific criteria for this were prepared on time (EFSA 2008). The stakeholders have obviously been quite effective in lobbying against the setting of nutrient profiles and there is little evidence of progress in this area since 2009. Currently, it is not even clear if nutrition profiles will be implemented at all (Cappuccio and Pravst 2011). Nevertheless, the producers of functional food must be aware that the overall composition of a final product should provide health benefits. Particular care should be directed towards those nutrients with the greatest public health importance for EU populations, such as saturated fatty acids, sodium and sugar, intakes of which generally do not comply with nutrient intake recommendations in many Member States (EFSA 2008).

8. Quality of functional foods

The quality and safety of functional foods is entirely the responsibility of the producer but can be controlled by national authorities. In practice, such controls mainly focus on assuring adequate safety by controlling for contaminants and additives (Pravst and Žmitek 2011). Nutritional composition is usually not considered a health risk and is therefore less controlled. In fact, while labelling requirements have been in existence in many countries for more than a decade, analyses of many food constituents are still challenging. The EU legislation currently concentrates on regulating the use of vitamins and minerals, while the use of other substances with a nutritional or physiological effect is not regulated in detail. When discussing the safety and quality control assessment of foods containing particular ingredients with biological activity we must distinguish products on the basis of their active ingredients; i.e. chemically stable dietary minerals, less stable vitamins, chemical compounds other than vitamins and minerals, living microorganisms (i.e. probiotics) etc. (Pravst and Žmitek 2011). The appropriate content of these ingredients in final products must be achieved using suitable production standards (including quality control of both raw materials and the final product) and stability during the manufacturing process and shelf life. The low content of an ingredient in a final product is often connected with either improper manufacturing (inappropriate purity or insufficient ingredients used in the production, uncontrolled manufacturing conditions, and inappropriate formulation) or its decomposition during shelf life. In situations where decomposition occurs either during manufacturing or shelf life this not only misleads the consumer but might also create increased health risks due to the possibility of the uncontrolled formation of by-products. In contrast, when not enough ingredients are used during manufacturing the primary concern is about misleading the consumer. In some cases, such scenarios can also pose a risk to human health, i.e. in instances of adulteration.

A significant problem that arises in the evaluation of the quality of functional foods is that there are no generally accepted tolerances for the declaration of nutrients and other active ingredients in the EU (DG SANCO 2006), although guidelines on this issue have been accepted in some countries (Table 11). The task of setting tolerable margins was identified as a priority 10 years ago during the discussion that led to the adoption of Directive 2002/46/EC on food supplements, but this goal has not yet been achieved (Pravst and Žmitek 2011). Nevertheless, there is a general agreement that such tolerances should be defined at the Community level in order to avoid trade barriers and ensure consumer protection (DG SANCO 2006).

Country	Tolerances for added nutrients [1]		
	Minerals	Water-soluble vitamins	Fat-soluble vitamins
Belgium	90% - 120%	90% - 120%	90% - 120%
Denmark	80% - 150%	80% - 150%	80% - 150%
France	80% - 200%	80% - 200%[2]	80% - 200%[2]
Italy	75% - 100%	80% - 130%[2]	80% - 130%[2]
Slovenia	80% - 150%	80% - 150%[2]	80% - 130%
Netherlands	80% - 150%	80% - 200%	80% - 200%
United Kingdom	50% - 200%	50% - 200%	70% - 130%

Notes: [1] if legislation prescribes minimum and maximum values for the addition of nutrients, the analysed amount must not exceed these limits; [2] with exceptions

Table 11. Tolerance values accepted or practiced in some EU member states (CIAA 2007; DG SANCO 2006; IVZ 2009).

9. Conclusions and future issues

In the last few years the regulation of nutrition and health claims has been one of the top food-related themes discussed in Europe. Regulation covering these areas was indeed required. Protecting consumers against misleading claims, along with the harmonisation of the European market, were the key issues that needed to be addressed. The regulation targets functional foods, a concept which emerged in Japan about 20 years ago to reduce the escalating health care costs with a category of foods offering potential health benefits, although from a different perspective. At the time, the USA and some EU-member states were also at the frontier of developments, but the European Union as a whole was lagging far behind. It was decided that the use of pre-approved evidence-based health claims on food labels would serve us best and in the ensuing time there has been a focus on creating a list of approved claims. In such a system, functional foods are basically defined by the limitations and the opportunities for the use of claims.

Essential nutrients are clearly the winner of the evaluation process. In cases where a well-established consensus among scientists exists on the biological role of a nutrient, the EFSA relied on that consensus and confirmed the cause-and-effect relationship without reviewing the primary scientific studies. In most cases, the proposed condition of use is to include at least 15% of the RDA of the nutrient per 100g of final product, to enable the use of health claims for such a nutrient. This will enable products which are a source of at least one such nutrient to communicate health claims, even in cases where there is no deficiency in the population. The consumer will recognise such a nutrient as a health added value and there are concerns that such claims might flood the market and enable consumers to be legally misled. While the authorisation of such health claims may pose a risk of misleading the consumer, there are also cases where concerns related to public health arise. Such an example is the claim concerning phosphorus and its role in the maintenance of normal bones. The intake of phosphorus easily exceeds the recommendations and a bigger intake might have adverse effects on bone health (Pravst 2011b). Therefore, both health and ethical concerns arise as to whether such claims should be allowed, even though science is not yet clear on this issue. A useful solution in such cases would be to authorise claims with more specific conditions of use.

Foods promoted with claims may be perceived by consumers as having a health advantage over other foods and this may encourage consumers to make choices which directly influence their total intake of individual nutrients in a way which would run counter to scientific advice. The regulation aims to avoid a situation where claims mask the overall nutritional status of a food product and confuse consumers when trying to make healthy choices in the context of a balanced diet with the introduction of nutrient profiles. However, these profiles have not yet been established and it is not even clear if they will be implemented at all. In a situation in which food producers have the power to stimulate the consumption of foods with a poor nutritional status we must count on their commitment to serving consumers (Pravst 2011a).

All of the above issues suggest that we are still far from the target. The scientific substantiation of health claims for non-essential ingredients is very important and substantial additional research will be needed to get new claims approved. A detailed examination of all the concerns raised by the EFSA in its published opinions, together with some additional advice about expectations related to the scientific substantiation of health claims, should result in the improved quality of clinical testing for bioactive components and functional foods.

10. Acknowledgments

I gratefully acknowledge Tobin Bales for providing help with the language. The work was financially supported by Ministry of Agriculture, Forestry and Food of the Republic of Slovenia, and Slovenian Research Agency and (Contract 1000-11-282007, Research project V7-1107).

11. References

AbuMweis, S.S., Jew, S. & Ames, N.P. (2010). Beta-glucan from barley and its lipid-lowering capacity: a meta-analysis of randomized, controlled trials. *Eur J Clin Nutr.* 64(12): 1472-1480.

Aggett, P.J., Antoine, J.M., Asp, N.G., Bellisle, F., Contor, L. et al. (2005). PASSCLAIM - Consensus on criteria. *European Journal of Nutrition.* 44 5-30.

Arvanitoyannis, I.S. & Houwelingen-Koukaliaroglou, M. (2005). Functional foods: A survey of health claims, pros and cons, and current legislation. *Crit. Rev. Food Sci. Nutr.* 45(5): 385-404.

Ashwell, M. (2002) *Concepts of Functional Foods*. ILSI - International LIfe Sciences Institute, Brussels.

Bech-Larsen, T. & Scholderer, J. (2007). Functional foods in Europe: consumer research, market experiences and regulatory aspects. *Trends in Food Science & Technology*. 18(4): 231-234.

Bitzios, M., Fraser, I. & Haddock-Fraser, J. (2011). Functional ingredients and food choice: Results from a dual-mode study employing means-end-chain analysis and a choice experiment. *Food Policy*. 36(5): 715-725.

Bjelakovic, G., Nikolova, D., Gluud, L.L., Simonetti, R.G. & Gluud, C. (2008). Antioxidant supplements for prevention of mortality in healthy participants and patients with various diseases. *Cochrane database of systematic reviews*.(2): CD007176.

Bonsmann, S.S.G., Celemin, L.F., Larranaga, A., Egger, S., Wills, J.M. et al. (2010). Penetration of nutrition information on food labels across the EU-27 plus Turkey. *European Journal of Clinical Nutrition*. 64(12): 1379-1385.

Burton, R.A. & Fincher, G.B. (2009). (1,3;1,4)-beta-D-Glucans in Cell Walls of the Poaceae, Lower Plants, and Fungi: A Tale of Two Linkages. *Molecular Plant*. 2(5): 873-882.

Cappuccio, F.P. & Pravst, I. (2011). Health claims on foods: Promoting healthy food choices or high salt intake? *British Journal of Nutrition*. In press (doi:10.1017/S0007114511002856)

Cazaubiel, M. & Bard, J.M. (2008). Use of biomarkers for the optimization of clinical trials in nutrition. *Agro Food Industry Hi-Tech*. 19(5): 22-24.

Celemin, L.F., Bonsmann, S.S.G., Carlsson, E., Larranaga, A. & Egger, S. (2011). Mapping public health nutrition awareness campaigns across Europe. *Agro Food Industry Hi-Tech*. 22(1): 38-40.

CIAA (2007) *CIAA manual on nutrition labelling (URL: http://www.ibec.ie/Sectors/FDII/FDII.nsf/vPages/Regulatory_Affairs~Consumer_Informati on~ciaa-nutrition-manual/$file/CIAA%20Nutrition%20Labelling%20Manual.doc, Accessed: June 2010)*. CIAA, Brussels.

Colombo, M.L. (2010). Conventional and new foods: health and nutritional claims The new functional role of food. *Agro Food Industry Hi-Tech*. 21(1): 42-44.

Covas, M.I., de la Torre, K., Farre-Albaladejo, M., Kaikkonen, J., Fito, M. et al. (2006a). Postprandial LDL phenolic content and LDL oxidation are modulated by olive oil phenolic compounds in humans. *Free Radical Biology and Medicine*. 40(4): 608-616.

Covas, M.I., Nyyssonen, K., Poulsen, H.E., Kaikkonen, J., Zunft, H.J.F. et al. (2006b). The effect of polyphenols in olive oil on heart disease risk factors - A randomized trial. *Annals of Internal Medicine*. 145(5): 333-341.

D-A-CH (2002) *Reference Values for Nutrient Intake*. German Nutrition Society, Bonn.

Demonty, I., Ras, R.T., van der Knaap, H.C.M., Duchateau, G.S.M.J., Meijer, L. et al. (2008). Continuous dose-response relationship of the LDL-cholesterol lowering effect of phytosterol intake. *The Journal of Nutrition*. 139 271-284.

DG SANCO. 2006. Directive 90/496/EEC on Nutrition Labelling for Foodstuffs: Discussion Paper on Revision of Technical Issues (URL: http://ec.europa.eu/food/food/labellingnutrition/nutritionlabel/discussion_pap er_rev_tech_issues.pdf, Accessed: June 2010). Brussels, European Commission, DG SANCO.

Earl, S., Cole, Z.A., Holroyd, C., Cooper, C. & Harvey, N.C. (2010). Dietary management of osteoporosis throughout the life course. *Proceedings of the Nutrition Society*. 69(01): 25-33.

EC. 2006. Regulation (EC) No 1924/2006 on nutrition and health claims made on foods.

EFSA. (2005). Opinion of the Scientific Panel on Dietetic Products, Nutrition and Allergies on a request from the Commission related to an application concerning the use of betaine as a novel food in the EU. *The EFSA Journal*. 191 1-17. (doi:10.2903/j.efsa.2005.191)

EFSA. (2008). The setting of nutrient profiles for foods bearing nutrition and health claims pursuant to Article 4 of the Regulation (EC) No 1924/2006 - Scientific Opinion of the Panel on Dietetic Products, Nutrition and Allergies. *The EFSA Journal*. 191 1-17. (doi:10.2903/j.efsa.2005.191)

EFSA. (2009a). Bimuno™ and help to maintain a healthy gastro-intestinal function - Scientific substantiation of a health claim related to BimunoTM and help to maintain a healthy gastro-intestinal function pursuant to Article 13(5) of Regulation (EC) No 1924/2006. *The EFSA Journal*. 1107 1-10. (doi:10.2903/j.efsa.2009.1107)

EFSA. (2009b). Opinion on the substantiation of health claims related to alpha linolenic acid and maintenance of normal blood cholesterol concentrations (ID 493) and maintenance of normal blood pressure (ID 625) pursuant to Article 13(1) of Regulation (EC) No 1924/2006. *EFSA Journal*. 7(9): 1252. (doi:10.2903/j.efsa.2009.1252)

EFSA. (2009c). Scientific Opinion on substantiation of health claims related to thiamine and energy-yielding metabolism (ID 21, 24, 28), cardiac function (ID 20), function of the nervous system (ID 22, 27), maintenance of bone (ID 25), maintenance of teeth (ID 25), maintenance of hair (ID 25), maintenance of nails (ID 25), maintenance of skin (ID 25) pursuant to Article 13(1) of Regulation (EC) No 1924/2006. *EFSA Journal*. 7(9): 1222. (doi:10.2903/j.efsa.2009.1222)

EFSA. (2009d). Scientific Opinion on the substantiation of health claims related to beta glucans and maintenance of normal blood cholesterol concentrations (ID 754, 755, 757, 801, 1465, 2934) and maintenance or achievement of a normal body weight (ID 820, 823) pursuant to Article 13(1) of Regulation (EC) No 1924/2006. *EFSA Journal*. 7(9): 1524. (doi:10.2903/j.efsa.2009.1254)

EFSA. (2009e). Scientific Opinion on the substantiation of health claims related to biotin and energy-yielding metabolism (ID 114, 117), macronutrient metabolism (ID 113, 114, 117), maintenance of skin and mucous membranes (ID 115), maintenance of hair (ID 118, 2876) and function of the nervous system (ID 116) pursuant to Article 13(1) of Regulation (EC) No 1924/2006. *EFSA Journal*. 7(9): 1209. (doi:10.2903/j.efsa.2009.1209)

EFSA. (2009f). Scientific Opinion on the substantiation of health claims related to calcium and maintenance of bones and teeth (ID 224, 230, 231, 354, 3099), muscle function and neurotransmission (ID 226, 227, 230, 235), blood coagulation (ID 230, 236), energy-yielding metabolism (ID 234), function of digestive enzymes (ID 355), and maintenance of normal blood pressure (ID 225, 385, 1419) pursuant to Article 13(1) of Regulation (EC) No 1924/2006. *EFSA Journal*. 7(9): 1210. (doi:10.2903/j.efsa.2009.1210)

EFSA. (2009g). Scientific Opinion on the substantiation of health claims related to calcium and vitamin D and maintenance of bone (ID 350) pursuant to Article 13(1) of Regulation (EC) No 1924/2006. *EFSA Journal.* 7(9): 1272. (doi:10.2903/j.efsa.2009.1272)

EFSA. (2009h). Scientific Opinion on the substantiation of health claims related to copper and protection of DNA, proteins and lipids from oxidative damage (ID 263, 1726), function of the immune system (ID 264), maintenance of connective tissues (ID 265, 271, 1722), energy-yielding metabolism (ID 266), function of the nervous system (ID 267), maintenance of skin and hair pigmentation (ID 268, 1724), iron transport (ID 269, 270, 1727), cholesterol metabolism (ID 369), and glucose metabolism (ID 369) pursuant to Article 13(1) of Regulation (EC) No 1924/2006. *EFSA Journal.* 7(9): 1211. (doi:10.2903/j.efsa.2009.1211)

EFSA. (2009i). Scientific Opinion on the substantiation of health claims related to EPA, DHA, DPA and maintenance of normal blood pressure (ID 502), maintenance of normal HDL-cholesterol concentrations (ID 515), maintenance of normal (fasting) blood concentrations of triglycerides (ID 517), maintenance of normal LDL-cholesterol concentrations (ID 528, 698) and maintenance of joints (ID 503, 505, 507, 511, 518, 524, 526, 535, 537) pursuant to Article 13(1) of Regulation (EC) No 1924/2006. *EFSA Journal.* 7(9): 1263. (doi:10.2903/j.efsa.2009.1263)

EFSA. (2009j). Scientific Opinion on the substantiation of health claims related to fluoride and maintenance of tooth mineralisation (ID 275, 276) and maintenance of bone (ID 371) pursuant to Article 13(1) of Regulation (EC) No 1924/2006. *EFSA Journal.* 7(9): 1212. (doi:10.2903/j.efsa.2009.1212)

EFSA. (2009k). Scientific Opinion on the substantiation of health claims related to folate and blood formation (ID 79), homocysteine metabolism (ID 80), energy-yielding metabolism (ID 90), function of the immune system (ID 91), function of blood vessels (ID 94, 175, 192), cell division (ID 193), and maternal tissue growth during pregnancy (ID 2882) pursuant to Article 13(1) of Regulation (EC) No 1924/2006. *EFSA Journal.* 7(9): 1213. (doi:10.2903/j.efsa.2009.1213)

EFSA. (2009l). Scientific Opinion on the substantiation of health claims related to glucomannan and maintenance of normal blood cholesterol concentrations (ID 836, 1560) pursuant to Article 13(1) of Regulation (EC) No 1924/2006. *EFSA Journal.* 7(9): 1258. (doi:10.2903/j.efsa.2009.1258)

EFSA. (2009m). Scientific Opinion on the substantiation of health claims related to iodine and thyroid function and production of thyroid hormones (ID 274), energy-yielding metabolism (ID 274), maintenance of vision (ID 356), maintenance of hair (ID 370), maintenance of nails (ID 370), and maintenance of skin (ID 370) pursuant to Article 13(1) of Regulation (EC) No 1924/2006. *EFSA Journal.* 7(9): 1214. (doi:10.2903/j.efsa.2009.1214)

EFSA. (2009n). Scientific Opinion on the substantiation of health claims related to iron and formation of red blood cells and haemoglobin (ID 249, ID 1589), oxygen transport (ID 250, ID 254, ID 256), energy-yielding metabolism (ID 251, ID 1589), function of the immune system (ID 252, ID 259), cognitive function (ID 253) and cell division (ID 368) pursuant to Article 13(1) of Regulation (EC) No 1924/2006. *EFSA Journal.* 7(9): 1215. (doi:10.2903/j.efsa.2009.1215)

EFSA. (2009o). Scientific Opinion on the substantiation of health claims related to lactase enzyme and breaking down lactose (ID 1697, 1818) pursuant to Article 13(1) of Regulation (EC) No 1924/2006. *EFSA Journal.* 7(9): 1236. (doi:10.2903/j.efsa.2009.1236)

EFSA. (2009p). Scientific Opinion on the substantiation of health claims related to linoleic acid and maintenance of normal blood cholesterol concentrations (ID 489) pursuant to Article 13(1) of Regulation (EC) No 1924/2006. *EFSA Journal.* 7(9): 1276. (doi:10.2903/j.efsa.2009.1276)

EFSA. (2009q). Scientific Opinion on the substantiation of health claims related to magnesium and electrolyte balance (ID 238), energy-yielding metabolism (ID 240, 247, 248), neurotransmission and muscle contraction including heart muscle (ID 241, 242), cell division (ID 365), maintenance of bone (ID 239), maintenance of teeth (ID 239), blood coagulation (ID 357) and protein synthesis (ID 364) pursuant to Article 13(1) of Regulation (EC) No 1924/2006. *EFSA Journal.* 7(9): 1216. (doi:10.2903/j.efsa.2009.1216)

EFSA. (2009r). Scientific Opinion on the substantiation of health claims related to manganese and protection of DNA, proteins and lipids from oxidative damage (ID 309), maintenance of bone (ID 310), energy-yielding metabolism (ID 311), and cognitive function (ID 340) pursuant to Article 13(1) of Regulation (EC) No 1924/2006. *EFSA Journal.* 7(9): 1217. (doi:10.2903/j.efsa.2009.1217)

EFSA. (2009s). Scientific Opinion on the substantiation of health claims related to niacin and energy-yielding metabolism (ID 43, 49, 54), function of the nervous system (ID 44, 53), maintenance of the skin and mucous membranes (ID 45, 48, 50, 52), maintenance of normal LDL-cholesterol, HDL cholesterol and triglyceride concentrations (ID 46), maintenance of bone (ID 50), maintenance of teeth (ID 50), maintenance of hair (ID 50, 2875) and maintenance of nails (ID 50, 2875) pursuant to Article 13(1) of Regulation (EC) No 1924/2006. *EFSA Journal.* 7(9): 1224. (doi:10.2903/j.efsa.2009.1224)

EFSA. (2009t). Scientific Opinion on the substantiation of health claims related to pantothenic acid and energy-yielding metabolism (ID 56, 59, 60, 64, 171, 172, 208), mental performance (ID 57), maintenance of bone (ID 61), maintenance of teeth (ID 61), maintenance of hair (ID 61), maintenance of skin (ID 61), maintenance of nails (ID 61) and synthesis and metabolism of steroid hormones, vitamin D and some neurotransmitters (ID 181) pursuant to Article 13(1) of Regulation (EC) No 1924/2006. *EFSA Journal.* 7(9): 1218. (doi:10.2903/j.efsa.2009.1218)

EFSA. (2009u). Scientific Opinion on the substantiation of health claims related to phosphorus and function of cell membranes (ID 328), energy-yielding metabolism (ID 329, 373) and maintenance of bone and teeth (ID 324, 327) pursuant to Article 13(1) of Regulation (EC) No 1924/2006. *EFSA Journal.* 7(9): 1219. (doi:10.2903/j.efsa.2009.1219)

EFSA. (2009v). Scientific Opinion on the substantiation of health claims related to selenium and protection of DNA, proteins and lipids from oxidative damage (ID 277, 283, 286, 1289, 1290, 1291, 1293, 1751), function of the immune system (ID 278), thyroid function (ID 279, 282, 286, 1289, 1290, 1291, 1293), function of the heart and blood vessels (ID 280), prostate function (ID 284), cognitive function (ID 285) and spermatogenesis (ID 396) pursuant to Article 13(1) of Regulation (EC) No 1924/2006. *EFSA Journal.* 7(9): 1220. (doi:10.2903/j.efsa.2009.1220)

EFSA. (2009w). Scientific Opinion on the substantiation of health claims related to sugar free chewing gum and dental and oral health, including gum and tooth protection and strength (ID 1149), plaque acid neutralisation (ID 1150), maintenance of tooth mineralisation (ID 1151), reduction of oral dryness (ID 1240), and maintenance of the normal body weight (ID 1152) pursuant to Article 13(1) of Regulation (EC) No 1924/2006. *EFSA Journal*. 7(9): 1271. (doi:10.2903/j.efsa.2009.1271)

EFSA. (2009x). Scientific Opinion on the substantiation of health claims related to vitamin A and cell differentiation (ID 14), function of the immune system (ID 14), maintenance of skin and mucous membranes (ID 15, 17), maintenance of vision (ID 16), maintenance of bone (ID 13, 17), maintenance of teeth (ID 13, 17), maintenance of hair (ID 17), maintenance of nails (ID 17), metabolism of iron (ID 206), and protection of DNA, proteins and lipids from oxidative damage (ID 209) pursuant to Article 13(1) of Regulation (EC) No 1924/2006. *EFSA Journal*. 7(9): 1221. (doi:10.2903/j.efsa.2009.1221)

EFSA. (2009y). Scientific Opinion on the substantiation of health claims related to vitamin B12 and red blood cell formation (ID 92, 101), cell division (ID 93), energy-yielding metabolism (ID 99, 190) and function of the immune system (ID 107) pursuant to Article 13(1) of Regulation (EC) No 1924/2006. *EFSA Journal*. 7(9): 1223. (doi:10.2903/j.efsa.2009.1223)

EFSA. (2009z). Scientific Opinion on the substantiation of health claims related to vitamin B6 and protein and glycogen metabolism (ID 65, 70, 71), function of the nervous system (ID 66), red blood cell formation (ID 67, 72, 186), function of the immune system (ID 68), regulation of hormonal activity (ID 69) and mental performance (ID 185) pursuant to Article 13(1) of Regulation (EC) No 1924/2006. *EFSA Journal*. 7(9): 1225. (doi:10.2903/j.efsa.2009.1225)

EFSA. (2009{). Scientific Opinion on the substantiation of health claims related to vitamin C and protection of DNA, proteins and lipids from oxidative damage (ID 129, 138, 143, 148), antioxidant function of lutein (ID 146), maintenance of vision (ID 141, 142), collagen formation (ID 130, 131, 136, 137, 149), function of the nervous system (ID 133), function of the immune system (ID 134), function of the immune system during and after extreme physical exercise (ID 144), non-haem iron absorption (ID 132, 147), energy-yielding metabolism (ID 135), and relief in case of irritation in the upper respiratory tract (ID 1714, 1715) pursuant to Article 13(1) of Regulation (EC) No 1924/2006. *EFSA Journal*. 7(9): 1226. (doi:10.2903/j.efsa.2009.1226)

EFSA. (2009|). Scientific Opinion on the substantiation of health claims related to vitamin D and maintenance of bone and teeth (ID 150, 151, 158), absorption and utilisation of calcium and phosphorus and maintenance of normal blood calcium concentrations (ID 152, 157), cell division (ID 153), and thyroid function (ID 156) pursuant to Article 13(1) of Regulation (EC) No 1924/2006. *EFSA Journal*. 7(9): 1227. (doi:10.2903/j.efsa.2009.1227)

EFSA. (2009}). Scientific Opinion on the substantiation of health claims related to vitamin K and maintenance of bone (ID 123, 127, 128, and 2879), blood coagulation (ID 124 and 126), and function of the heart and blood vessels (ID 124, 125 and 2880) pursuant to Article 13(1) of Regulation (EC) No 1924/2006. *EFSA Journal*. 7(9): 1228. (doi:10.2903/j.efsa.2009.1228)

EFSA. (2009~). Scientific Opinion on the substantiation of health claims related to zinc and function of the immune system (ID 291, 1757), DNA synthesis and cell division (ID 292, 1759), protection of DNA, proteins and lipids from oxidative damage (ID 294, 1758), maintenance of bone (ID 295, 1756), cognitive function (ID 296), fertility and reproduction (ID 297, 300), reproductive development (ID 298), muscle function (ID 299), metabolism of fatty acids (ID 302), maintenance of joints (ID 305), function of the heart and blood vessels (ID 306), prostate function (ID 307), thyroid function (ID 308), acid-base metabolism (ID 360), vitamin A metabolism (ID 361) and maintenance of vision (ID 361) pursuant to Article 13(1) of Regulation (EC) No 1924/2006. *EFSA Journal*. 7(9): 1229. (doi:10.2903/j.efsa.2009.1229)

EFSA. (2009). Water-soluble tomato concentrate (WSTC I and II) and platelet aggregation. *The EFSA Journal*. 1101 1-15. (doi:10.2903/j.efsa.2009.1101)

EFSA. (2010a). Scientific Opinion on Dietary Reference Values for carbohydrates and dietary fibre. *EFSA Journal*. 8(3): 1462. (doi:10.2903/j.efsa.2010.1462)

EFSA. (2010b). Scientific Opinion on Dietary Reference Values for fats, including saturated fatty acids, polyunsaturated fatty acids, monounsaturated fatty acids, *trans* fatty acids, and cholesterol. *EFSA Journal*. 8(3): 1461. (doi:10.2903/j.efsa.2010.1461)

EFSA. (2010c). Scientific Opinion on the substantiation of a health claim related to ethanol-water extract of *Caralluma fimbriata* (Slimaluma®) and helps to reduce waist circumference pursuant to Article 13(5) of Regulation (EC) No 1924/2006. *EFSA Journal*. 8(5): 1602. (doi:10.2903/j.efsa.2010.1602)

EFSA. (2010d). Scientific Opinion on the substantiation of health claims related to biotin and maintenance of normal skin and mucous membranes (ID 121), maintenance of normal hair (ID 121), maintenance of normal bone (ID 121), maintenance of normal teeth (ID 121), maintenance of normal nails (ID 121, 2877), reduction of tiredness and fatigue (ID 119), contribution to normal psychological functions (ID 120) and contribution to normal macronutrient metabolism (ID 4661) pursuant to Article 13(1) of Regulation (EC) No 1924/2006. *EFSA Journal*. 8(10): 1728. (doi:10.2903/j.efsa.2010.1728)

EFSA. (2010e). Scientific Opinion on the substantiation of health claims related to calcium and maintenance of normal bone and teeth (ID 2731, 3155, 4311, 4312, 4703), maintenance of normal hair and nails (ID 399, 3155), maintenance of normal blood LDL-cholesterol concentrations (ID 349, 1893), maintenance of normal blood HDL-cholesterol concentrations (ID 349, 1893), reduction in the severity of symptoms related to the premenstrual syndrome (ID 348, 1892), "cell membrane permeability" (ID 363), reduction of tiredness and fatigue (ID 232), contribution to normal psychological functions (ID 233), contribution to the maintenance or achievement of a normal body weight (ID 228, 229) and regulation of normal cell division and differentiation (ID 237) pursuant to Article 13(1) of Regulation (EC) No 1924/2006. *EFSA Journal*. 8(10): 1725. (doi:10.2903/j.efsa.2010.1725)

EFSA. (2010f). Scientific Opinion on the substantiation of health claims related to chromium and contribution to normal macronutrient metabolism (ID 260, 401, 4665, 4666, 4667), maintenance of normal blood glucose concentrations (ID 262, 4667), contribution to the maintenance or achievement of a normal body weight (ID 339, 4665, 4666), and reduction of tiredness and fatigue (ID 261) pursuant to Article 13(1) of Regulation (EC) No 1924/2006. *EFSA Journal*. 8(10): 1732. (doi:10.2903/j.efsa.2010.1732)

EFSA. (2010g). Scientific Opinion on the substantiation of health claims related to docosahexaenoic acid (DHA) and maintenance of normal (fasting) blood concentrations of triglycerides (ID 533, 691, 3150), protection of blood lipids from oxidative damage (ID 630), contribution to the maintenance or achievement of a normal body weight (ID 629), brain, eye and nerve development (ID 627, 689, 704, 742, 3148, 3151), maintenance of normal brain function (ID 565, 626, 631, 689, 690, 704, 742, 3148, 3151), maintenance of normal vision (ID 627, 632, 743, 3149) and maintenance of normal spermatozoa motility (ID 628) pursuant to Article 13(1) of Regulation (EC) No 1924/2006. *EFSA Journal.* 8(10): 1734. (doi:10.2903/j.efsa.2010.1734)

EFSA. (2010h). Scientific Opinion on the substantiation of health claims related to dried plums of 'prune' cultivars (Prunus domestica L.) and maintenance of normal bowel function (ID 1164) pursuant to Article 13(1) of Regulation (EC) No 1924/2006. *EFSA Journal.* 8(2): 1486. (doi:10.2903/j.efsa.2010.1486)

EFSA. (2010i). Scientific Opinion on the substantiation of health claims related to eicosapentaenoic acid (EPA), docosahexaenoic acid (DHA), docosapentaenoic acid (DPA) and maintenance of normal cardiac function (ID 504, 506, 516, 527, 538, 703, 1128, 1317, 1324, 1325), maintenance of normal blood glucose concentrations (ID 566), maintenance of normal blood pressure (ID 506, 516, 703, 1317, 1324), maintenance of normal blood HDL-cholesterol concentrations (ID 506), maintenance of normal (fasting) blood concentrations of triglycerides (ID 506, 527, 538, 1317, 1324, 1325), maintenance of normal blood LDL-cholesterol concentrations (ID 527, 538, 1317, 1325, 4689), protection of the skin from photo-oxidative (UV-induced) damage (ID 530), improved absorption of EPA and DHA (ID 522, 523), contribution to the normal function of the immune system by decreasing the levels of eicosanoids, arachidonic acid-derived mediators and pro-inflammatory cytokines (ID 520, 2914), and "immunomodulating agent" (4690) pursuant to Article 13(1) of Regulation (EC) No 1924/2006. *EFSA Journal.* 8(10): 1796. (doi:10.2903/j.efsa.2010.1796)

EFSA. (2010j). Scientific Opinion on the substantiation of health claims related to folate and contribution to normal psychological functions (ID 81, 85, 86, 88), maintenance of normal vision (ID 83, 87), reduction of tiredness and fatigue (ID 84), cell division (ID 195, 2881) and contribution to normal amino acid synthesis (ID 195, 2881) pursuant to Article 13(1) of Regulation (EC) No 1924/2006. *EFSA Journal.* 8(10): 1760. (doi:10.2903/j.efsa.2010.1760)

EFSA. (2010k). Scientific Opinion on the substantiation of health claims related to guar gum and maintenance of normal blood glucose concentrations (ID 794), increase in satiety (ID 795) and maintenance of normal blood cholesterol concentrations (ID 808) pursuant to Article 13(1) of Regulation (EC) No 1924/2006. *EFSA Journal.* 8(2): 1464. (doi:10.2903/j.efsa.2010.1464)

EFSA. (2010l). Scientific Opinion on the substantiation of health claims related to hydroxypropyl methylcellulose (HPMC) and maintenance of normal bowel function (ID 812), reduction of post-prandial glycaemic responses (ID 814), maintenance of normal blood cholesterol concentrations (ID 815) and increase in satiety leading to a reduction in energy intake (ID 2933) pursuant to Article 13(1) of Regulation (EC) No 1924/2006. *EFSA Journal.* 8(10): 1739.

(doi:10.2903/j.efsa.2010.1739)

EFSA. (2010m). Scientific Opinion on the substantiation of health claims related to iodine and contribution to normal cognitive and neurological function (ID 273), contribution to normal energy-yielding metabolism (ID 402), and contribution to normal thyroid function and production of thyroid hormones (ID 1237) pursuant to Article 13(1) of Regulation (EC) No 1924/2006. *EFSA Journal.* 8(10): 1800. (doi:10.2903/j.efsa.2010.1800)

EFSA. (2010n). Scientific Opinion on the substantiation of health claims related to iron and formation of red blood cells and haemoglobin (ID 374, 2889), oxygen transport (ID 255), contribution to normal energy-yielding metabolism (ID 255), reduction of tiredness and fatigue (ID 255, 374, 2889), biotransformation of xenobiotic substances (ID 258), and "activity of heart, liver and muscles" (ID 397) pursuant to Article 13(1) of Regulation (EC) No 1924/2006. *EFSA Journal.* 8(10): 1740. (doi:10.2903/j.efsa.2010.1740)

EFSA. (2010o). Scientific Opinion on the substantiation of health claims related to konjac mannan (glucomannan) and reduction of body weight (ID 854, 1556, 3725), reduction of post-prandial glycaemic responses (ID 1559), maintenance of normal blood glucose concentrations (ID 835, 3724), maintenance of normal (fasting) blood concentrations of triglycerides (ID 3217), maintenance of normal blood cholesterol concentrations (ID 3100, 3217), maintenance of normal bowel function (ID 834, 1557, 3901) and decreasing potentially pathogenic gastro-intestinal microorganisms (ID 1558) pursuant to Article 13(1) of Regulation (EC) No 1924/2006. *EFSA Journal.* 8(10): 1798. (doi:10.2903/j.efsa.2010.1798)

EFSA. (2010p). Scientific Opinion on the substantiation of health claims related to lactulose and decreasing potentially pathogenic gastro-intestinal microorganisms (ID 806) and reduction in intestinal transit time (ID 807) pursuant to Article 13(1) of Regulation (EC) No 1924/2006. *EFSA Journal.* 8(10): 1806. (doi:10.2903/j.efsa.2010.1806)

EFSA. (2010q). Scientific Opinion on the substantiation of health claims related to live yoghurt cultures and improved lactose digestion (ID 1143, 2976) pursuant to Article 13(1) of Regulation (EC) No 1924/2006. *EFSA Journal.* 8(10): 1763. (doi:10.2903/j.efsa.2010.1763)

EFSA. (2010r). Scientific Opinion on the substantiation of health claims related to magnesium and "hormonal health" (ID 243), reduction of tiredness and fatigue (ID 244), contribution to normal psychological functions (ID 245, 246), maintenance of normal blood glucose concentrations (ID 342), maintenance of normal blood pressure (ID 344, 366, 379), protection of DNA, proteins and lipids from oxidative damage (ID 351), maintenance of the normal function of the immune system (ID 352), maintenance of normal blood pressure during pregnancy (ID 367), resistance to mental stress (ID 375, 381), reduction of gastric acid levels (ID 376), maintenance of normal fat metabolism (ID 378) and maintenance of normal muscle contraction (ID 380, ID 3083) pursuant to Article 13(1) of Regulation (EC) No 1924/2006. *EFSA Journal.* 8(10): 1807. (doi:10.2903/j.efsa.2010.1807)

EFSA. (2010s). Scientific Opinion on the substantiation of health claims related to manganese and reduction of tiredness and fatigue (ID 312), contribution to normal formation of connective tissue (ID 404) and contribution to normal energy yielding metabolism

(ID 405) pursuant to Article 13(1) of Regulation (EC) No 1924/2006. *EFSA Journal.* 8(10): 1808. (doi:10.2903/j.efsa.2010.1808)

EFSA. (2010t). Scientific Opinion on the substantiation of health claims related to meal replacements for weight control (as defined in Directive 96/8/EC on energy restricted diets for weight loss) and reduction in body weight (ID 1417), and maintenance of body weight after weight loss (ID 1418) pursuant to Article 13(1) of Regulation (EC) No 1924/2006. *EFSA Journal.* 8(2): 1466. (doi:10.2903/j.efsa.2010.1466)

EFSA. (2010u). Scientific Opinion on the substantiation of health claims related to melatonin and alleviation of subjective feelings of jet lag (ID 1953), and reduction of sleep onset latency, and improvement of sleep quality (ID 1953) pursuant to Article 13(1) of Regulation (EC) No 1924/2006. *EFSA Journal.* 8(2): 1467. (doi:10.2903/j.efsa.2010.1467)

EFSA. (2010v). Scientific Opinion on the substantiation of health claims related to molybdenum and contribution to normal amino acid metabolism (ID 313) and protection of DNA, proteins and lipids from oxidative damage (ID 341) pursuant to Article 13(1) of Regulation (EC) No 1924/2006. *EFSA Journal.* 8(10): 1745. (doi:10.2903/j.efsa.2010.1745)

EFSA. (2010w). Scientific Opinion on the substantiation of health claims related to niacin and reduction of tiredness and fatigue (ID 47), contribution to normal energy-yielding metabolism (ID 51), contribution to normal psychological functions (ID 55), maintenance of normal blood flow (ID 211), and maintenance of normal skin and mucous membranes (ID 4700) pursuant to Article 13(1) of Regulation (EC) No 1924/2006. *EFSA Journal.* 8(10): 1757. (doi:10.2903/j.efsa.2010.1757)

EFSA. (2010x). Scientific Opinion on the substantiation of health claims related to pantothenic acid and mental performance (ID 58), reduction of tiredness and fatigue (ID 63), adrenal function (ID 204) and maintenance of normal skin (ID 2878) pursuant to Article 13(1) of Regulation (EC) No 1924/2006. *EFSA Journal.* 8(10): 1758. (doi:10.2903/j.efsa.2010.1758)

EFSA. (2010y). Scientific Opinion on the substantiation of health claims related to pectins and reduction of post-prandial glycaemic responses (ID 786), maintenance of normal blood cholesterol concentrations (ID 818) and increase in satiety leading to a reduction in energy intake (ID 4692) pursuant to Article 13(1) of Regulation (EC) No 1924/2006. *EFSA Journal.* 8(10): 1747. (doi:10.2903/j.efsa.2010.1747)

EFSA. (2010z). Scientific Opinion on the substantiation of health claims related to plant sterols and plant stanols and maintenance of normal blood cholesterol concentrations (ID 549, 550, 567, 713, 1234, 1235, 1466, 1634, 1984, 2909, 3140), and maintenance of normal prostate size and normal urination (ID 714, 1467, 1635) pursuant to Article 13(1) of Regulation (EC) No 1924/2006. *EFSA Journal.* 8(10): 1813. (doi:10.2903/j.efsa.2010.1813)

EFSA. (2010{). Scientific Opinion on the substantiation of health claims related to potassium and maintenance of normal muscular and neurological function (ID 320, 386) and maintenance of normal blood pressure (ID 321) pursuant to Article 13(1) of Regulation (EC) No 1924/2006. *EFSA Journal.* 8(10): 1469. (doi:10.2903/j.efsa.2010.1469)

EFSA. (2010|). Scientific Opinion on the substantiation of health claims related to protein and increase in satiety leading to a reduction in energy intake (ID 414, 616, 730), contribution to the maintenance or achievement of a normal body weight (ID 414, 616, 730), maintenance of normal bone (ID 416) and growth or maintenance of muscle mass (ID 415, 417, 593, 594, 595, 715) pursuant to Article 13(1) of Regulation (EC) No 1924/2006. *EFSA Journal.* 8(10): 1811. (doi:10.2903/j.efsa.2010.1811)

EFSA. (2010}). Scientific Opinion on the substantiation of health claims related to riboflavin (vitamin B2) and contribution to normal energy-yielding metabolism (ID 29, 35, 36, 42), contribution to normal metabolism of iron (ID 30, 37), maintenance of normal skin and mucous membranes (ID 31, 33), contribution to normal psychological functions (ID 32), maintenance of normal bone (ID 33), maintenance of normal teeth (ID 33), maintenance of normal hair (ID 33), maintenance of normal nails (ID 33), maintenance of normal vision (ID 39), maintenance of normal red blood cells (ID 40), reduction of tiredness and fatigue (ID 41), protection of DNA, proteins and lipids from oxidative damage (ID 207), and maintenance of the normal function of the nervous system (ID 213) pursuant to Article 13(1) of Regulation (EC) No 1924/2006. *EFSA Journal.* 8(10): 1814. (doi:10.2903/j.efsa.2010.1814)

EFSA. (2010~). Scientific Opinion on the substantiation of health claims related to selenium and maintenance of normal hair (ID 281), maintenance of normal nails (ID 281), protection against heavy metals (ID 383), maintenance of normal joints (ID 409), maintenance of normal thyroid function (ID 410, 1292), protection of DNA, proteins and lipids from oxidative damage (ID 410, 1292), and maintenance of the normal function of the immune system (ID 1750) pursuant to Article 13(1) of Regulation (EC) No 1924/2006. *EFSA Journal.* 8(10): 1727. (doi:10.2903/j.efsa.2010.1727)

EFSA. (2010). Scientific Opinion on the substantiation of health claims related to thiamin and reduction of tiredness and fatigue (ID 23) and contribution to normal psychological functions (ID 205) pursuant to Article 13(1) of Regulation (EC) No 1924/2006. *EFSA Journal.* 8(10): 1755. (doi:10.2903/j.efsa.2010.1755)

EFSA. (2010€). Scientific Opinion on the substantiation of health claims related to vitamin B12 and contribution to normal neurological and psychological functions (ID 95, 97, 98, 100, 102, 109), contribution to normal homocysteine metabolism (ID 96, 103, 106), maintenance of normal bone (ID 104), maintenance of normal teeth (ID 104), maintenance of normal hair (ID 104), maintenance of normal skin (ID 104), maintenance of normal nails (ID 104), reduction of tiredness and fatigue (ID 108), and cell division (ID 212) pursuant to Article 13(1) of Regulation (EC) No 1924/2006. *EFSA Journal.* 8(10): 1756. (doi:10.2903/j.efsa.2010.1756)

EFSA. (2010). Scientific Opinion on the substantiation of health claims related to vitamin B6 and contribution to normal homocysteine metabolism (ID 73, 76, 199), maintenance of normal bone (ID 74), maintenance of normal teeth (ID 74), maintenance of normal hair (ID 74), maintenance of normal skin (ID 74), maintenance of normal nails (ID 74), contribution to normal energy-yielding metabolism (ID 75, 214), contribution to normal psychological functions (ID 77), reduction of tiredness and fatigue (ID 78), and contribution to normal cysteine synthesis (ID 4283) pursuant to Article 13(1) of Regulation (EC) No 1924/2006. *EFSA Journal.* 8(10): 1759. (doi:10.2903/j.efsa.2010.1759)

EFSA. (2010,). Scientific Opinion on the substantiation of health claims related to vitamin C and reduction of tiredness and fatigue (ID 139, 2622), contribution to normal psychological functions (ID 140), regeneration of the reduced form of vitamin E (ID 202), contribution to normal energy-yielding metabolism (ID 2334, 3196), maintenance of the normal function of the immune system (ID 4321) and protection of DNA, proteins and lipids from oxidative damage (ID 3331) pursuant to Article 13(1) of Regulation (EC) No 1924/2006. *EFSA Journal.* 8(10): 1815. (doi:10.2903/j.efsa.2010.1815)

EFSA. (2010ƒ). Scientific Opinion on the substantiation of health claims related to vitamin D and normal function of the immune system and inflammatory response (ID 154, 159), maintenance of normal muscle function (ID 155) and maintenance of normal cardiovascular function (ID 159) pursuant to Article 13(1) of Regulation (EC) No 1924/2006. *EFSA Journal.* 8(2): 1468. (doi:10.2903/j.efsa.2010.1468)

EFSA. (2010„). Scientific Opinion on the substantiation of health claims related to vitamin E and protection of DNA, proteins and lipids from oxidative damage (ID 160, 162, 1947), maintenance of the normal function of the immune system (ID 161, 163), maintenance of normal bone (ID 164), maintenance of normal teeth (ID 164), maintenance of normal hair (ID 164), maintenance of normal skin (ID 164), maintenance of normal nails (ID 164), maintenance of normal cardiac function (ID 166), maintenance of normal vision by protection of the lens of the eye (ID 167), contribution to normal cognitive function (ID 182, 183), regeneration of the reduced form of vitamin C (ID 203), maintenance of normal blood circulation (ID 216) and maintenance of normal a scalp (ID 2873) pursuant to Article 13(1) of Regulation (EC) No 1924/2006. *EFSA Journal.* 8(10): 1816. (doi:10.2903/j.efsa.2010.1816)

EFSA. (2010…). Scientific Opinion on the substantiation of health claims related to wheat bran fibre and increase in faecal bulk (ID 3066), reduction in intestinal transit time (ID 828, 839, 3067, 4699) and contribution to the maintenance or achievement of a normal body weight (ID 829) pursuant to Article 13(1) of Regulation (EC) No 1924/2006. *EFSA Journal.* 8(10): 1817. (doi:10.2903/j.efsa.2010.1817)

EFSA. (2010†). Scientific Opinion on the substantiation of health claims related to zinc and maintenance of normal skin (ID 293), DNA synthesis and cell division (ID 293), contribution to normal protein synthesis (ID 293, 4293), maintenance of normal serum testosterone concentrations (ID 301), "normal growth" (ID 303), reduction of tiredness and fatigue (ID 304), contribution to normal carbohydrate metabolism (ID 382), maintenance of normal hair (ID 412), maintenance of normal nails (ID 412) and contribution to normal macronutrient metabolism (ID 2890) pursuant to Article 13(1) of Regulation (EC) No 1924/2006. *EFSA Journal.* 8(10): 1819. (doi:10.2903/j.efsa.2010.1819)

EFSA. (2011a). General guidance for stakeholders on the evaluation of Article 13.1, 13.5 and 14 health claims. *EFSA Journal.* 9(4): 2135. (doi:10.2903/j.efsa.2011.2135)

EFSA. (2011b). Guidance on the scientific requirements for health claims related to gut and immune function. *EFSA Journal.* 9(4): 1984. (doi:10.2903/j.efsa.2011.1984)

EFSA. (2011c). Scientific and technical guidance for the preparation and presentation of an application for authorisation of a health claim (revision 1). *EFSA Journal.* 9(5): 2170. (doi:10.2903/j.efsa.2011.2170)

EFSA. (2011d). Scientific Opinion on the substantiation of a health claim related to melatonin and reduction of sleep onset latency (ID 1698, 1780, 4080) pursuant to Article 13(1) of Regulation (EC) No 1924/2006. *EFSA Journal*. 9(6): 2241. (doi:10.2903/j.efsa.2011.2241)

EFSA. (2011e). Scientific Opinion on the substantiation of health claims related to arabinoxylan produced from wheat endosperm and reduction of post-prandial glycaemic responses (ID 830) pursuant to Article 13(1) of Regulation (EC) No 1924/2006. *EFSA Journal*. 9(6): 2205. (doi:10.2903/j.efsa.2011.2205)

EFSA. (2011f). Scientific Opinion on the substantiation of health claims related to beta-glucans from oats and barley and maintenance of normal blood LDL-cholesterol concentrations (ID 1236, 1299), increase in satiety leading to a reduction in energy intake (ID 851, 852), reduction of post-prandial glycaemic responses (ID 821, 824), and "digestive function" (ID 850) pursuant to Article 13(1) of Regulation (EC) No 1924/2006. *EFSA Journal*. 9(6): 2207. (doi:10.2903/j.efsa.2011.2207)

EFSA. (2011g). Scientific Opinion on the substantiation of health claims related to betaine and contribution to normal homocysteine metabolism (ID 4325) pursuant to Article 13(1) of Regulation (EC) No 1924/2006. *EFSA Journal*. 9(4): 2052. (doi:10.2903/j.efsa.2011.2052)

EFSA. (2011h). Scientific Opinion on the substantiation of health claims related to caffeine and increase in physical performance during short-term high-intensity exercise (ID 737, 1486, 1489), increase in endurance performance (ID 737, 1486), increase in endurance capacity (ID 1488) and reduction in the rated perceived exertion/effort during exercise (ID 1488, 1490) pursuant to Article 13(1) of Regulation (EC) No 1924/2006. *EFSA Journal*. 9(4): 2053. (doi:10.2903/j.efsa.2011.2053)

EFSA. (2011i). Scientific Opinion on the substantiation of health claims related to caffeine and increased fat oxidation leading to a reduction in body fat mass (ID 735, 1484), increased energy expenditure leading to a reduction in body weight (ID 1487), increased alertness (ID 736, 1101, 1187, 1485, 1491, 2063, 2103) and increased attention (ID 736, 1485, 1491, 2375) pursuant to Article 13(1) of Regulation (EC) No 1924/2006. *EFSA Journal*. 9(4): 2054. (doi:10.2903/j.efsa.2011.2054)

EFSA. (2011j). Scientific Opinion on the substantiation of health claims related to carbohydrate-electrolyte solutions and reduction in rated perceived exertion/effort during exercise (ID 460, 466, 467, 468), enhancement of water absorption during exercise (ID 314, 315, 316, 317, 319, 322, 325, 332, 408, 465, 473, 1168, 1574, 1593, 1618, 4302, 4309), and maintenance of endurance performance (ID 466, 469) pursuant to Article 13(1) of Regulation (EC) No 1924/2006. *EFSA Journal*. 9(6): 2211. (doi:10.2903/j.efsa.2011.2211)

EFSA. (2011k). Scientific Opinion on the substantiation of health claims related to chitosan and reduction in body weight (ID 679, 1499), maintenance of normal blood LDL-cholesterol concentrations (ID 4663), reduction of intestinal transit time (ID 4664) and reduction of inflammation (ID 1985) pursuant to Article 13(1) of Regulation (EC) No 1924/2006. *EFSA Journal*. 9(6): 2214. (doi:10.2903/j.efsa.2011.2214)

EFSA. (2011l). Scientific Opinion on the substantiation of health claims related to choline and contribution to normal lipid metabolism (ID 3186), maintenance of normal liver function (ID 1501), contribution to normal homocysteine metabolism (ID 3090), maintenance of normal neurological function (ID 1502), contribution to normal cognitive function (ID 1502), and brain and neurological development (ID 1503) pursuant to Article 13(1) of Regulation (EC) No 1924/2006. *EFSA Journal.* 9(4): 2056. (doi:10.2903/j.efsa.2011.2056)

EFSA. (2011m). Scientific Opinion on the substantiation of health claims related to creatine and increase in physical performance during short-term, high intensity, repeated exercise bouts (ID 739, 1520, 1521, 1522, 1523, 1525, 1526, 1531, 1532, 1533, 1534, 1922, 1923, 1924), increase in endurance capacity (ID 1527, 1535), and increase in endurance performance (ID 1521, 1963) pursuant to Article 13(1) of Regulation (EC) No 1924/2006. *EFSA Journal.* 9(7): 2303. (doi:10.2903/j.efsa.2011.2303)

EFSA. (2011n). Scientific Opinion on the substantiation of health claims related to fats and "function of the cell membrane" (ID 622, 2900, 2911) and normal absorption of fat-soluble vitamins (ID 670, 2902) pursuant to Article 13(1) of Regulation (EC) No 1924/2006. *EFSA Journal.* 9(6): 2220. (doi:10.2903/j.efsa.2011.2220)

EFSA. (2011o). Scientific Opinion on the substantiation of health claims related to foods with reduced amounts of saturated fatty acids (SFAs) and maintenance of normal blood LDL cholesterol concentrations (ID 620, 671, 4332) pursuant to Article 13(1) of Regulation (EC) No 1924/2006. *EFSA Journal.* 9(4): 2062. (doi:10.2903/j.efsa.2011.2062)

EFSA. (2011p). Scientific Opinion on the substantiation of health claims related to foods with reduced amounts of sodium and maintenance of normal blood pressure (ID 336, 705, 1148, 1178, 1185, 1420) pursuant to Article 13(1) of Regulation (EC) No 1924/2006. *EFSA Journal.* 9(6): 2237. (doi:10.2903/j.efsa.2011.2237)

EFSA. (2011q). Scientific Opinion on the substantiation of health claims related to foods with reduced lactose content and decreasing gastro-intestinal discomfort caused by lactose intake in lactose intolerant individuals (ID 646, 1224, 1238, 1339) pursuant to Article 13(1) of Regulation (EC) No 1924/2006. *EFSA Journal.* 9(6): 2236. (doi:10.2903/j.efsa.2011.2236)

EFSA. (2011r). Scientific Opinion on the substantiation of health claims related to glycaemic carbohydrates and maintenance of normal brain function (ID 603, 653) pursuant to Article 13(1) of Regulation (EC) No 1924/2006. *EFSA Journal.* 9(6): 2226. (doi:10.2903/j.efsa.2011.2226)

EFSA. (2011s). Scientific Opinion on the substantiation of health claims related to intense sweeteners and contribution to the maintenance or achievement of a normal body weight (ID 1136, 1444, 4299), reduction of post-prandial glycaemic responses (ID 4298), maintenance of normal blood glucose concentrations (ID 1221, 4298), and maintenance of tooth mineralisation by decreasing tooth demineralisation (ID 1134, 1167, 1283) pursuant to Article 13(1) of Regulation (EC) No 1924/2006. *EFSA Journal.* 9(6): 2229. (doi:10.2903/j.efsa.2011.2229)

EFSA. (2011t). Scientific Opinion on the substantiation of health claims related to monacolin K from red yeast rice and maintenance of normal blood LDL cholesterol concentrations (ID 1648, 1700) pursuant to Article 13(1) of Regulation (EC) No 1924/2006. *EFSA Journal.* 9(7): 2304. (doi:10.2903/j.efsa.2011.2304)

EFSA. (2011u). Scientific Opinion on the substantiation of health claims related to oat and barley grain fibre and increase in faecal bulk (ID 819, 822) pursuant to Article 13(1) of Regulation (EC) No 1924/2006. *EFSA Journal.* 9(6): 2249. (doi:10.2903/j.efsa.2011.2249)

EFSA. (2011v). Scientific Opinion on the substantiation of health claims related to polyphenols in olive and protection of LDL particles from oxidative damage (ID 1333, 1638, 1639, 1696, 2865), maintenance of normal blood HDL cholesterol concentrations (ID 1639), maintenance of normal blood pressure (ID 3781), "anti-inflammatory properties" (ID 1882), "contributes to the upper respiratory tract health" (ID 3468), "can help to maintain a normal function of gastrointestinal tract" (3779), and "contributes to body defences against external agents" (ID 3467) pursuant to Article 13(1) of Regulation (EC) No 1924/2006. *EFSA Journal*. 9(4): 2033. (doi:10.2903/j.efsa.2011.2033)

EFSA. (2011w). Scientific Opinion on the substantiation of health claims related to resistant starch and reduction of post-prandial glycaemic responses (ID 681), "digestive health benefits" (ID 682) and "favours a normal colon metabolism" (ID 783) pursuant to Article 13(1) of Regulation (EC) No 1924/2006. *EFSA Journal*. 9(4): 2024. (doi:10.2903/j.efsa.2011.2024)

EFSA. (2011x). Scientific Opinion on the substantiation of health claims related to rye fibre and changes in bowel function (ID 825), reduction of post prandial glycaemic responses (ID 826) and maintenance of normal blood LDL-cholesterol concentrations (ID 827) pursuant to Article 13(1) of Regulation (EC) No 1924/2006. *EFSA Journal*. 9(6): 2258. (doi:10.2903/j.efsa.2011.2258)

EFSA. (2011y). Scientific Opinion on the substantiation of health claims related to sodium and maintenance of normal muscle function (ID 359) pursuant to Article 13(1) of Regulation (EC) No 1924/2006. *EFSA Journal*. 9(6): 2260. (doi:10.2903/j.efsa.2011.2260)

EFSA. (2011z). Scientific Opinion on the substantiation of health claims related to sugar-free chewing gum with carbamide and plaque acid neutralisation (ID 1153) pursuant to Article 13(1) of Regulation (EC) No 1924/2006. *EFSA Journal*. 9(4): 2071. (doi:10.2903/j.efsa.2011.2071)

EFSA. (2011{). Scientific Opinion on the substantiation of health claims related to the replacement of mixtures of saturated fatty acids (SFAs) as present in foods or diets with mixtures of monounsaturated fatty acids (MUFAs) and/or mixtures of polyunsaturated fatty acids (PUFAs), and maintenance of normal blood LDL cholesterol concentrations (ID 621, 1190, 1203, 2906, 2910, 3065) pursuant to Article 13(1) of Regulation (EC) No 1924/2006. *EFSA Journal*. 9(4): 2069. (doi:10.2903/j.efsa.2011.2069)

EFSA. (2011 |). Scientific Opinion on the substantiation of health claims related to the sugar replacers xylitol, sorbitol, mannitol, maltitol, lactitol, isomalt, erythritol, D-tagatose, isomaltulose, sucralose and polydextrose and maintenance of tooth mineralisation by decreasing tooth demineralisation (ID 463, 464, 563, 618, 647, 1182, 1591, 2907, 2921, 4300), and reduction of post-prandial glycaemic responses (ID 617, 619, 669, 1590, 1762, 2903, 2908, 2920) pursuant to Article 13(1) of Regulation (EC) No 1924/2006. *EFSA Journal*. 9(4): 2076. (doi:10.2903/j.efsa.2011.2076)

EFSA. (2011}). Scientific Opinion on the substantiation of health claims related to very low calorie diets (VLCDs) and reduction in body weight (ID 1410), reduction in the sense of hunger (ID 1411), reduction in body fat mass while maintaining lean body mass (ID 1412), reduction of post-prandial glycaemic responses (ID 1414), and maintenance of normal blood lipid profile (1421) pursuant to Article 13(1) of Regulation (EC) No 1924/2006. *EFSA Journal*. 9(6): 2271.

(doi:10.2903/j.efsa.2011.2271)

EFSA. (2011~). Scientific Opinion on the substantiation of health claims related to water and maintenance of normal physical and cognitive functions (ID 1102, 1209, 1294, 1331), maintenance of normal thermoregulation (ID 1208) and "basic requirement of all living things" (ID 1207) pursuant to Article 13(1) of Regulation (EC) No 1924/2006. *EFSA Journal.* 9(4): 2075. (doi:10.2903/j.efsa.2011.2075)

Geissler, C. and Powers, H. Geissler, C. and Powers, H. Eds.(2005) *Human Nutrition.* Elsevier, Edinburgh.

Gibson, G.R. & Roberfroid, M.B. (1995). Dietary Modulation of the Human Colonic Microbiota: Introducing the Concept of Prebiotics. *The Journal of Nutrition.* 125(6): 1401-1412.

Grossklaus, R. (2009). Codex recommendations on the scientific basis of health claims. *European Journal of Nutrition.* 48 S15-S22.

He, F.J. & MacGregor, G.A. (2003). How far should salt intake be reduced? Hypertension. *Hypertension.* 42 1093-1099.

He, F.J. & MacGregor, G.A. (2010). Reducing population salt intake worldwide: from evidence to implementation. *Prog Cardiovasc Dis.* 52 363-382.

Hoefkens, C., Verbeke, W. & Van Camp, J. (2011). European consumers' perceived importance of qualifying and disqualifying nutrients in food choices. *Food Quality and Preference.* 22(6): 550-558.

Howlett, J. (2008) *Functional Foods: From science to health and claims.* ILSI Europe, Brussels.

Ikeda, Y., Iki, M., Morita, A., Kajita, E., Kagamimori, S. et al. (2006). Intake of fermented soybeans, natto, is associated with reduced bone loss in postmenopausal women: Japanese population-based osteoporosis (JPOS) study. *J. Nutr.* 136(5): 1323-1328.

IVZ (2009) *Tolerance values for labeling of nutritional composition of foods (URL: http://ivz.arhiv.over.net/javne_datoteke/datoteke/2068-HVtolerance_27102009.doc, Accessed: June 2010).* Slovenian Institute of Public Health, Ljubljana.

Jadhav, S.J., Lutz, S.E., Ghorpade, V.M. & Salunkhe, D.K. (1998). Barley: Chemistry and value-added processing. *Critical Reviews in Food Science and Nutrition.* 38(2): 123-171.

Katsuyama, H., Ideguchi, S., Fukunaga, M., Fukunaga, T., Saijoh, K. et al. (2004). Promotion of bone formation by fermented soybean (Natto) intake in premenopausal women. *Journal of Nutritional Science and Vitaminology.* 50(2): 114-120.

Krutulyte, R., Grunert, K.G., Scholderer, J., Lahteenmaki, L., Hagemann, K.S. et al. (2011). Perceived fit of different combinations of carriers and functional ingredients and its effect on purchase intention. *Food Quality and Preference.* 22(1): 11-16.

Lalor, F., Kennedy, J., Flynn, M.A. & Wall, P.G. (2010). A study of nutrition and health claims? a snapshot of what's on the Irish market. *Public Health Nutrition.* 13(05): 704-711.

Ling, W.H. & Jones, P.J.H. (1995). Dietary phytosterols: A review of metabolism, benefits and side effects. *Life Sciences.* 57(3): 195-206.

Marrugat, J., Covas, M.I., Fito, M., Schroder, H., Miro-Casas, E. et al. (2004). Effects of differing phenolic content in dietary olive oils on lipids and LDL oxidation - A randomized controlled trial. *European Journal of Nutrition.* 43(3): 140-147.

Palacios, C. (2006). The role of nutrients in bone health, from A to Z. *Crit. Rev. Food Sci. Nutr.* 46(8): 621-628.

Pothoulaki, M. & Chryssochoidis, G. (2009). Health claims: Consumers' matters. *Journal of Functional Foods.* 1(2): 222-228.

Pravst, I. (2010). The evaluation of health claims in Europe: What have we learned? *AgroFOOD industry hi-tech.* 21(4): 4-6.

Pravst, I. (2011a). Health claims: Where are we now and where are we going? (editorial) *Agro Food Industry Hi-Tech.* 22(4): 2-3.

Pravst, I. (2011b). Risking public health by approving some health claims? – The case of phosphorus. *Food Policy.* 36: 725-727.

Pravst, I. & Žmitek, K. (2011). The coenzyme Q10 content of food supplements. *Journal fur Verbraucherschutz und Lebensmittelsicherheit-Journal of Consumer Protection and Food Safety.* 6: 457-463.

Pravst, I., Žmitek, K. & Žmitek, J. (2010). Coenzyme Q10 contents in foods and fortification strategies. *Crit. Rev. Food Sci. Nutr.* 50(4): 269-280.

Raspor, P. 2011. Definition of functional foods and safety aspects of functional nutrition, Functional Foods Seminar, ZKZP/GZS: Ljubljana, Slovenia.

Sioen, I., Matthys, C., Huybrechts, I., Van Camp, J. & De Henauw, S. (2011). Consumption of plant sterols in Belgium: consumption patterns of plant sterol-enriched foods in Flanders, Belgium. *British Journal of Nutrition.* 105(06): 911-918.

St Onge, M.P. & Jones, P.J.H. (2003). Phytosterols and human lipid metabolism: efficacy, safety, and novel foods. *Lipids.* 38(4): 367-375.

Strazzullo, P., D'Elia, L., Kandala, N.-B. & Cappuccio, F.P. (2009). Salt intake, stroke and cardiovascular disease: a meta-analysis of prospective studies. *Br Med J.* 339 b4567.

Talati, R., Baker, W.L., Pabilonia, M.S., White, C.M. & Coleman, C.I. (2009). The Effects of Barley-Derived Soluble Fiber on Serum Lipids. *Ann Fam Med.* 7(2): 157-163.

Turner, N.D. & Lupton, J.R. (2011). Dietary Fiber. *Advances in Nutrition: An International Review Journal.* 2(2): 151-152.

Urala, N. & Lahteenmaki, L. (2004). Attitudes behind consumers' willingness to use functional foods. *Food Quality and Preference.* 15(7-8): 793-803.

Verbeke, W. (2005). Consumer acceptance of functional foods: socio-demographic, cognitive and attitudinal determinants. *Food Quality and Preference.* 16(1): 45-57.

Verbeke, W. (2006). Functional foods: Consumer willingness to compromise on taste for health? *Food Quality and Preference.* 17(1-2): 126-131.

Verbeke, W. (2010). Consumer reactions to foods with nutrition and health claims. *Agro Food Industry Hi-Tech.* 21(6): 5-8.

Verbeke, W., Scholderer, J. & Lahteenmaki, L. (2009). Consumer appeal of nutrition and health claims in three existing product concepts. *Appetite.* 52(3): 684-692.

Vermeer, C., Shearer, M.J., Zittermann, A., Bolton-Smith, C., Szulc, P. et al. (2004). Beyond deficiency: Potential benefits of increased intakes of vitamin K for bone and vascular health. *Eur. J. Nutr.* 43(6): 325-335.

Weinbrenner, T., Fito, M., de la Torre, R., Saez, G.T., Rijken, P. et al. (2004). Olive oils high in phenolic compounds modulate oxidative/antioxidative status in men. Journal of Nutrition. 134(9): 2314-2321.

WHO (2003). Recomendations for preventing cardiovascular diseases. In: *WHO Technical Report Series.* pp. 81-94. World Health Organization, Geneva.

Wills, J.M., Schmidt, D.B., Pillo-Blocka, F. & Cairns, G. (2009). Exploring global consumer attitudes toward nutrition information on food labels. *Nutrition Reviews.* 67(5): S102-S106.

Zeisel, S.H. & Caudill, M.A. (2010). Choline. *Advances in Nutrition: An International Review Journal.* 1(1): 46-48.

12

Organic Food Preference: An Empirical Study on the Profile and Loyalty of Organic Food Customers

Pelin Özgen
Atılım University
Turkey

1. Introduction

Eating habits in the world have shown some phases over time. In the last century, the sole aim was to feed oneself, however, later in time, food industry has been affected from industrialization trend, and agriculture turned into a sector in which large-scale food products are produced and consumed primarily based on their being cost effective. In parallel with this trend, one of the major challenges in agriculture is to increase efficiency. As a result of this challenge, two types of food production have aroused- one being the *conventional method* and the other method utilizing *genetic engineering* which emerged in the last decade.

Conventional method is the oldest and the most widely used technique. According to Knorr and Watkins (1984) conventional agriculture is defined as "capital-intensive, large-scale, highly mechanized agriculture with monocultures of crops and extensive use of artificial fertilizers, herbicides and pesticides, with intensive animal husbandry". However, heavy reliance on synthetic chemical fertilizers and pesticides is said to have serious impacts on public health and the environment (Pimentel et al. 2005). Therefore, as people are becoming more environmental conscious, this method is being questioned and tried to be developed during the last decades.

The second technique, called as genetically modified foods (GM foods or GMO foods), were first put on the market in the early 1990s (wikipedia.org). These food products are derived from genetically modified organisms, (GMOs), which are obtained by using advanced techniques of genetic engineering. Currently, genetic modification is mostly applied to soybean, corn, canola, cotton seed and sugar beet and the application area is observed to expand everyday.

Naturally, genetically modified foods are not without advantages and disadvantages. The biggest advantage of using genetically modified foods is the ability to grow faster and bigger crops. In addition to that, weaknesses against certain types of disease and insects might be eliminated by genetic modification (hubpages.com(1)). Moreover, higher crop yields are thought to make food prices decrease and therefore lead to less starvation in the world. Besides these advantages, GDOs have also disadvantages, such as their tendency to make harm to other organisms (such as in the case of monarch butterflies which are poisoned by GMO corns), possible damages to environment in the long run, possible health problems in humans and unforeseen risks and dangers due to the complexity of nature (hubpages.com(2)).

Apart from the technological developments and the increased need for food, beginning with the 1970s, consciousness about health and environmental issues has aroused. This awakening led to many changes in both production and consumption patterns of food. In parallel with these, environmental friendly agriculture, which is also called as ecological or organic agriculture, started to be employed and supported by the governments.

Detailed information about organic foods, discussion on customer loyalty in food sector and empirical study are given next.

2. Organic foods

The first formal organization to promote and regulate organic agriculture is the International Federation of Organic Agriculture Movement (IFOAM), which was established in 1972. According to IFOAM, organic agriculture is "a production system that sustains the health of soil, ecosystem and people. It relies on ecological processes, biodiversity and cycles adapted to local conditions, rather than the use of inputs with adverse effects." In other words, organic foods are foods which are produced by using organic farming techniques, in which use of synthetic fertilizers, pesticides, fungicides, growth regulators and livestock feed additives and antibiotics as well as genetic modification are strictly prohibited (Lohr, 2001). Because of the naturalness of the production, organic foods are said be superior over food products that are produced with other techniques. Therefore, due to this perceived superiority, there is an increased attention towards organic practices.

According to a study conducted by Hau and Joaris (2000), certified organic products make up about 2% of the world food market. Despite its small market share, organic food products is the fastest growing segment of the food industry especially in the developed countries such as USA, Japan and EU countries (Raynolds, 2004). In the USA, the demand for organic food market was reported to increase at a rate of 18.5% (Klonsky, 2007), and in France the demand increase rate was about 10% annually (Monier, et.al, 2009), which is about ten times the rate of demand for total food products. This also stands as an evidence that the organic food market is growing very rapidly and special attention should be given for this segment in the food industry. Consequently, some studies (for example Vindigni et. al, 2002, Thompson, 1998, Makatouni, 2002, Lohr, 2002, Davies et.al, 1995) have been addressed to this issue in developed countries, yet there exist very few studies addressed in developing countries such as Turkey.

According to Turkish Statistical Institute (TUIK), about 43 million people live in Turkey in 20-60 age group, who decide personally on their food and may be considered as possible organic food consumers. This large number is appealing for marketers of several products, including organic food. However, previous research reveals that organic food production in Turkey was started to be employed not until 1985 (Karakoc and Baykam, 2009)- about 15 years later than it was started to be encouraged in the international markets with the establishment of International Federation of Organic Agriculture Movement (IFOAM) in 1972 (www.ifoam.org). Official encouragement of organic food production in Turkey was started only a decade ago, with the establishment of Association of Ecological Agriculture Organization (ETO) (Yanmaz, 2005). Currently, in Europe, about 6% of the agricultural fields are allocated to organic farming, whereas in Turkey only 1% of the area is reserved for the same purpose (Deniz, 2007) and organic food products are still considered to be new products for Turkish consumers. Despite the short history of organic food in Turkey, currently about 250 different organic products are produced and almost all certified products are exported to developed countries, which are European Union countries, USA and Japan in particular. Also, it is reported that Turkey holds the market leader position in dry and dried organic fruits (www.eto.org.tr). However, it should not be overlooked that, logistics and certification process causes international trade to hold more problems than marketing domestically, especially in food products where freshness, food standards and reliability are the major concerns. However, despite its difficulties, as stated above, almost all of the organic food is produced for international markets and it is clear that the demand for organic food in domestic market needs to be promoted more. Considering that Turkey has a large population, who are getting more conscious in organic food production and consumption, reaching domestic customers might be more rewarding for businesses.

3. Customer loyalty in the food sector

Customer loyalty is another issue that is examined in this research. According to Jacoby and Kryner (1974), customer loyalty is defined as customer's repeat purchase which is resulted from a series of psychological processes. However, it should be noted that, if people only focus on the repeated purchase, than it might be misleading, and this repeated buying should not be treated as customer loyalty. As Dick and Basu (1994) point out, even a relatively important repeat purchase may not reflect true loyalty, but may merely be the result of situational conditions. Therefore, many studies (for example, Jacoby & Chestnut, 1978; Kahn & Meyer, 1991; Dick & Basu, 1994) are available in the literature suggesting that loyalty should be divided into 2 types of loyalty: behavioral loyalty and attitudinal loyalty. In order to make a satisfying and comprehensive definition, both attitudinal and behavioral components should be present (Kim et.al, 1994). In parallel with this, Samuelson and Sandvick (1997) state that, behavioral approach to loyalty is still valid as a component of loyalty, however, attitudinal approaches to loyalty should also be present to supplement the behavioral approach.

Dick and Basu (1994) have developed a Loyalty Model, in which loyalty is shown to have different levels, affected from different attitudinal levels. As seen in the figure below (Figure 1), true loyalty can only exist when there is a highly positive relative attitude accompanied with a behavioral measure, which is called here as repeated patronage.

		Repeated Patronage	
		High	Low
Relative	High	True Loyalty	Latent Loyalty
Attitude	Low	Spurious Loyalty	No Loyalty

Fig. 1. Dick and Basu's Loyalty Model (Garland and Gendall, 2004)

Looking at the financial perspective of a firm, one of the most important roles of marketing is to increase the market for a product and to create continous cash flows for the company. According to many researchers, (such as Gupta and Zeithaml (2006), Rust et.al (2000) Srinivasan et.al, (2005), Baloglu (2002)) creating loyal customers to the firm is the first step and is very essential for increasing the market for a product. Keeping in mind that the market for organic food products needs to be enlarged, customer loyalty concept should also be examined in depth. Therefore, results and implications of the empirical study will be discussed in the following sections of this study.

4. Methodology

Survey method is used in order to gather data in this research. The questionnaire was formed based on previous research (such as Sarikaya, 2007; Monier, 2009; McIver, 2004, etc.) in addition to questions formed by the researcher, parallel with the aim of the study. Since the questionnaire is newly formed, it should be tested for internal consistency before employing it as the actual data gathering instrument (Tabacknick and Fidell, 2001). Therefore the questionnaire was applied to a test group of 37 in order to see the internal consistency coefficient, known as Cronbach's Alfa. After running the test, the Cronbach's Alfa was found to be .72. Since the acceptable level of Cronbach Alfa is .70 for internal consistency (Nunnally and Bernstein, 1974), the created questionnaire was found to be suitable to be used in the current research and it is applied for the actual study.

The questionnaire consists of five parts. In the first part, attitudes towards organic food products and buying patterns are investigated with 22 questions. Based on author's personal experience and observation, there is a debate on the fact that people's choice of organic food show difference according to for whom they are buying the food. In other words, when people are shopping for their children, it is seen that tendency towards buying organic alternative increases. Whether this observation is true for Turkish customers is also tried to be answered with the help of questions in the first part.

In the second part, the accessibility of organic food products and the place where respondents get their organic foods are asked. In the third part, respondents are asked to rank the reasons for choosing organic foods, where in the fourth part, they are kindly asked to reveal their opinions about what needs to be developed in the organic food sector.

The final part of the questionnaire is devoted to demographic questions, consisting of educational level, marital status, family size, gender, monthly income and age.

All of the questions in the questionnaire are formed as structured and pre-coded questions; therefore reluctance towards participating in the study is minimized due to minimized effort required from respondents.

The questionnaire was applied on June 2011 in Turkey's two largest cities- Ankara and Istanbul, due to convenience and purposive reasons. Respondents were chosen via mall-intercept method and before applying the questionnaire, a filtering question whether or not they purchase organic products was asked and it is made sure that the sample group consists of only organic customers.

A number of 138 questionnaires were filled, however, after preliminary screening, only 122 of them were found to be useful. The data gathered was analyzed with respect to descriptive and univariate analysis such as frequency tables, t-test and ANOVA by using SPSS 15.0 software package.

5. Empirical results / findings

The current study was conducted in Ankara and Istanbul in June 2011. Respondents were selected via mall-intercept method among organic food purchasers, with respect to availability. According to the results of the study, majority of the respondents are found out to be female, have university degree and are married with children. Considering that the respondents are chosen among organic food buyers, these demographic findings might be considered as the general profile of organic food buyers segment. The characteristics of the respondents are given in the tables below.

Gender	Frequency	%	Cum. %
Male	35	29	29
Female	87	71	100
Total	122	100	

Table 1. Gender distribution of respondents

As seen from the table above, 71% of the respondents are female. This result might imply that, female are the major customers of organic food products.

Age	Frequency	%	Cum. %
< 25	8	6.6	6.6
26-30	16	13.1	19.7
31-35	38	31.1	50.8
36-40	28	23.0	73.8
> 41	32	26.2	100
Total	122	100	

Table 2. Age distribution of respondents

When the age distribution is examined, it is seen that about 30% of the respondents are between 31 and 35, and the smallest group with respect to age is composed of people who are younger than 25. This result might be interpreted as younger people are less likely to buy organic food products. However, this interpretation should be approached with caution, because the sample is not taken via random sampling method and therefore is susceptible to sampling errors (Malhotra, 2011).

Marital Status	Frequency	%	Cum. %
Single	38	31.1	31.1
Married	84	68.9	100
Total	122	100	

Table 3. Marital Status of the respondents

According to the results, 84 of the respondents are married and remaining 38 of respondents are single. The married group consists of about 70% of the total respondents. In order to increase the expressiveness of this result, family size is also measured, as shown below.

Family Size	Frequency	%	Cum. %
Spouse and myself	20	16.4	15.6
Spouse, myself, and kid(s) younger than 3 years old	46	37.7	54.1
Spouse, myself, and kid(s) older than 3 years old	41	33.6	87.7
Myself and my parents	5	4.1	91.8
Only myself	10	8.2	100
Total	122	100	

Table 4. The family size of the respondents

As shown in the table above (Table 4), 71.3% of the respondents are living with spouse and kid(s). The smallest group with respect to family size is the group who are living with his/her parents, composing only 4% of the respondents. This result may lead to a stereotyping that, there is an increased tendency towards buying organic food products if a person has a child. Whether this stereotype is true or not, in other words, whether there is a significant difference between buying food for kids and buying for adults will be tested as an hypothesis in the following parts of this study.

Education Level	Frequency	%	Cum. %
< High school	4	3.3	3
High school	22	18	21.3
University	67	54.9	76.2
Master's degree	20	16.4	92.6
PhD/Doctorate	9	7.4	100
Total	122	100	

Table 5. Education level of respondents

According to the results, it is seen that 97% of the respondents have at least high school degree. In addition to this, it is seen that 67 of the people have a university diploma, which corresponds to 55% of the respondents. Moreover, 29 of the respondents have graduate degree, corresponding to 23%. By looking at this education data, it can be concluded that, organic foods are preferred by mostly educated people.

Monthly Income	Frequency	%	Cum. %
1001- 2000 TL	20	16.4	16.4
2001- 3000 TL	39	32.0	48.4
3001- 4000 TL	38	31.1	79.5
> 4000 TL	25	20.5	100
Total	122	100	

Table 6. Monthly Income of respondents

As seen in the table above (Table 6), 48% of the respondents earn less than 3000 TL a month and 52% have monthly income higher than 3000 TL. According to TUIK (2011), the average monthly income for household in Turkey is 1750 TL for the year 2010. This result might be interpreted as that, the shoppers for organic food products have a higher income level than Turkish households.

5.1 Attitudes towards organic foods and organic food preference
Respondent's attitudes towards organic food products are measured with Likert- scaled items in the questionnaire. The items are scaled from 1 to 5, with "1" being "strongly disagree" and "5" being strongly agree" to given statements.
By looking at the results, it is seen that, the average score for the statement "organic foods are more delicious than traditional food products" is 4.26, implying that respondents strongly agree that there is a difference in taste in favor of organically produced food products. Similarly, the statement of "organic foods are more nutritious" has received 3.8 points and the statement "organic food products are healthier than traditional food products" have received an average of 4.12. These results show that, there is a general strong belief towards organic food's being healthier.
When they are asked for their buying behavior, it is seen that respondents are not reluctant to pay a premium price in order to buy an organic food product. Currently, organic foods are about twice more expensive than regular food products in Turkish market. Even though this 100% premium price is paid, either willingly or unwillingly, 79 of the respondents (64.7%) believe that, in order to increase the demand for organic food products, the price of the organic food products should be decreased. In addition to price, respondents state that, the barriers to purchase organic foods include availability and organic food product range. These factors should also be developed, if organic food market is wanted to expand.
In an attempt to find out the reasons why people do not complain much about premium price they pay for organic foods, an open ended question is placed in the questionnaire. The most frequent answer is related with health concerns. 83 of the respondents (68%) believe that, organic foods are worth a price premium due to being healthier and have more nutritive value. Other reasons for household's willingness to pay more are listed as quality, certification and environmental concerns, where environmental concerns being seen as the least important reason. This finding contradicts with the research made by Bellows et.al (2008), in which environmental concerns are seen as more important factor for demand for organic products.

As stated above, customer loyalty is very essential in order to expand a market of a product. Since nothing can be improved unless measured, two questions were placed in the questionnaire in order to find out loyalty levels of the respondents towards organic food products. In the first question, which was asked in order to identify the behavioral component of customer loyalty, respondents are asked "if they prefer to buy the organic alternative if there existed one". This question has received an average score of 3.74, implying that there is a tendency to buy the organic alternative. The second question about loyalty is asked to test for attitudinal loyalty. In this question respondents are asked "if they recommend organic foods to other people". This question has received an average score of 4.2, which may be translated as that they make recommendations to other people which leads to a result that they have attitudinal loyalty. By comparing these two loyalty scores, it might be concluded that respondents are both behaviorally and attitudinally loyal to organic food products. This result may imply that organic food product market might continue to expand in the following years as well.

5.2 Changes in behavior according to whom the food is bought for

Based on author's personal experience and observation, it is seen that people become selective in shopping if they are buying goods for other people, especially if they are buying foods for children. Eventhough less attention is paid to health concerns while buying for himself/herself, the picture changes when it comes to shop for the children. Considering that there is a positive attitude towards organic food products, it is expected that people would buy organic food products for their children more often than they buy for themselves. In order to answer the question whether this observation is statistically provable for Turkish customers, questions in the first part of the questionnaire should be analyzed.

As shown above in Table 4, 70% of the respondents are living with spouse and kid(s), leading to a thought that there is an increased tendency towards buying organic food products if a person has a child. However, if this stereotyping is statistically significant or not should be tested.

A one sample t-test is conducted to see if there is significant change between the intention to buy organic foods for respondent himself or for his children. Firstly, the general tendency towards organic food buying is tested. The sample mean for "tendency to buy organic food products for children" is 4.19, which is found to be significantly different from the mean for "tendency to buy organic food products for himself", which has an average of 3.15 ($t(121)=1.82$, $p=.04$) at 95% confidence level.

After t-test is employed for testing the difference in general buying tendencies, individual tests, with respect to some product groups are conducted. These product groups include fresh vegetables, fresh fruits, milk and dairy products and dried fruits. However, no significant change is observed in the tendency to buy organic alternative in any of the individual product groups cited above. This result shows that despite the observed significant difference in tendency to buy organic food for children and for adults, there is no significant relation between tendency to buy organic foods for these two groups, with respect to special product groups.

6. Conclusion

Nowadays the society is mainly concerned with topics such as global warming, ecological impacts, health issues and better nutrition. In the area of food production, organic production techniques are seen as the best alternative for these issues. In parallel with the increased consciousness on health and environmental issues, the demand for organic food products is rapidly increasing worldwide. It is reported that annual growth rate in demand for organic foods is about 10 times the rate of demand for total food products (Monier et.al.2009). Despite the increasing domestic demand in Turkey, it is seen that majority of the produced organic food is exported to European countries, USA and to Japan. In spite of the high potential for organic food production in Turkey, it is seen that only about 1% of the total agricultural area is devoted to organic farming. Considering that demand is increasing everyday, Turkey should act intelligently to utilize its potential to become a major local and international organic food supplier. Yet, certification process, issues in labeling and logistics of the organic foods constitute important barriers to exporting. Therefore, in order to increase the production of organic products, it is firstly essential to expand the domestic market for organic products.

For the purpose of increasing the domestic demand, the general attitude of Turkish consumers towards organic foods, the profile of organic buyers and customer loyalty in organic food products market is tried to be investigated in this study. A survey is applied to 122 respondents, of which the majority of the respondents are highly educated females, who are married and have children. In addition to that, respondents have about 3000 TL monthly income, which is seen to be higher than Turkey average income. This profile is parallel with other studies (such as Sarıkaya, 2007, Monier et.al, 2009, Yanmaz, 2005) concerning the buyers for organic food products.

According to the results, there is a strong belief that organic foods are more delicious than other foods, and they are believed to have more nutritious value. In an overall assessment, organic products are preferred over conventionally produced or genetically modified food, especially if people are buying for their children. The fact that preference for organic foods differs according to whom the food is bought for is also tested and is proved to have a statistical significance. However, no significant change is observed in the tendency to buy organic alternative for specific product groups. The reason for not observing a significant difference might be due to small sample size and signals that tests should be repeated in further studies. In addition to this, in general, it can be said that organic food products are preferred mainly due to health concerns. Nevertheless, trust is an important issue in customer's minds and it is believed that strict controls and procedures in both production labeling should be implemented.

Another topic investigated in this research is about customer loyalty. According to the results, it is seen that here is a high loyalty among organic customers both in attitudinal and behavioral dimensions. This result is especially important for organic food producers, because high attitudinal loyalty is considered as a signal that consumers are willing to buy the organic alternative if there exists one and they recommend organic products to their families and friends. Moreover, it is seen that people are not satisfied with the currently available product range. Therefore, one can conclude that organic demand is congruent and the market for organic food products are expected to expand provided that the industry and the retailers ensure regular and easy supply with a high product variability.

To sum it up, according to the results of the empirical study, it is seen that the domestic market for organic food products is eager for new products and there is a strong loyalty among organic customers towards these products. In order to utilize this market potential, availability should be increased via utilizing supermarkets and alternative marketing channels more effectively. Currently, marketing and distribution for organic foods are relatively inefficient due to small volumes, which leads to high costs. Provided that marketing channels are better organized for organic foods, then the prices will eventually decrease, which will lead to an increase in demand. Moreover, there is still a lack of information about organic food products in the domestic market. Both the producer and customers should be better informed. Considering the demographic profile of the organic customers, marketing of the organic food products should be targeted mostly to educated women who have children. Finally, the variability in the organic foods should be increased since current customers for organic products are eager to buy organic alternative if possible. Provided that these actions are taken, then it should be no surprise to see Turkey as the leader in organic food production and consumption.

7. Limitations and suggestions for further study

As in every study, this study has also its limitations. One of major limitations of the study is due to application of survey method. There might be some errors due to factors such as social desirability or affect of the interviewer. It is believed that more reliable results could be obtained if survey method could have been backed up with other research methods such as observation or even experimentation. Especially, market basket analysis is expected to lead to interesting findings in consumer behavior in food sector.

Even though 122 is a satisfying sample size, the results of the findings could be more reliable and maybe the statistical associations which could not be observed could be observed if the sample size was larger.

Therefore, in further studies, it is recommended to study with a larger sample size and with other techniques in order to increase the generalizability and reliability of the results.

8. References

Association of Ecological Agriculture http://www.eto.org.tr

Baloglu, S. (2002) Dimensions of Customer Loyalty: Seperating Friends From Well Wishers. *Cornell Hotel and Restaurant Administration Quarterly*, 43, (1), 47-59

Bellows, Anne C.; Onyango, Benjamin; Diamond, Adam; and Hallman, William K. (2008) Understanding Consumer Interest in Organics: Production Values vs. Purchasing Behavior. *Journal of Agricultural & Food Industrial Organization*. 6, 1, pp.1-31

Davies, A., Titterington, A.J. & Cochrane, C. (1995). Who buys organic food?: A profile of the purchasers of organic food in Northern Ireland, *British Food Journal*, 97, 10, pp. 17 – 23

Deniz, N. (2007) Turkish Export potential. Paper presented in 1st Organic Agriculture Congress, Bahcesehir University, October 19-20 2007 Istanbul, Turkey

Gupta, S.& Zeithaml, V. (2006). Customer Metrics and Their Impact on Financial Performance. *Marketing Science*, Nov-Dec. 718-739.

Hau and Joaris, (2000) Organic farming, EU report, The European Commission in Vindigni, G.,Janssen, M.A., Jager, W. (2002). Organic Food Consumption. A multi-theoretical framework of consumer decision making. *British Food Journal*, 104, 8, pp. 624-642

Hubpages.com (1): http://benjimester.hubpages.com/hub/Genetically-Modified-foods-Pros-and-Cons

Hubpages.com (2): http://hubpages.com/hub/GMO-advantages-and-disadvantages

International Federation of Organic Agriculture Movement (IFOAM) www.ifoam.org

Jacoby, J. & Chestnut, R.W. (1978). *Brand Loyalty: Measurement and Management*. Wiley, ISBN:0471028452 New York

Jacoby, J.& Kryner, D.B. (1973). Brand Loyalty vs. Repeat Purchasing Behaviour. *Journal of Marketing Research*, Vol. 10, pp. 1-9

Karakoc, U. & Baykam, B.G. (2009). Türkiye'de Organik Tarım Gelisiyor. Betam, Arastirma Notu, 09-35. Retrieved on April 6, 2011 from www.betam.bahcesehir.edu.tr

Klonsky, K. (2007) Organic Agriculture and the US Farm Bill. University of Carolina, Agricultural Issues Center As cited in Monier, S., Hassan, D., Nichèle, V., Simioni, M. (2009). Organic Food Consumption Patterns, *Journal of Agricultural & Food Industrial Organization*, 7, 2. Article 12.

Knorr, D., Watkins, T.R (1984) *Alterations in Food Production. New York* As cited in Beus, C.E., Dunlap, R.E. (1990), Conventional versus Alternative Agriculture: The Paradigmatic Roots of the Debate, *Rural Sociology*, 55, 4, pp.590-616

Lohr, L. (2001) Factors Affecting International Demand and Trade in Organic Food Products. As cited in Regmi, A. (Editor) *Changing Structure of Global Food Consumption and Trade*. Market and Trade Economics Division, Economic Research Service / WRS-01-1

Makatouni, A. (2002) What Motivates Consumers to Buy Organic Food in the UK?: Results from a Qualitative Study", *British Food Journal*, 104, 3/4/5, pp. 345 – 352

Malhotra, N. (2011) *Marketing Research: An Applied Orientation*, Pearson International Press.

McIver, H. (2004). Organic hip: Popular Picks at Health Food Stores. *Better Nutrition*, 66, 2

Monier, S., Hassan, D., Nichèle, V. & Simioni, M. (2009) "Organic Food Consumption Patterns" *Journal of Agricultural & Food Industrial Organization*, 7, 2. Article 12.

Nunnally, J.C.& Berstein, I.H. (1994). *Psychometric Theory*. 3rd Ed., New York: McGraw-Hill.

Pimentel, D., Hepperly, P., Hanson, J., Douds, D. & Seidel, R. (2005). *"Environmental, Energetic, and Economic Comparisons of Organic and Conventional Farming Systems"*, Bioscience, 55(July), 7, pp.573-582

Raynolds, L.T. (2004). The Globalization of Organic Agro-Food Networks. *World Development* 32, 5, pp.725-743

Rust, R.T., Zeithaml,V.A. & Lemon, K.N. (2000). *Driving Customer Equity.*The Free Press, ISBN: 0684864665 New York

Sarıkaya, N. (2007) Organik Ürün Tüketimini Etkileyen Faktörler ve Tutumlar Üzerine bir Saha Çalısması (A fieldwork on Factors Affecting the Consumption and Attitudes Towards Organic Foods). *Kocaeli Universitesi Sosyal Bilimler Enstitusu Dergisi*, 14,2, pp. 110- 125.

Srinivasan, V., Park, C.S. & Chang, D.R. (2005). An Approach to the Measurement, Analysis and Prediction of Brand Equity and Its Sources. *Management Science*, 51, 9,pp. 1433-1448

Thompson, G.D.(1998). Consumer Demand for Organic Foods: What We Know and What We Need to Know. *American Journal of Agricultural Economics*, 80, pp. 1113-1118

Turkish Statistical Institute (TUIK) www.tuik.gov.tr

Vindigni, G., Janssen, M.A., Jager, W. (2002). Organic Food Consumption. A multi-theoretical Framework of Consumer Decision Making. *British Food Journal*, 104, 8, pp. 624-642

Genetically Modified Food. Retrieved from
    http://en.wikipedia.org/wiki/ Genetically_modified_food on 22.07.2011

Yanmaz, R. (2005) Organik Ürünlerin Pazarlanması ve Ticareti. (Marketing and Commercial Trade of Organic Products) *Symposium on Food Safety and Reliability*, Ankara-Turkey Conference Proceedings, Full Text, pp. 349-365

Permissions

The contributors of this book come from diverse backgrounds, making this book a truly international effort. This book will bring forth new frontiers with its revolutionizing research information and detailed analysis of the nascent developments around the world.

We would like to thank Benjamin Valdez, Michael Schorr and Roumen Zlatev, for lending their expertise to make the book truly unique. They have played a crucial role in the development of this book. Without their invaluable contribution this book wouldn't have been possible. They have made vital efforts to compile up to date information on the varied aspects of this subject to make this book a valuable addition to the collection of many professionals and students.

This book was conceptualized with the vision of imparting up-to-date information and advanced data in this field. To ensure the same, a matchless editorial board was set up. Every individual on the board went through rigorous rounds of assessment to prove their worth. After which they invested a large part of their time researching and compiling the most relevant data for our readers. Conferences and sessions were held from time to time between the editorial board and the contributing authors to present the data in the most comprehensible form. The editorial team has worked tirelessly to provide valuable and valid information to help people across the globe.

Every chapter published in this book has been scrutinized by our experts. Their significance has been extensively debated. The topics covered herein carry significant findings which will fuel the growth of the discipline. They may even be implemented as practical applications or may be referred to as a beginning point for another development. Chapters in this book were first published by InTech; hereby published with permission under the Creative Commons Attribution License or equivalent.

The editorial board has been involved in producing this book since its inception. They have spent rigorous hours researching and exploring the diverse topics which have resulted in the successful publishing of this book. They have passed on their knowledge of decades through this book. To expedite this challenging task, the publisher supported the team at every step. A small team of assistant editors was also appointed to further simplify the editing procedure and attain best results for the readers.

Our editorial team has been hand-picked from every corner of the world. Their multi-ethnicity adds dynamic inputs to the discussions which result in innovative outcomes. These outcomes are then further discussed with the researchers and contributors who give their valuable feedback and opinion regarding the same. The feedback is then collaborated with the researches and they are edited in a comprehensive manner to aid the understanding of the subject.

Apart from the editorial board, the designing team has also invested a significant amount of their time in understanding the subject and creating the most relevant covers. They scrutinized every image to scout for the most suitable representation of the subject and create an appropriate cover for the book.

The publishing team has been involved in this book since its early stages. They were actively engaged in every process, be it collecting the data, connecting with the contributors or procuring relevant information. The team has been an ardent support to the editorial, designing and production team. Their endless efforts to recruit the best for this project, has resulted in the accomplishment of this book. They are a veteran in the field of academics and their pool of knowledge is as vast as their experience in printing. Their expertise and guidance has proved useful at every step. Their uncompromising quality standards have made this book an exceptional effort. Their encouragement from time to time has been an inspiration for everyone.

The publisher and the editorial board hope that this book will prove to be a valuable piece of knowledge for researchers, students, practitioners and scholars across the globe.

List of Contributors

A. Derossi, T. De Pilli and C. Severini
Department of Food Science, University of Foggia, Italy

D. B. Luiz
Federal University of Santa Catarina, Brazil
Embrapa Fisheries and Aquaculture, Brazil

H. J. Jose and R. F. P. M. Moreira
Federal University of Santa Catarina, Brazil

Izabel Soares, Zacarias Távora, Rodrigo Patera Barcelos and Suzymeire Baroni
Federal University of the Bahia Reconcavo / Center for Health Sciences, Brazil

Ebrahim Alizadeh Doughikollaee
University Of Zabol, Iran

Yukihiro Yamamoto and Setsuko Hara
Seikei University, Japan

Benjamin Valdez Salas and Michael Schorr Wiener
Instituto de Ingenieria, Departamento de Materiales, Minerales y Corrosion, Universidad Autonoma de Baja California, Mexico

Gustavo Lopez Badilla
UNIVER, Plantel Oriente, Mexicali, Baja California, Mexico

Francisco Javier Gutiérrez, Mª Luisa Mussons, Paloma Gatón and Ruth Rojo
Centro Tecnológico CARTIF, Parque Tecnológico de Boecillo, Valladolid, España

Makoto Kanauchi
Miyagi University, Japan

Ladislav Mura
Dubnica Institute of Technology, Department of Specialised Subjects, Slovak Republic

Håkan Jönsson and Hans Knutsson
Lund University, Sweden

Carl-Otto Frykfors
Linköping University, Sweden

Igor Pravst
Nutrition Institute, Ljubljana, Slovenia

Pelin Özgen
Atılım University, Turkey